Arsenic:

INDUSTRIAL, BIOMEDICAL, ENVIRONMENTAL PERSPECTIVES

Arsenic:

INDUSTRIAL, BIOMEDICAL, ENVIRONMENTAL PERSPECTIVES

Proceedings of the Arsenic Symposium, Gaithersburg, Maryland, Sponsored by the Arsenic Panel of The Chemical Manufacturer's Association and the National Bureau of Standards

edited by

William H. Lederer

and

Robert J. Fensterheim

VAN NOSTRAND REINHOLD ENVIRONMENTAL ENGINEERING SERIES

VNR VAN NOSTRAND REINHOLD COMPANY
NEW YORK CINCINNATI TORONTO LONDON MELBOURNE

Library of Congress Catalog Card Number: 82-8339
ISBN: 0-442-21496-0

Manufactured in the United States of America

Published by Van Nostrand Reinhold Company Inc.
135 West 50th Street, New York, N.Y. 10020

Van Nostrand Reinhold Publishing
1410 Birchmount Road
Scarborough, Ontario M1P 2E7, Canada

Van Nostrand Reinhold
480 Latrobe Street
Melbourne, Victoria 3000, Australia

Van Nostrand Reinhold Company Limited
Molly Millars Lane
Wokingham, Berkshire, England

15 14 13 12 11 10 9 8 7 6 5 4 3 2 1

Library of Congress Cataloging in Publication Data

Main entry under title:

Arsenic–industrial, biomedical, environmental
perspectives.

(Van Nostrand Reinhold environmental engineering series)
"Proceedings of an Arsenic Symposium sponsored by the Chemical Manufacturers Association and the National Bureau of Standards, November 4–6, 1981 at Gaithersburg, Maryland"–Pref.
Includes index.
1. Arsenic poisoning–Congresses. 2. Arsenic–Environmental aspects–Congresses. 3. Arsenic–Industrial applications–Congresses. 4. Arsenic–Congresses.
I. Lederer, William H. II. Fensterheim, Robert J.
III. Arsenic Symposium (1981: Gaithersburg, Md.)
IV. Chemical Manufacturers Association (U.S.) V. United States. National Bureau of Standards.
RA1231.A7A77 1982 363.1'79 82-8339
ISBN 0-442-21496-0 AACR2

Van Nostrand Reinhold Environmental Engineering Series

THE VAN NOSTRAND REINHOLD ENVIRONMENTAL ENGINEERING SERIES is dedicated to the presentation of current and vital information relative to the engineering aspects of controlling man's physical environment. Systems and subsystems available to exercise control of both the indoor and outdoor environment continue to become more sophisticated and to involve a number of engineering disciplines. The aim of the series is to provide books which, though often concerned with the life cycle—design, installation, and operation and maintenance—of a specific system or subsystem, are complementary when viewed in their relationship to the total environment.

The Van Nostrand Reinhold Environmental Engineering Series includes books concerned with the engineering of mechanical systems designed (1) to control the environmental within structures, including those in which manufacturing processes are carried out, and (2) to control the exterior environment through control of waste products expelled by inhabitants of structures and from manufacturing processes. The series includes books on heating, air conditioning and ventilation, control of air and water pollution, control of the acoustic environment, sanitary engineering and waste disposal, illumination, and piping systems for transporting media of all kinds.

Van Nostrand Reinhold Environmental Engineering Series

ADVANCED WASTEWATER TREATMENT, by Russell L. Culp and Gordon L. Culp

ARCHITECTURAL INTERIOR SYSTEMS—Lighting, Air Conditioning, Acoustics, John E. Flynn and Arthur W. Segil

SOLID WASTE MANAGEMENT, by D. Joseph Hagerty, Joseph L. Pavoni and John E. Heer, Jr.

THERMAL INSULATION, by John F. Malloy

AIR POLLUTION AND INDUSTRY, edited by Richard D. Ross

INDUSTRIAL WASTE DISPOSAL, edited by Richard D. Ross

MICROBIAL CONTAMINATION CONTROL FACILITIES, by Robert S. Rurkle and G. Briggs Phillips

SOUND, NOISE, AND VIBRATION CONTROL (Second Edition), by Lyle, F. Yerges

NEW CONCEPTS IN WATER PURIFICATION, by Gordon L. Culp and Russell L. Culp

HANDBOOK OF SOLID WASTE DISPOSAL: MATERIALS AND ENERGY RECOVERY, by Joseph L. Pavoni, John E. Heer, Jr., and D. Joseph Hagerty

ENVIRONMENTAL ASSESSMENTS AND STATEMENTS, by John E. Heer, Jr. and D. Joseph Hagerty

ENVIRONMENTAL IMPACT ANALYSIS: A New Dimension in Decision Making by R. K. Jain, L. V. Urban and G. S. Stacey

CONTROL SYSTEMS FOR HEATING, VENTILATING, AND AIR CONDITIONING (Second Edition), by Roger W. Haines

WATER QUALITY MANAGEMENT PLANNING, edited by Joseph L. Pavoni

HANDBOOK OF ADVANCED WASTEWATER TREATMENT (Second Edition), by Russell L. Culp, George Mack Wesner and Gordon L. Culp

HANDBOOK OF NOISE ASSESSMENT, edited by Daryl N. May

NOISE CONTROL: HANDBOOK OF PRINCIPLES AND PRACTICES, edited by David M. Lipscomb and Arthur C. Taylor

AIR POLLUTION CONTROL TECHNOLOGY, by Robert M. Bethea

POWER PLANT SITING, by John V. Winter and David A. Conner

DISINFECTION OF WASTEWATER AND WATER FOR REUSE, by Geo. Clifford White

LAND USE PLANNING: Techniques of Implementation, by T. William Patterson

BIOLOGICAL PATHS TO SELF-RELIANCE, by Russell E. Anderson

HANDBOOK OF INDUSTRIAL WASTE DISPOSAL, by Richard A. Conway and Richard D. Ross

HANDBOOK OF ORGANIC WASTE CONVERSION, edited by Michael W. M. Bewick

LAND APPLICATIONS OF WASTE (Volume 1), by Raymond C. Loehr, William J. Jewell, Joseph D. Novak, William W. Clarkson and Gerald S. Friedman

LAND APPLICATIONS OF WASTE (Volume 2), by Raymond C. Loehr, William J. Jewell, Joseph D. Novak, William W. Clarkson and Gerald S. Friedman

STRUCTURAL DYNAMICS: Theory and Computation, by Mario Paz

HANDBOOK OF MUNICIPAL WASTE MANAGEMENT SYSTEMS: Planning and practice, by Barbara J. Stevens

INDUSTRIAL POLLUTION CONTROL: Issues and Techniques, by Nancy J. Sell

WASTE RECYCLING AND POLLUTION CONTROL HANDBOOK, by A. V. Bridgwater and C. J. Mumford

WATER CLARIFICATION PROCESSES: Practical Design and Evaluation, by Herbert E. Hudson, Jr.

HANDBOOK OF NONPOINT POLLUTION: Sources and Management, by Vladimir Novotny and Gordon Chesters

ENVIRONMENTAL RISK ANALYSIS FOR CHEMICALS, edited by Richard A. Conway

NATURAL SYSTEMS FOR WATER POLLUTION CONTROL, by Ray Dinges

ARSENIC–INDUSTRIAL, BIOMEDICAL, ENVIRONMENTAL PERSPECTIVES, edited by William H. Lederer and Robert J. Fensterheim

PREFACE

This book contains the proceedings of an Arsenic Symposium sponsored by the Chemical Manufacturers Association and the National Bureau of Standards, November 4-6, 1981 at Gaithersburg, Maryland. The Symposium was divided into five sessions: (1) industrial sources, (2) industrial uses, (3) biomedical perspectives, (4) epidemiology and (5) environmental perspectives. Each session had a chairman who made some brief introductory remarks followed by the speakers. Generally, after each speaker there was a discussion except in session IV, epidemiology, in which the discussion for the entire session occurred after the last presentation. In addition to the table of contents, presentations and discussions, this book contains a list of registered attendees, author index and subject index.

The purpose of the Symposium was to serve as a forum for the presentation of information on arsenic compounds. The Symposium consisted of invited lectures by experts in the production and use of arsenic and its compounds, in their environmental and biomedical properties, and in epidemiology. The Symposium also provided a means whereby industry and government agencies could reach an understanding for cost effective regulation of arsenic as a hazardous material through knowledge of production and use patterns, toxicology properties, and the presence of arsenic in the environment.

Arsenic is a ubiquitous metal naturally occurring in fossil fuels, ores, and some drinking water supplies. Arsenic containing mineral deposits include: copper, lead, zinc, phosphates, uranium, gold, cobalt and silver. Smelting, mining or leaching of these deposits produces commercial products used extensively in industrial and agricultural applications. Arsenic may be present as an industrial byproduct in air and water emissions and as a solid waste.

The properties of arsenic, as measured in environmental and biological systems, are related to its valence state, ability to form a variety of chemical complexes, and its interactions with other elements or compounds. Scientifically sound development of air and water standards, regulation of pesticides and solid waste, and health and safety issues must necessarily take these factors into consideration. The Symposium brought the current state of knowledge concerning the properties of arsenic compounds into an open forum for discussion. It also served as a guide for future biomedical and environmental research on arsenic.

The Chemical Manufacturers Association (CMA) Arsenic Program Panel provided support for this Symposium. The Arsenic Program was established under CMA in January, 1981 to provide a means whereby industrial operations involved with arsenic could collectively sponsor research and educational programs. The objectives of this Program Panel include:

1. Educate scientific, industrial and government communities concerning arsenic production, the presence of arsenic in the environment, important applications, and state of the scientific data base.

2. Provide a forum for discussion of issues associated with production and use of arsenic.

3. Gather information and assess the need for further research. Provide funds for future studies, if needed.

4. Maintain a liaison with other trade associations and agencies having an interest in arsenic.

Membership of the Arsenic Program Panel consists of the following nine companies which provide support for the program.

Amax Lead & Zinc, Inc.
Anaconda Copper Company
ASARCO, Inc.
Diamond Shamrock Corporation
Koppers Company, Inc.
Osmose Wood Preserving Company
Pennwalt Corporation
Salsbury Laboratories
Sunshine Mining Company

ACKNOWLEDGMENTS

The editors wish to express their appreciation to the other members of the Arsenic Symposium Organization Committee and their staffs who were instrumental in organizing the Symposium. Specifically, William J. Baldwin, John P. McCarthy, Hasmukh C. Shah, Carol R. Stack, William H. Kirchhoff and Kathy C. Stang warrant a special acknowledgement for their outstanding contributions. Robert D. Arsenault was instrumental in initiating this Symposium and appreciation for his efforts are deserving.

The editors wish to thank Mary B. Churilla for her outstanding administrative and secretarial assistance. Additionally, a special note of appreciation is in order to Margaret S. D'Andrea for her excellent technical assistance in the development of this book.

The outstanding facilities at the National Bureau of Standards and the cooperation of its staff and the Chemical Manufacturers Association staff made the atmosphere of the Symposium especially conducive to scientific and social discourse. Appreciation is also expressed to the Chemical Manufacturers Association Arsenic Program Panel for support of the Symposium. Finally, and most important, the outstanding presentations by the speakers, the fine leadership by the session chairmen, and the enthusiastic participation by the attendees made the Symposium an educational and scientifically stimulating experience.

TABLE OF CONTENTS

SESSION I - INDUSTRIAL SOURCES

Session Chairman: Ken Nelson
ASARCO Incorporated
3422 South-700 West
Salt Lake City, UT 84106

INTRODUCTION

I believe this is the first open conference to be held in this country devoted exclusively to arsenic, which is rather surprising since arsenic has been around for a long, long time. It occurred to me that it's interesting also that we don't have the equivalent of a trade association dealing with arsenic. We have, for example, the Lead Industry Association, Zinc Institute, Silver Institute, and then we have product associations such as The American Petroleum Institute and the Bituminous Coal Institute, but we have no organization devoted to arsenic that I am aware of. Perhaps this conference is a beginning.

Arsenic is reported to have been first isolated by Albertus Magnus in the year 1250. In the time since then, someone discovered that arsenic was an effective poison--that arsenic trioxide could be inserted in food or drink in small amounts and would effectively dispatch the consumer. So arsenic acquired a reputation because of its toxicity and is known to the public perhaps mostly by its toxicity. But, it has many other interesting properties.

Arsenic is rarely found as an element in nature, but occurs principally as a trisulfide or in arsenopyrites and is found usually in association with lead, zinc, copper, and gold-bearing minerals. If the 92 natural elements are ranked in decending order of their percentages in the earth's crust, arsenic would be about 20th on the list. The average crustal content of arsenic is estimated to be between 1.5 and three parts per million. It's found literally everywhere in traces--it's present in the air, the oceans and fresh waters, in soil, and in all living things. Its ubiquitousness, along with results from certain animal-feeding experiments, suggests that it may indeed be an element essential to life.

Since we first have to recover the metal from the earth before we can use it, our first session will be devoted to the means of gathering the element from its sources.

1

ARSENIC SOURCES, PRODUCTION AND APPLICATIONS IN THE 1980's

LOUIS D. FITZGERALD
MANAGER, BY-PRODUCT SALES
ASARCO Incorporated
120 Broadway
New York, New York 10271

Large changes, which have evolved through the last decade, have drastically changed the future picture of arsenic trioxide. These changes include:

1. Environmental pressures, which reduce production.

2. Application changes, which in a period of shortage, disrupt market patterns.

3. New sources at high prices, which tend to alleviate shortages in the most profitable applications.

No major changes in production technics have evolved during the past decade. At the present time, nearly all arsenic production results as a by-product of copper, lead, zinc, gold or cobalt smelting.

The economic viability of producing arsenic as a prime product has yet to be demonstrated. Data on sources and production for 1970 and 1980 are presented. Application data is presented to show the shifts in use patterns. These data are discussed and future directions in application growth are suggested.

1. Introduction

During the last few years, arsenic has been under attack from almost every environmental or safety standpoint. The production of this material and its uses have gone through agonizing scrutiny. At the same time, many compounds which compete with arsenic have gone through a similar scrutiny; and it has been apparent that arsenic has been found to be, in the overall picture, more acceptable than the competing chemicals.

First of all, let's keep arsenic in its true perspective. This material is a by-product of non-ferrous metal production. At the present time, there is no producing facility in the world where arsenic is the prime product being produced from arsenical ores.

As a by-product, arsenic is produced whether or not the producing plant wants to make it or whether or not the market is willing to accept it. Conversely the producing plant cannot crank up to turn out more arsenic should the demand increase nor can an increased demand be expected to evolve large additional production of arsenic. This is best illustrated by the current market situation. There is a 7-10,000 ton worldwide shortfall in arsenic production at the present time. There are several producers of arsenic residues that have accumulated stockpiles of these residues. The economics of converting their current production as well as the accumulated residues to saleable arsenic trioxide involves a large capital commitment which in turn demands high arsenic prices. Historically, these kind of prices have not been obtainable for arsenic on a long term basis. Sooner or later the current imbalance will be adjusted. The producers who have been long established in by-product arsenic sales will continue to sell their product.

This is the nature of by-products operations and it should not be ignored while being dazzled by near term large potential profit.

2. Arsenic Production and Consumption Through the Past Decade.

In 1970 the environmental problems for arsenic were just beginning. The major application was in agriculture, using approximately 70% of arsenic trioxide production. The other applications including glass, wood preserving and other industrial uses were much smaller. It was about this time, however that some herbicidal uses for sodium arsenite were under attack and were being phased out. The sale of arsenicals for household applications declined rapidly. At the same time extensive environmental studies were put in motion to determine if arsenic were safe both in the work place and in applications which dispersed it into the environment. Other regulations regarding sulfur dioxide emissions were already coming into force. The combination of these regulations and studies gave all arsenic producers considerable pause for reflection. In the early 1970's there were large stockpiles of arsenic trioxide and a consumption level that indicated that such stockpiles would continue to grow. By the mid 70's, however arsenic trioxide production had begun to drop appreciably. Some producers simply phased out because of low arsenical values in their intake materials. Others, such as ASARCO restricted their smelter capacities because of SO_2 emissions thus reducing the availability of by-product arsenic. World arsenic production had peaked out in 1970 at about 77,000 tons of arsenic trioxide. By 1973 the production was below 70,000 tons and production has continued to fall since that time with 1980 production estimated at less than 40,000 tons (fig. 1, table 1).

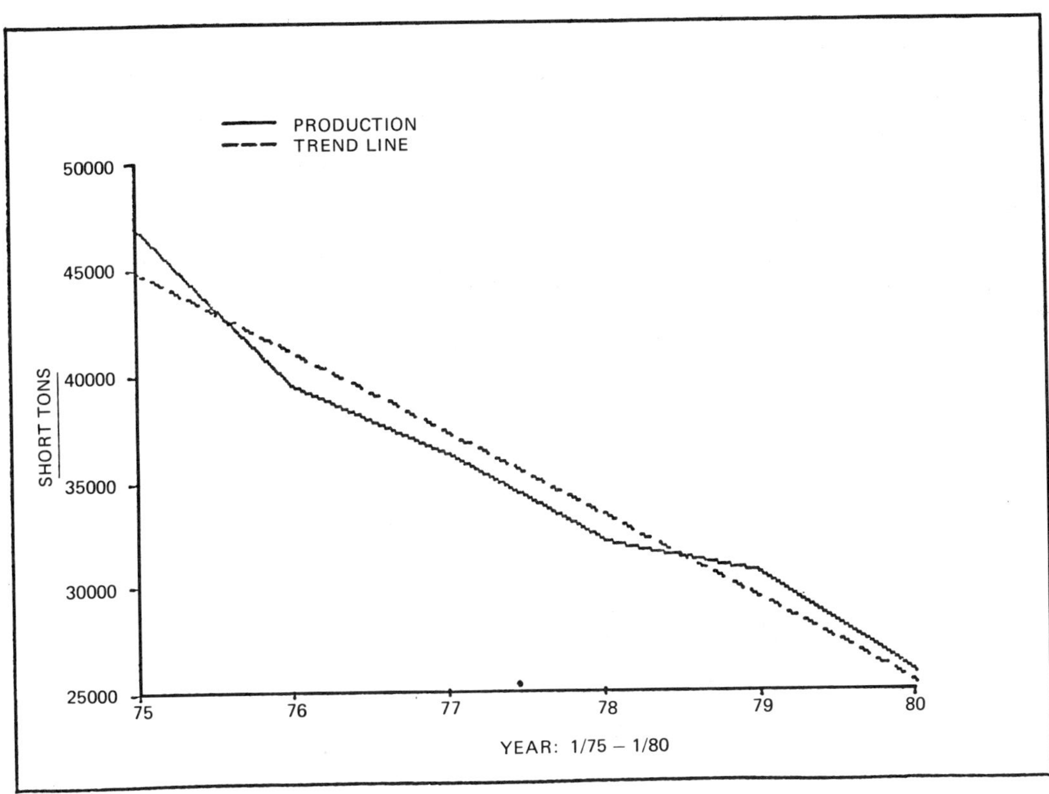

Figure 1. Free world arsenic trioxide production.

Table 1. World supply of arsenic trioxide by smelter location, and country of origin of arsenic values.

Smelter Location	Origin	1970 (Short Term)	1980 (Est.)
Sweden	Sweden	18,100	6,000
U.S.A.	Philippines	9,900	4,000
	Peru	2,800	1,000
	U.S.A./Canada	1,400	--
France	France/Morocco	15,400	5,000
S.W. Africa	S.W. Africa	4,500	1,000
U.S.S.R.	U.S.S.R.	7,900	4,000
Peru (Oroya)	Peru	5,000	2,000
Mexico	Mexico	10,100	4,000
Brazil	Brazil	400	--
Japan	Japan	400	--
Portugal	Portugal	1,000	--
Spain	Spain	200	--
Other	Other	--	2,000
		77,100	29,000

During the entire decade no new arsenic producing facilities were constructed although there were two plants under construction at the end of the decade which should go on stream in 1982.

During the same ten years, the applications consuming arsenic have changed as well. Sodium arsenite and lead arsenite are no longer of commercial importance. Arsenic for the glass industry has largely been replaced and is no longer a major market. Much of the arsenic originally sold for metal alloying has been shifted to arsenic metal rather than the trioxide. It is, however, in the field of agricultural chemicals and wood preservatives that the changes have been the greatest (fig. 2).

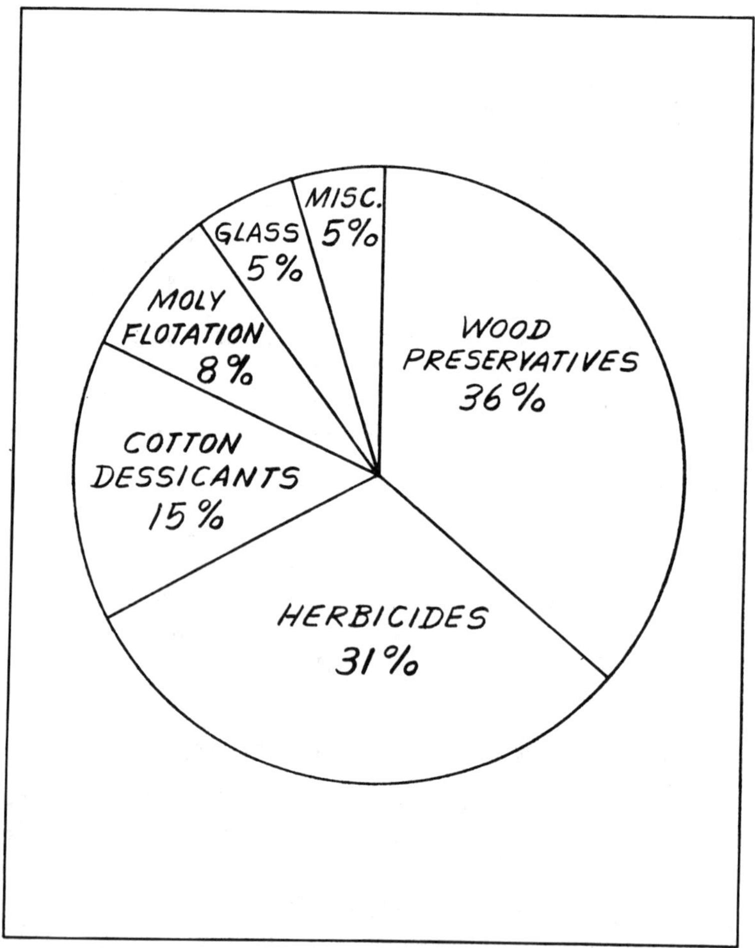

Figure 2. 1981 U.S. markets for As_2O_3 estimated by ASARCO.

Along with environmental attacks on arsenic, came environmental attacks on many organic chemicals which were used for weed control. In the late 1970's 2-4-5-t usage was severely restricted. As a result of this, arsenicals were purchased for the removal of undesirable vegetation on railroad and highway rights of way. At about the same time, the wood preservative markets began to shift drastically. The growth of copper chrome arsenate (CCA) wood preservatives was phenomenal. Forty percent growth rates for this product were common in the last few years of the decade. Now we enter the 1980's with arsenic production down and a large, unsatisfied and growing demand for this product. We currently estimate the supply shortfall to be in the vicinity of 7-10,000 tons per year of arsenic trioxide (fig. 3).

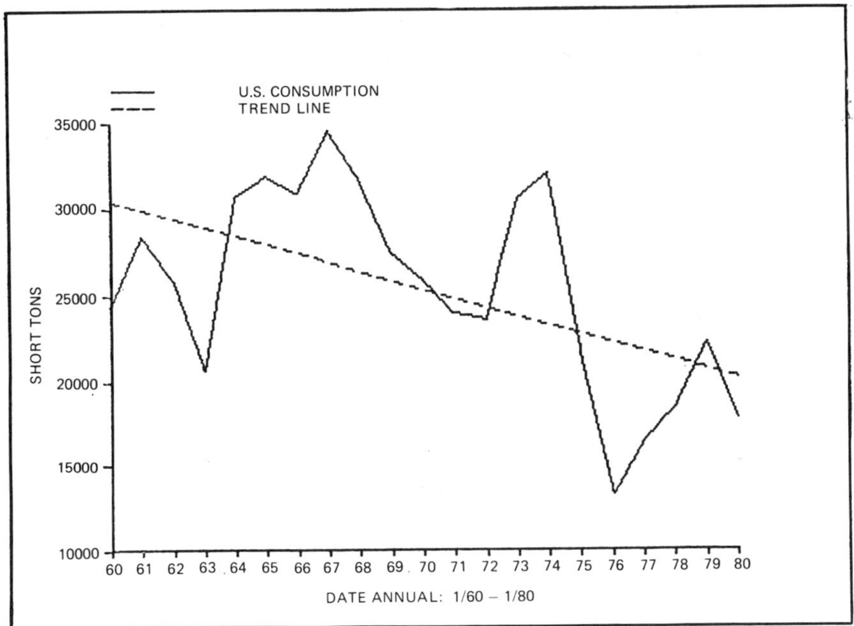

Figure 3. United States apparent consumption of arsenic trioxide.

3. Occurrence, Recovery and Production Arsenic Trioxide

I'm going to briefly review where the arsenic comes from and how it is made since all of this information has been published in OSHA and EPA documents. The principal production of arsenic is from copper ores. This covers the material produced by Sweden and Asarco's Tacoma Plant based on Philippine ores. In addition, Mexican, Peruvian, Russian, Chinese and African production is based on arsenic containing copper or cobalt ores. Appreciable arsenic is found in zinc-lead deposits and in silver, nickel and gold deposits. For the most part, recovery of arsenic trioxide from these latter ores has not been economical and the material has either been stockpiled or processed through an arsenic trioxide production facility tied to a copper or cobalt source.

The arsenic trioxide production begins with the roasting and smelting of the ore concentrate at which time the arsenic trioxide is sublimed and trapped in dust collectors. These arsenical dusts are then further processed either by selective resublimation or by hydrometallurgical leaching followed by crystallization to a marketable arsenic trioxide product. The principal source of environmental arsenic is not as one might suppose at the arsenic plant itself where the dusts are processed to finished material, but in the smelters where the original roasting is done. These facilities are so large and require such a large throughput of material including the air for roasting that is is infeasible to gather in all the potential sources of dust and vapor.

Once the arsenical dusts have been collected from the smelter, they can be handled in a much more tightly sealed system and arsenic evolution and loss is kept to a minimum. Hydrometallurgical processes for producing arsenic trioxide dust used as starting material has a high arsenic content. The volumes of liquid to be moved and the subsequent crystalization costs make the process both capital and energy intensive.

On the other hand these hydrometallurgial facilities produce a very pure arsenic trioxide (99% or better) and this product can, in current markets be sold at a premium to compensate for the extra plant and operating costs.

4. Conclusions

We have attempted to briefly review the production and consumption of arsenic trioxide. This material has gone from a large surplus in the early 1970's to an allocation position in the past 4 years. As an allocated material the price of arsenic trioxide has been rising rapidly but particular applications in herbicides and wood preservatives, have continued to grow in spite of rising costs.

The current shortage appears to be long term and even the advent of newer producers will see little easing of pressure on demand. Current arsenic prices as high as $1.00 to $1.25 per pound on the free market should entice some companies who have been burying or otherwise disposing of arsenical residues to recover arsenic trioxide if it can be done on a short term, economical basis.

DISCUSSION

P. J. Neff: You indicated that you expect this shortage to end. Do you have any idea when that will happen?

L. D. Fitzgerald: I am amazed that the shortage has lasted as long as it has. It has been here now for four years plus. We all basically agree, shortages eventually come to an end. Things must balance in the long term. The position where balance occurs is highly dependent on the relationship of suppliers and consumers. For the last few years, we've had consumers running off selling material like they were going out of business and the suppliers have been hard put to just barely keep up. We do not, in fact, see an instant end to this situation. I think we, as the major by-product suppliers, have a responsibility to maintain applications, because a lost application tends to eliminate a market on a permanent basis. A market just doesn't come and go--a market needs to be developed, it matures, and if that market fails to be supplied or is priced out of existence, it disappears. It is a basic objective of by-product suppliers to have as many viable applications as possible. From a practical point of view, maintaining this balance is especially important, particularly for companies who are talking about small-scale production (secondary level production). Some balance is a requisite because the by-products producers are always going to sell their material. I do not have a projection as to what year or even what decade the shortage may end. It could very well be that we could go into 1990 and I'd stand here and say, "Gee, you know what? There's a 10,000-ton shortage of arsenic." But at the moment, all I can really do is give you the numbers and let you conjure with your own crystal ball.

2

THE HYDROMETALLURGICAL RECOVERY OF ARSENIC FROM SMELTER FLUE DUSTS

Dr. James R. Wolfe
Nedlog Technology Group, Inc.
Rt. 3, Box 668
Laramie, WY 82070

Lead, zinc, and copper ores often contain appreciable quantities of arsenic. During smelting operations, highly volatile arsenic concentrates in the smelter flue dusts and is collected to avoid both polluting the countryside and the loss of valuable metal contents. Often the flue dust is recycled back into the smelting process, but eventually the impurity content (primarily the arsenic) in the recycled dust reaches levels which preclude further recycling. Nedlog has developed two hydrometallurgical processes for recovering the metal content, including the arsenic, from smelter flue dusts and other arsenical materials. The processes are similar in that both require oxidative autoclave leaching using sulfuric acid. They differ in the manner in which the arsenic is treated and eventually recovered. Flow sheets of both processes are discussed and a description of a semi-commercial plant now in operation by Nedlog is described.

1. Introduction

Arsenic occurs in the earth's crust at an average concentration of about two parts per million or 0.0002%. Nature, however, has selectively increased the concentration of arsenic in numerous ore bodies throughout the world to the point that during the winning of the metals of interest, arsenic concentrations on the order of 1% to 10% are not uncommon. The higher arsenic concentrations of 10% or more are usually found in the bag house dusts that are collected from various stages of copper, lead and zinc smelting operations, when concentrates containing appreciable amounts of arsenic are used as a feed. The higher vapor pressure of arsenic and many of its compounds is responsible for this concentration and the vapor loss of arsenic acts as a bleed stream for this impurity.

In addition to arsenic, a typical bag house smelter dust often contains relatively high concentrations of the base metals being smelted and a concentration of minor metals of commercial value, such as indium, germanium, tellurium, bismuth, and occasionally silver and gold. A typical analysis of a lead smelter dust is shown in figure 1. A similar analysis for a typical copper smelter dust is shown in figure 2. It is the obviously high metal content and the large stockpiles of accumulated

Metallic Element	Concentration % or (troy oz/ton)
Lead	49.4
Zinc	8.5
Arsenic	10.7
Antimony	4.0
Cadmium	1.0
Copper	0.2
Indium	?
Silver	(3)

Figure 1. Metal content in a typical lead smelter flue dust

Metallic Element	Concentration % or (troy oz/ton)
Copper	15.7
Lead	10.0
Zinc	2.4
Arsenic	10.5
Iron	5.1
Bismuth	1.5
Tin	1.0
Cadmium	0.2
Germanium	0.03
Silver	(6)
Gold	(0.1)

Figure 2. Metal content in a typical copper smelter flue dust

dust in the western U. S. that first attracted Nedlog's attention to the possibilities of economic metal recovery. For the past several years Nedlog has devoted an appreciable portion of its resources to developing technology for the recovery of metal values from arsenical smelter flue dusts. The recovery of arsenic in marketable form has been a key element in Nedlog's technical and commercial development.

This paper is divided into two sections: Namely, (1) the recovery of sodium arsenate from copper smelter dusts and (2) the recovery of arsenic acid from lead smelter dusts.

2. The Recovery of Sodium Arsenate From Copper Smelter Dusts

This process was developed by scientists at Hazen Research, Inc. in Golden, Colorado under contract to Nedlog Technology. The basic technology is covered by two patents issued during 1981. [1,2][1] This process has received extensive bench scale technical development and preliminary economics for a 60 ton-per-day plant appear favorable. Nedlog has scheduled pilot plant tests for 1982.

---

[1]Figures in brackets indicate the literature references at the end of this paper.

The process consists of three separate selective leaches in which the various metal values are removed and recovered. The key to the process is the relatively high iron content in the copper smelter dusts (figure 2) which allows the arsenic to be fixed early in the process as ferric arsenate and prevents leaching during the recovery of the metals of interest.

Figure 3 is a schematic of the copper smelter flue dust recovery process. The material is transported in steel containers and slurried in ten-ton batches. The slurried flue dust is then batch leached in a pressure autoclave, using sulfuric acid and an oxygen or air overpressure. A leach time of three hours at 100°C and 50 psi is sufficient to solubilize the copper and several other metals of interest. The leach slurry is filtered with the leach liquors going to metal recovery operations and the filter cake containing the arsenic and several metals, insoluble in H_2SO_4, going to the next leach step. The second leach is a hot brine leach that leads to the recovery of lead and the precious metal content. The ferric arsenate remains insoluble and proceeds to the final leach step.

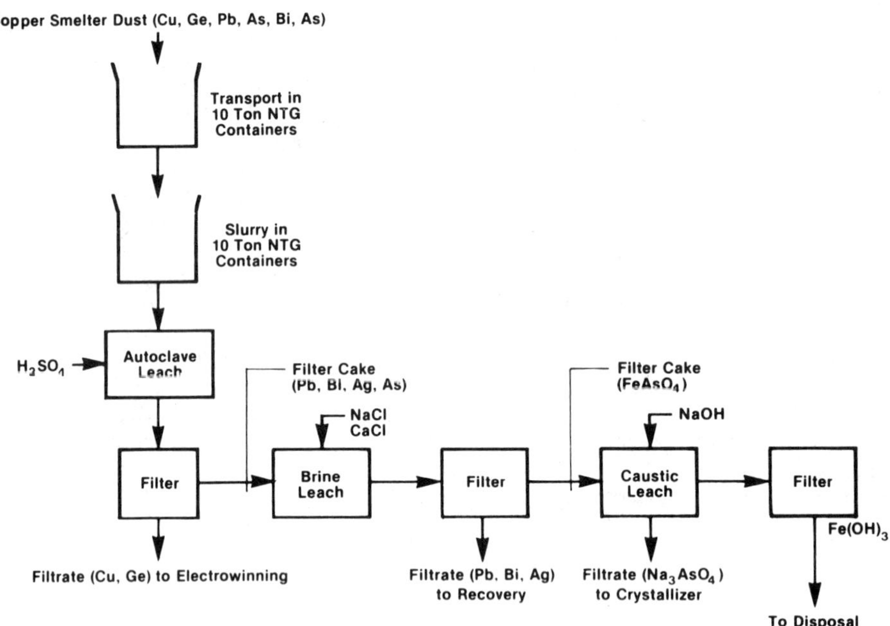

Figure 3. Copper smelter dust metal recovery flowsheet

Recovery of the arsenic is made by leaching the ferric arsenate in a hot caustic solution according to the reaction:

$$3\ NaOH + FeAsO_4 = Na_3AsO_4 + Fe(OH)_3$$

A detailed schematic of this process is shown in figure 4. The final product is a high-purity arsenate in dry crystalline form. The $Fe(OH)_3$ residue has a small amount of unreacted $FeAsO_4$. Leachate tests have shown that this material is extremely stable and can be landfilled under existing EPA regulations.

3. The Recovery of Arsenic Acid From Lead Smelter Dusts

This process was developed jointly by scientists at Hazen Research, Inc., and Nedlog Technology. The process is covered by two patent applications that have not yet been acted on by the U. S. Patent Office. Therefore, certain proprietary aspects of the technology will not be discussed in detail. The process has recently been put into commercial application at Nedlog's Laramie, Wyoming plant.

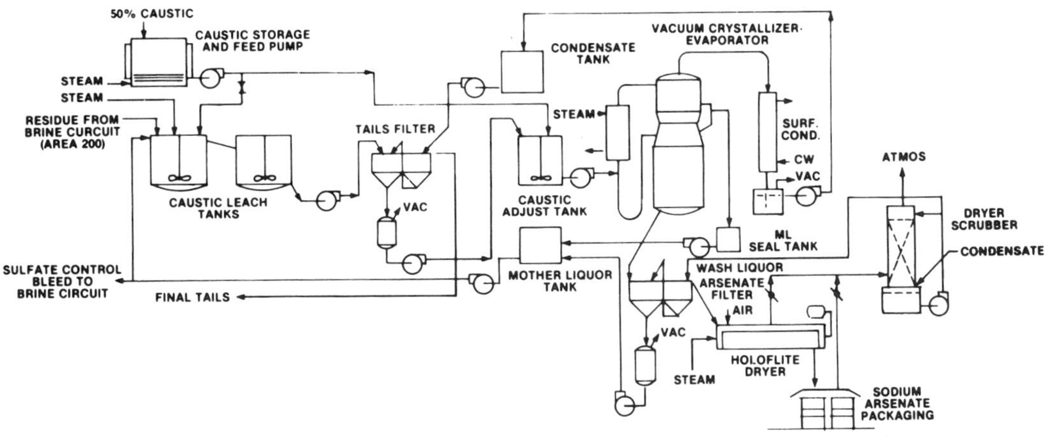

Figure 4. Copper smelter dust arsenic recovery circuit schematic

Lead smelter dust contains about the same concentrations (10%) of arsenic as the copper smelter dust, but the main recoverable metals are quite different as can be seen by comparing figures 1 and 2. The primary leach is again a sulfuric acid pressure leach, under nominally more agressive conditions. Leach conditions are four hours at 150°C and 200 psi under oxygen. The lack of iron in the system allows 90% of the arsenic to solubilize and report to the leach liquor. The filter cake is essentially pure $PbSO_4$, which can be recycled to the smelter as a relatively high grade lead concentrate.

The leach liquor contains several metals that can be recovered economically in addition to the arsenic. The arsenic in the original flue dust was primarily in its trivalent state as As_2O_3. The oxidation leach converts the trivalent arsenic to pentavalent arsenic and makes direct recovery of arsenic acid possible. Nedlog has developed a process for recovering the pentavalent arsenic utilizing solvent extraction (liquid/liquid) technology. Nedlog utilized pulse columns for both extraction and subsequent stripping. The advantage of pulsed columns is that numerous stages of extraction and stripping can be accomodated in a relatively small area. The arsenic is stripped from the loaded organic as arsenic acid, which is fed to an evaporator for final preparation of 75% arsenic acid. The process flowsheet is presented in figure 5.

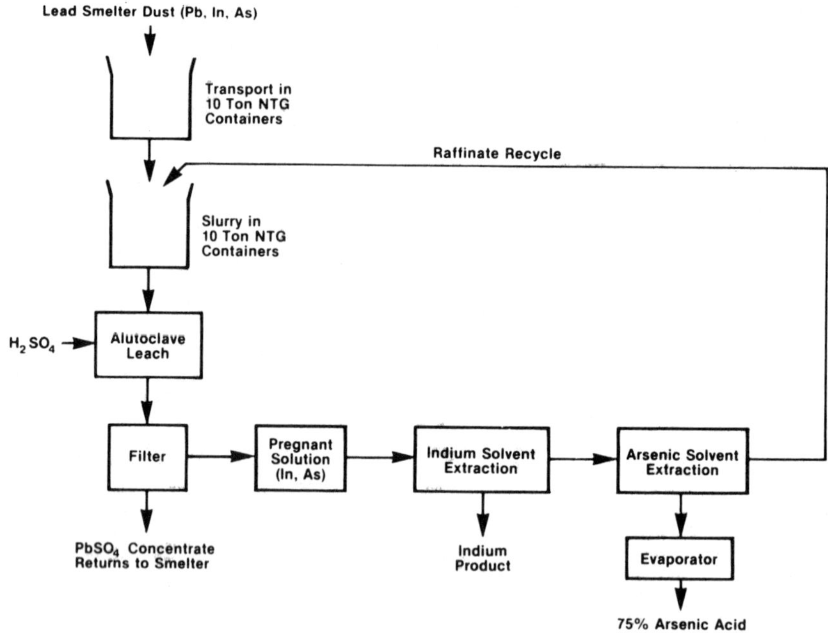

Figure 5. Lead smelter dust metal recovery flowsheet

4. Summary

Nedlog has developed two processes for the recovery of arsenic products from smelter flue dusts. In one case (copper dusts) the arsenic is fixed early in the process and finally recovered after all other metals of interest have been recovered. In the other case (lead dusts), the arsenic is leached during the first step and recovered as arsenic acid by solvent extraction techniques.

5. References

[1] Reynolds, J. E. and Coltrinari, E. L., Process for recovering metal values from materials containing arsenic, United States Patent No. 4,244,734, (Jan. 13, 1981)

[2] Reynolds, J. E. and Coltrinari, E. L., Process for recovering arsenic compounds by sodium hydroxide leaching, United States Patent No. 4,244,927, (Jan. 13, 1981)

DISCUSSION

E. A. Woolson: In your copper smelter flue dust, I noticed there was no antimony, whereas antimony was present in the lead smelter flue dust. Is that true in all cases or just in the facts you happen to have?

J. R. Wolfe: No, there is antimony in that material, too. I just didn't put in on the list because it was not of commercial interest.

E. A. Woolson: Do you know the approximate percentage?

J. R. Wolfe: I think it's on the order of one percent. Antimony will follow along in those circuits just as well. In fact, you practically have the whole periodic table in these materials and I just cut it off at some level.

E. A. Woolson: I raise this point because antimony trioxide has been shown to cause cancer in animals.

3

ARSENIC IN ENERGY SOURCES:

A Future Supply or an Environmental Problem?

R. A. Schraufnagel
ARCO Coal Company
Research and Development
P. O. Box 2819
Dallas, Texas 75221

Today, essentially all the domestic arsenic production is from the processing of flue dust associated with copper smelting. However, with increased environmental control devices on all types of processes and the advent of new processing technologies, an expanded supply of arsenical residues is anticipated in the future. Current and expected future technology was surveyed to determine the occurrence of arsenic in sources of energy, the possible production of arsenic from these sources and the environmental impacts associated with the processing of these sources. These sources include petroleum, geothermal sources, coal, and shale oil production.

There are only trace amounts of arsenic in crude oils. Arsenic precipitates along with a number of other heavy metals from geothermal sources. If economic separations exist, this would be a small source of arsenic. Full development of geothermal sources will necessitate environmentally sound practices for disposal of precipitates.

Both inorganic and organic forms of arsenic are present in coal and oil shale. Thermal processing causes the organic arsenic compounds and some inorganic arsenic compounds to be distributed throughout the boiling point range in liquid products from shale retorting. In coal inorganic arsenic is primarily in the ash and "heavy ends." Significant arsenic removal from coal can be achieved either through stack gas scrubbing or deashing prior to processing.

Major emphasis in this paper is the arsenic in shale oil production. Removal is necessary because arsenic rapidly deactivates the catalysts used to upgrade the raw shale oil to a syncrude product. Proprietary technology concentrates the arsenic on a guard bed material for disposal or sale to produce arsenic and provide a byproduct credit. Proper disposal requires that future leaching be minimized. This technology is also available.

1. Introduction

Essentially all the domestic arsenic production results as a by-product of copper smelting. Domestic demand is met by this source and by importation (primarily as arsenic trioxide). United States demand has ranged between 20-25,000 tons per year in the early 70's while supplies over the same period have ranged between 25 and 35,000 tons [1][1]. The fluxuating supply is based primarily on the economics of copper processing and the fluxuating demand primarily on the need for herbicides and pesticides [2]. The domestic outlook for the year 2000 is on the order of 26,200 tons but could be affected by the stringency of new environmental and safety standards [1,2]. Though the use of arsenic is not limited by supply, arsenic is considered an inelastic commodity because its availability depends on the demands of lead and copper ore [2].

Advances in smelting, especially those for minimizing air pollution, are expected to increase the capability for arsenic recovery. Increased environmental control devices on other types of processes and new processing technologies, such as the development of a domestic shale oil industry, also expect to expand the supply of arsenic.

In this paper the recent literature is surveyed to determine if current and expected future energy technology will be important to supplying future arsenic needs. Energy sources surveyed include conventional petroleum production, geothermal sources, coal, and shale oil production. The occurrence of arsenic in these sources and process streams associated with these sources, possible production of arsenic from these sources and environmental impacts associated with arsenic in these sources is discussed. More detailed information in all of these areas can be obtained from the numerous references which are cited.

2. Natural Occurrence of Arsenic

Arsenic is ubiquitous. It is found in trace quantities in all living things, the atmosphere, water and geologic formations. The crustal average arsenic content is 1.8 ppm. Typical average concentrations of arsenic in geologic formations and soils are shown in table 1 [3,4,5]. Highest concentrations are found in shales and soils formed from carbonate residue. These higher concentrations may be due to the fact that these materials originated from microscopic plant or animal organisms that biologically concentrated trace metals. Little difference is noted between concentrations observed in cultivated and uncultivated soils despite the common usage of arsenic containing herbicides.

The average occurrence of arsenic in sea water is 0.003 ppm [3,5]. Except near mineral wells where concentrations in excess of 4 ppm are common, arsenic in groundwater is generally much less than .01 ppm [6].

---

[1]Figures in brackets indicate literature references at the end of this paper.

The mean concentration of arsenic in fresh water of the United States is 0.064 ppm with a range of 0.005 to .336 ppm [6,7]. Arsenic in air is usually less than 0.04 μg/m^3 [7]. Though coal burning can contribute to some atmospheric arsenic, significant quantities result from crustal weathering and blowing soil [8].

Table 1

Concentration of Arsenic in Geologic Formations and Soil

Formation	Concentration ppm	
Geologic Formations	Mean	Range
Granite	.8	<1-19
Basalt	2.4	-
Shales	13	1.4-27
Sandstone	1	<1-25
Carbonates	1	-
Soils		
U. S. Average	5	1-30
Carbonate Residuum	15	3.7-42
Cultivated	5.5	1.8-27
Uncultivated	6.7	2.7-20

Arsenic is found in varying concentrations in plants and animals ranging from several tenths of a ppm in land plants and mammals to in excess of 30 ppm in brown algae and shrimp [7]. Plants and animals will biologically concentrate arsenic. Concentration factors range from 40-333 for fresh water animals, 3,300 for sea animals, 1,670-3,300 for fresh water plants and 6,000 for brown algae.

The typical human ingests between 29 and 169 μg of arsenic per day [9]. Two recent studies indicated that trace amounts of arsenic are nutritionally beneficial [10,11].

Now that we have some knowledge of normal natural background levels of arsenic in our environment we can access the significance of arsenic in energy sources. Let's start by briefly examining the occurrence of arsenic in crude and refined petroleum, determine if enough arsenic is present to be considered a future source and access the environmental impact of arsenic in this source.

3. Arsenic in Petroleum

Arsenic is a minor constituent of crude petroleum [12]. Table 2 shows typical trace metal concentrations in crude oil. Studies of crudes that we normally process have shown that arsenic is present in the parts per billion range. Arsenic in crude oils and refined products can be detected using neutron activation analysis. Concentrations averaging .263 ppm have been reported using this method. Though reported arsenic concentrations will vary with type of crude oil and possibly with the testing procedure employed the reported values are always low when compared to iron, nickel vanadium and the other trace metals normally present [13,14].

Though the arsenic compounds present are oil soluble, approximately one-third can be extracted when mixed with distilled water [15]. Arsenic is present in all fractions when crude oil is separated into a methanol soluble fraction, resins, and asphaltenes; however, the highest concentration, 2.25 ppm, is found in the asphaltenes. Within the methanol soluble and asphaltene fractions arsenic concentration increases with increased molecular weight [15].

Table 2
Metal Content of Crude Oils

Metal	Average Concentration ppm
Arsenic	.015
Cobalt	1.71
Copper	1.32
Iron	40.67
Manganese	1.17
Nickel	16.58
Selenium	.53
Vanadium	88.55
Zinc	29.80

A study of trace metal emissions from a number of oil fired power plants was recently completed. Arsenic was always found to be below the detection limit in the particulate emissions [16]. Environmental hazards due to arsenic or other trace metals in petroleum are not anticipated. If necessary residual wastes can be economically disposed of in a secured landfill [17].

In summary, since arsenic is present in crude oil and refined fractions in very small quantities it will not be a potential source of the element nor cause undue environmental problems. However, more stringent rules regarding the disposal of heavy metal wastes could increase refining costs.

4. Arsenic in Geothermal Sources

Geothermal source is a general term used to include four types of systems: vapor dominated, hot-water, hot dry rock, and geopressured. To date processing has centered around vapor dominated and hot-water systems.

4.1. Occurrence of Arsenic

Arsenic is present in geothermal brines in varying concentrations and is usually mixed with a number of other metals and minerals, including gold, silver, bismuth, antimony, quartz, calcite, and sulfides of iron, copper, lead, and zinc [5]. This is because geothermal sources commonly occur near fractures surrounding hydrothermal veins. An example of a geothermal brine is shown in table 3 [18].

Table 3

Chemical Analysis of a Geothermal Brine

Species	Concentration (Mg/l)
SiO_2 (total)	518
SiO_2 (dissolved)	100
Cl^-	11,000
$SO_4^=$	32
F	2
$CO_3^=$	0
HCO_3	17
K^+	890
Ca^{++}	510
Na^+	6,400
Mg^{++}	.4
As^{+6} ?	11
B	125

Arsenic is present in trace quantities in almost all geothermal brines. Though removal may be required for environmental reasons, this is not expected to be a future service of arsenic (in part, due to the limited future production). However, processing of arsenic from hydrothermal veins surrounding geothermal deposits might occur separately if the economics are favorable and concentrated arsenic sources are found.

There are a number of recent studies on chemical species present in brine, including studies of the Imperial Valley in California, Klamath Falls, Oregon, Yellowstone Park and the State of Wyoming [19,20,21,22]. Arsenic can be aerially distributed around centers of geothermal systems making it useful in exploration and drilling of new resources [23].

4.2. Environmental Affects of Arsenic

Recently a detailed report was published which surveyed the environmental regulations applying to the geothermal resource [24].

Since the geothermal resource is new, specific source performance standards have not been promulgated as yet. Some feel that in areas such as The Geysers in California strict interpretation of existing environmental law and promulgation of new laws will severely limit the ultimate development of this resource [25].

The major impact that arsenic may have on the environment will be related to water quality and brine disposal. Elevated levels of arsenic in the Madison River System of Wyoming is attributed to geothermal source waters in Yellowstone Park [26]. Seventy-five percent of the arsenic in Lake Ohakuri in New Zealand is due to discharge of spent water from the Wairakei geothermal plant [27]. To minimize environmental effects reinjection of wastes was proposed though this method may ultimately reduce the efficiency of the heat recovery.

4.3 Recovery of Arsenic from Waste Streams

Recent literature discussing methods of removing arsenic from water are too numerous to mention. A smaller but significant number of articles specifically address removal of arsenic from geothermal sources. These include several Japanese patents [28, 29]. The desalination of geothermal fluids has been demonstrated on a pilot scale [30]. Though methods are available arsenic removal from brines is not cheap. A recent study showed that a process based on lime-soda addition could effectively remove arsenic but would require a capital expenditure in excess of $1.5 million dollars. This is based on 1976 costs for a 4.4 million gallons per day plant to be added to an already operating facility. Operating costs were estimated to be $2.29/1,000 gallons [18].

4.4 Geothermal Brines as a Future Arsenic Source

Though arsenic removal processes are discussed in the literature, the removal is to satisfy environmental constraints and not to profitably extract arsenic. A review of the literature already cited, the transaction of the Geothermal Resources Council for 1977, 1978, and 1979, and the proceedings of "The Second United Nations Symposium on the Development and Use of Geochemical Resource" in 1975 shows that there is no interest expressed by the geothermal community to extract minerals from brines.

5. Arsenic in Coal

5.1. Occurrence

Coal has always been known to contain a number of trace elements. Arsenic content can vary with the coal type, sulfur and ash content and location of the reserve. Texas lignites contain between 1-5.5 ppm arsenic [31]. Some Montana coals contain between 4 and 5 ppm arsenic, bituminous coals may have 3.0 to 20 ppm arsenic [32, 33]. Concentrations from other seams are also cited in the recent literature [34-36]. A high sulfur Czechoslovakian brown coal is known to contain arsenic levels as high as 330-460 ppm [37]. The concentration of arsenic in American coals is shown by region in table 4 [7].

Table 4

Concentration of Arsenic in American Coal

	As, ppm
Powder River Basin	3
Western Interior	16
Eastern Interior	14
Appalachian Region	18

Despite the varation of arsenic content of coals, the National Bureau of Standards laboratory has chosen a standard coal and fly ash for testing purposes. This is shown in table 5. Arsenic is 5 ppm in the coal and concentrated to 62 ppm fly ash and is representative of an "average" coal that is burned in power plants [38].

Table 5

Occurrence of Trace Metals in Coal and Fly Ash

	Concentration ppm	
Element	NBS Coal	NBS Fly Ash
Manganese	16	400
Nickel	14	98
Copper	16	106
Zinc	18	185
Arsenic	5	62
Selenium	2.6	8
Rubidium	19	130
Strontium	152	1,490
Ytterbium	7.5	66
Molybdenum	1.7	23
Lead	6	66

It is not certain why metals are present in the coal formations. Possible sources of minerals in coal ash are [5]:

1) accumulation by plants during growth;
2) transportation into the coal swamp as a component of inorganic sediments;
3) absorption in the formation during or following coalification.

All the mechanisms could play a role in the presence of metals though experts tend to favor the last [5].

Recent coal cleaning studies shed additional light on the source of trace metals. Arsenic is readily removed by float sink methods.

There is a strong correlation between arsenic and the pyritic sulfur and iron content as well as a weaker correlation between arsenic and total sulfur in the coal [33, 39]. The majority of the arsenic in the coal is inorganic in nature.

5.2. Coal Processing--Combustion

The most common coal process is combustion. The greatest concern related to arsenic in the combustion process is its transport in fly ash.

A number of studies have been done to characterize the fly ash from power plants, compare alternative removal techniques and optimize the pollution control and energy conversion efficiency of power plants. Arsenic tends to be preferentially concentrated in the fly ash when compared to the original coal or bottom ash [40]. When fly ash fractions are broken down by particle size, enrichment of As increases with decreasing particle size in the 1-10μ size range and becomes independent of particle size for submicron particles [41, 42]. Arsenic is also associated with particles less than 0.4μ in diameter [43]. Though arsenic can form volatile compounds at high temperature, detailed material balance studies have shown that more than 95 percent of the arsenic is accounted for when the particles of flue gas are collected [44-46].

Two common methods of removing particulates from stack gases are the venturi scrubber and electrostatic precipitator. Several recent articles compare the efficiency of removing particulates and trace metals with these two methods [47-50]. Venturi scrubbers are much more efficient at removing total particulates but electrostatic precipitators are more efficient in removal of the small particles containing trace elements.

A detailed study of trace metals in the soil and plants surrounding the Four Corners power plant in New Mexico was recently completed [40]. The conventional particulate removal technology employed at the site was adequate in removing trace metals including arsenic. Trace metals were not elevated above normal concentrations even in vegetation grown within 3 km of the plant.

5.3. Environmental Impacts of Trace Metals in Combustion

When fly ash is removed from a stack with a scrubber, it becomes a solid waste. The most common technique of disposing of fly ash is in settling ponds. Proper disposal requires knowledge of the leachability of the sludge and possible interaction with groundwater [51-53]. Ash can be landfilled in an environmentally acceptable manner but monitoring of adjacent water sources is required [54]. Disposal methods should be studied on an individual basis. In the case of one power plant, it was determined that plume dispersion was satisfactory. A few more particulates in the stack gas had less of an environmental impact than ponding the fly ash and was the recommended control measure [55].

A number of choices are available in controlling trace metal emmissions from coal combustion sources. One needs to be aware not only of the atmospheric impacts of fly ash but the impacts associated with disposal of the recovered solid waste. In a number of instances it appears that methods used to control total particulate are adequate in controlling arsenic and other concentrations of trace metal. Precleaning the coal using conventional heavy media (float/sink) processing shows promise in removing arsenic prior to combustion [39].

Optimizing the energy conversion efficiency and pollution control efficiency of a given procss is complex. A recent paper outlines the required steps to consider in this optimization [56]. The interpretation of existing environmental laws and the passage of new laws can have a significant impact on this optimization. One needs to be sure that environmental laws 1) realistically assess the risk, 2) are not so rigid that they eliminate possible control options, 3) do not overly constrain the system by over controlling one stream, and 4) are cost effective. In particular, laws regarding the handling of solid wastes from power plants will have to be carefully formulated to allow for optimization of combustion processes.

5.4. Other Coal Processes

Other coal processes that may be impacted by trace metals are coal liquefaction, coal gasification, and coal storage. Information is available on the distribution of arsenic in the SRC I and SRC II solvent refined coal liquefaction process [57, 58]. The feed coal contains 18 ppm arsenic. Arsenic is undetectable in the process solvent and naphtha. The coal liquid contained 2 ppm arsenic and the mineral residue 77 ppm arsenic. Even considering the high concentrations of As in the process water which is later treated, closure on the arsenic balance is not complete. It is believed that some of the arsenic is volatilized in the off gases but this was never measured.

Coal gasification can be done aboveground or in situ. Arsenic can be present in the produced particulates, tar, unreacted char, or process water. Gasifier char is the major repository of trace elements though arsenic has been found in coal tars [59, 60]. However, leachates of the chars contain only low concentrations of toxic metals [61]. Similar conclusions are noted for in situ coal gasification tests. Arsenic, due to particulates in the product gas stream, ranged in concentration from 6.6-88 ppb (V/v) at the Hanna test [62]. The concentration of arsenic on the tars and particulates of the product stream at Hanna IV and Hoe Creek II tests done on two types of Wyoming coal varied from 0.008 to .3 weight percent [63]. The wide variability is due in part to a very small sample and may not be statistically significant. Tests on the groundwater several years after completion of the test showed a minimal change in baseline levels of arsenic [64].

There is a possibility that trace elements can be leached from coal storage piles. A number of recent studies have been done relating to this problem [65, 66]. Though some elevation in metals concentration has been noted, this does not appear to be a problem. One would have to contend with increased acidity and total solids content of the water before contamination with trace metals became important.

5.5. Coal as a Potential Source of Arsenic

Over 400 million metric tons of coal is annually shipped to utilities in the United States. If this coal contained on the average 10 ppm of arsenic, 4,400 tons of arsenic or about 20 percent of current United States demand would be processed in coal combustion. However, the arsenic is always present in very low concentrations in these processing streams. It seldom exceeds 100 ppm in the most concentrated of fly ash. Therefore, it is not economical to recover arsenic from today's coal processing.

Despite a very intense search of the recent literature, no articles discussing the use of coal processing streams as a potential source of arsenic were found. Though no estimates can be made at this time, coal might supply some arsenic in the future. As pointed out earlier, the burning of coal concentrates significant amounts of trace metals in the small particle size fly ash. With further study it may be possible to collect and more completely separate and concentrate this fraction.

6. Arsenic in Shale Oil Production

There is in excess of 300 billion barrels of shale oil reserves in the western United States. Commercializaton of this resource is just starting to begin with an estimated 200,000 bbl/day production by 1990. By 2000 production could be 800,000 bbl/day and be in the 1 to 2.5 million bbl/day range by 2025 [67].

6.1. Occurrence

As was presented earlier, rocks geologically classified as shales have an arsenic content of 13 ppm. On the other hand oil shale, which geologically is actually a kerogen containing marlstone, contains arsenic in concentrations more than ten times the earth's crustal concentration. Shale was formed over 50 million years ago with the accumulation of dying microscopic plant and animal material at the bottom of a huge fresh water lake. Besides supplying a source of organic material, these organisms likely also acted biologically to concentrate trace metals much the way shrimp or algae do today. Table 6 shows a typical trace metal analysis of raw shale [38]. The trace metal concentration can be a function of the oil in the shale rock. One 15 gal/ton assay shale contained 81 ppm arsenic and a 40 gal/ton assay sample contained 108 ppm [58].

Table 6

Typical Trace Element Analysis of Raw Shale Rock

Element	Concentration ppm
Fluorine	1070
Manganese	196
Nickel	24
Copper	44
Zinc	70
Arsenic	70
Selenium	3.2
Rubidium	60
Strontium	580
Molybdenium	27
Lead	27
Uranium	5.4

There are two methods of procesing the shale, in situ or in an aboveground retort. Simulated in situ retort work done by Lawrence Livermore National Laboratory and Laramie Energy Technical Center indicate that the arsenic ranges from 28 to 51 ppm in the raw shale, 26 to 76 ppm in the spent oil shale, 2.5 to 7 ppm in the oils, and .23 to 6.5 ppm in whole retort waters. About 75 percent of the arsenic present in retort water is particulate in nature [58].

Arsenic is associated with pyrite in raw shale. However, there is also reason to believe some arsenic is present as an organometallic compound [68]. There has been difficulty in obtaining closure of the arsenic material balance which can cause variability of arsenic cited in various literature articles [68]. In particular, it was found that unstable organic arsenic compounds in shale oils need to be properly handled to assure accurate analysis [69]. Careful analytical techniques have shown that total arsenic present in TOSCO II shale oil is 40 ppm, somewhat higher than reported for the simulated in situ retort [70].

Table 7 shows the arsenic distribution in a raw shale oil [69]. The gas oil fraction contains the highest average arsenic concentration, about 52 ppm. A substantial arsenic peak occurs in the 204-260°C (400-500°F) boiling range. This may be due in part to the presence of arsenic trioxide. Following this peak, the arsenic concentration decreases. A signficiant arsenic level is observed in the residuum.

Table 7

Arsenic Distribution in Shale Oil

Boiling Range	Fraction Volume Percent	Arsenic Content ppm
*IBP-204°C (IBP-400°F)	18	10
204-482°C (400-900°F)	58	52
>482°C (>900°F)	24	38

*Initial Boiling Point

6.2. Removal from Process Streams

Arsenic has been shown to rapidly and permanently deactivate commercial hydrotreating catalysts. Even where hydrotreating is not required, arsenic removal may be desirable. In the recent years a number of patents have been issued for the removal of arsenic from shale oil products [71-84]. Total arsenic removal varies from process to process, final arsenic concentration may be 10-15 ppm, 1-5 ppm, or even less [71-76]. One especially promising removal process involves the use of one or more guard beds which removes arsenic prior to hydrotreating [72,74,75,78,79]. The guard bed contains a proprietary catalyst that contains a metal oxide such as Fe_2O_3. After subjecting the oil to temperatures in excess of 300°F and pressures in excess of 500 psig in the presence of hydrogen, the arsenic is deposited on the catalyst [72]. The manner in which the contaminants are removed from the oil is not entirely clear but could involve the breakdown of the organo-arsenic compounds by the catalyst. Analysis of the catalyst shows the presence of iron arsenide compounds [74]. The process has been demonstrated to effectively remove arsenic after more than 1,000 hours of operation [69].

6.3. Potential Source of Arsenic

One attractive benefit of the process, especially if one were considering using this catalyst material as a source of arsenic, is that the guard bed will contain in excess of 8 weight percent arsenic before new catalyst is required [69]. Depending on process conditions, the catalyst might even contain 20-30 weight percent arsenic before it becomes unsatisfactory for dearsenation of the raw oil [79]. When compared to naturally occurring arsenic ore or the 17.5 pounds of arsenic per ton of copper commonly found in present day smelters, this is indeed a concentrated source.

When the catalyst can no longer remove enough arsenic from the oil, there are three things that could be done: 1) discard the catalyst in an environmentally accceptable manner, 2) sell the catalyst to an arsenic smelter, or 3) regenerate the catalyst and sell the arsenic as a by-product. It would be preferred to sell the arsenic-rich catalyst. If a market is not found before the first shale plants are operating, the spent shale catalyst will have to be disposed of properly. The patent literature also indicates this catalyst could be regenerated to remove the arsenic [79]. In this regeneration process, arsenic is removed as a sulfur compound but in a second step could be oxidized to arsenic trioxide. The patent indicates that the process is more efficient in removing arsenic from the guard bed than conventional roasting.

Since it appears that the guard bed material could be a source of arsenic as a by-product of shale oil production, let us now estimate what quantity of arsenic might be available from this source. As a basis let's assume that 40 ppm of arsenic is removed from a 50,000 bbl/day raw shale oil stream. This represents 609 lbs/day or 111 tons/year of arsenic from a single shale plant. For a projected one million bbl/day production in the early 21st century, this would amount to 2,222 tons/year of arsenic. This represents about 10 percent of the projected arsenic demand and could be a significant future source. Since arsenic would be a by-product of shale production, the guard bed material or arsenic compound could be provided at a low cost.

6.4. Environmental Concerns of Shale Oil Production

Environmental emphasis in shale oil production has centered around the disposal of processed shale and, in the case of in situ retorts, the migration of products in groundwater after the oil has been extracted.

Shale oil production will generate large volumes of processed shale. A 50,000 bpd plant will produce 73,000 tons/day of processed shale. When this is scaled up to a million bbl/day industry, one year of operation would deposit processed shale containing 16,000 tons of As [85]. Unfortunately, the concentration of arsenic, though greater than ten times the crustal average, is uneconomical to recover.

Surface disposal of processed shale in canyons is an approved disposal method. The processed shale is compacted, watered, and fertilized. After a number of man years of research specific plant species have been selected to reseed the processed shale. After about three years of cultivation, the propagation of the plant species becomes self-sustaining.

The disposed process shale contains about 80 ppm of arsenic. Studies have shown that some of this arsenic is leachable with water concentrations at 0.04-0.05 mg/1 arsenic being fairly common [86]. To minimize any adverse impact of water runoff increasing arsenic concentrations downstream of the plant, runoff is collected and recycled at the plant site.

Consideration is also being given to return some of the procesed material back to the mine. Because of a volume increase on processing, not all the material could be returned to the mine. Before this method is viable a method of strengthing the material to allow further mining of the unprocessed shale pillars and examination of long term leaching of metals needs to be examined. The recent literature shows a number of concepts to minimize leaching of arsenic from solid wastes [87-93]. Some of the techniques use cement to bind the trace metals in place and may prove beneficial in this application.

7. Environmental Regulations Concerning Arsenic In Energy Sources

In order to better access possible impacts of more stringent environmental regulations let us first examine current regulations concerning arsenic in energy sources. Though it is not the purpose of the paper to present specific costs of increased control, some general statements concerning the costs and benefit balance associated with various levels of regulations will be briefly discussed.

7.1. Current Regulations

Current environmental rules pertaining to arsenic are summarized with the understanding that additional regulations are possible. Originally the Occupational Safety and Health Administration (OSHA) standard for workplace exposure to arsenic dusts, fumes or mists during an eight-hour shift was 0.5 mg/m^3 and maximum permitted exposure for arsine gas was 0.05 ppm. As of August 1, 1978 OSHA's permanent standard was reduced to 10 $\mu g/m^3$ [2].

Environmental Protection Agency regulations cover water, air, and solid wastes [94]. National Interim drinking water regulations require As^{+3} be less than 50 μg/l. Ambient water quality standards for the protection of wildlife require As^{+3} be less than 440 μg/l in fresh water and 506 μg/l in sea water. Inorganic arsenic is considered a hazardous waste, but there is no ambient air standard or current regulation pertinent to its presence in energy sources. Regulations pertaining to smelters do exist, however. The Clean Air Act defines lead and particulates as hazardous air pollutants and these regulations, especially those pertaining to particulates, will provide some degree of control over atmospheric emissions of arsenic.

The greatest uncertainity in federal regulations is in the disposal of solids wastes. There are specific regulations regarding the disposal of arsenic acid, arsenic pentoxide, and arsenic trioxide. Wastes which produce leachate at pH 5 containing more than 5 mg/l arsenic are considered hazardous. Clarification of rules regarding disposal of power plant wastes, solids from stack gas scrubbers, or other wastes containing arsenic mixed with other metals is still being considered [94].

The Resource Conservation and Recovery Act of 1976 (RCRA) will have a wide-ranging impact on the handling of solid wastes from energy sources in the following ways [7]:

1. possible occurrence of time delays in bringing new facilities on stream;

2. need to characterize in detail the solid wastes from all processes;
3. require the demonstration of pollution control technology on new processes so their feasibility can be determined;
4. require the recycling and reprocessing of more of a resource.

If RCRA is implemented in a reasonable manner, it could be very beneficial. However, care must be taken that RCRA realistically assesses environmental risks so as not to act as an expensive time consuming constraint that provides minimum benefit.

7.2. Optimization of Regulations

In establishing limits on arsenic in various process streams one has to make sure that the probability and cost of the risk is balanced by the cost of control. One has a limited number of processing options. Since arsenic is an element, it is conserved. Removing arsenic from water or air will cause an increase in the arsenic content to be disposed of as a solid waste. In processing, one can either dilute the process stream so the concentration is below the accceptable limit or remove and concentrate the material. In setting regulations one needs to look at the entire system to optimize the costs of control. Unfortunately, this is a difficult process, especially when human emotions and politics are involved. Sometimes, when regulations have been set in the past if any optimization occurred it tended to only be with one variable; e.g., the concentration in water. Impacts such as increased quantities of solid wastes caused by increased removal from the water usually were not fully considered.

It is hoped that, at a symposium such as this, one can start to realistically assess the costs and risks associated with various levels of control so the entire system can be optimized. We will be evaluating in detail the risks and benefits associated with arsenic. When we consider proposing specific standards we should remember what Paracelsus, a scientist sometimes known as the father of industrial hygiene, said in the early 1500's, "There is no substance which is not a poison, the right dose differentiates a poison and a remedy. Dose alone makes poison." Our task is to realistically assess the risks and benefits of small quantities of arsenic, present in process stream associated with energy sources.

8. Summary--Energy Sources as a Future Arsenic Supply

The recent literature was surveyed to determine if petroleum, geothermal, coal, or shale oil production would be important as a future supply of domestic arsenic. The occurrence of arsenic in these sources and process streams associated with these sources was determined. Environmental impacts associated with the process streams and possible production of arsenic from environmental control equipment attached to the process was also examined.

In only one case, shale oil production, was arsenic ever present in a process stream at a sufficiently high concentration that economic recovery appears feasible. The guard bed catalyst, which removes arsenic from shale oil, concentrates the arsenic adequately for further processing. At estimated shale oil production rates in the early 21st century, 2-3,000 tons of arsenic per year could be produced as a by-product or about 10 percent of domestic United States demand. If economic ways are found to remove arsenic from processed shale which contains 80 ppm arsenic, there would be almost inexhaustible supply of As, >16,000 tons/year. This, however, appears very unlikely in the near future.

It may also be worthwhile to examine more closely the removal of arsenic from geothermal brines and from fine particulates associated with fly ash. At some point in the future economical processing might be developed to concentrate and separate the arsenic in these streams. At no point in the future will crude oil be considered a source of arsenic.

References

[1] Greenspoon, Gertrude N., Arsenic, pp. 99-106, Mineral Facts And Problems, (United States Bureau of Mines Bulletin 667 1975).

[2] Carapella, S. C., Arsenic and Arsenic Alloys, Kirk-Othmer Encyclopedia Of Chemical Technology, 3rd Edition 3, pp. 243-250, (John Wiley & Sons, New York 1978).

[3] Bowen, H. J. M., Trace Elements In Biochemistry, pp. 16 (Academic Press, New York, 1966).

[4] Connor, J. J. and Shacklette, H. T., Background geochemistry of some rocks, soils, plants and vegetables in the conterminous United States, U. S. Geological Survey Professional paper 547-F (U. S. Government Printing Office Washington, D. C., 1975).

[5] Arsenic, Encyclopedia Brittanicia, 15th edition I pp.549, (Helen Hemingway Benton, Chicago 1977).

[6] Kopp, J. F. and Kroner, R. C., Trace metals in waters of the United States. A five year summary of trace metals in rivers and lakes of the United States, U. S. Department of Interior Federal Water Pollution Control Administration (1967).

[7] Baird, J. N. et al, Environmental and health aspects of disposal of solid wastes from coal conversion: an information assessment, ORNL-5361, Oak Ridge National laboratory (Sept. 1978).

[8] Crecelius, Eric A. et al, Background air particulate chemistry near coldstrip Montana, Environ. Sci. Technol. 14 [4] 422-8 (April 1980).

References (continued)

[9] Shunichi, Hariguchi et al, An attempt at comparative daily intake of several metals (arsenic, copper, lead, manganese zinc) from foods in thirty countries in the world, Osaka City Med. J. 24 [2] 237-42 (1978) in Chemical Abstracts 92: 16530h, (1980).

[10] Nielsen, F. H., Myron D. R., Uthus E. O., Newer trace elements - vanadium and arsenic deficiency signs and possible metabolic roles, Trace Elem. Metab. Man Anin., Proc. Int. Symp. 3rd (1977) 244-247 in Chemical Abstracts 90: 150662g (1979).

[11] The contribution of drinking water to mineral nutrition in humans, National Research Council Report PB80-114184 244pp in Chemical Abstracts 93: 79473h (1980).

[12] Smith, Ivan C., Ferguson, Thomas L. and Carson, Bonnie L, Metals in new and used petroleum products and by-products quantities and consequences, Chapter 7, The Role Of Trace Metals In Petroleum, pp. 123-148 (Ann Arbor Science, Ann Arbor, Michigan 1975).

[13] Bergervioux, C., Galinier, J. L., Zikousky, L., Determination of trace element pathways in a petroleum distillation unit by instrumental neutron activation analysis, J. Radioanal. Chem. 54 [1-2] 255-265 (1979).

[14] Block, C., Dams, R., Concentration - data of elements in liquid fuel oils as obtained by neutron activation analysis, J. Radioanal. Chem. 46 [1] 137-144 (1978).

[15] Filby, R. H., The Nature of Metals in Petroleum, Chapter 2, The Role Of Trace Metals In Petroleum pp.31-58 (Ann Arbor Science, Ann Arbor, Michigan 1975).

[16] Bennett, Ray L., Knapp, Kenneth T., Particulate sulfur and trace metal emissions from oil fired power plants, AICHE Symp. Ser. 75 [188] 174-180 (1979).

[17] Sanjour, William, Weisberg, Eugene, Crvse, Henry, Assessment of industrial hazardous waste practices in the petroleum refining industry, National Pet. Refiners Assoc. Washington, D. C. paper AM-76-33 (1976).

[18] Christensen, D. C., McNeese, J. A., Removal of arsenic and boron from geothermal brines, p 242-251 in Proc. 32nd Ind. Waste Conf. Purdue University (May 10-12, 1977).

[19] Pimental, K. D., Ireland, R. R., Tompkins, G. A., Chemical fingerprints to assess the effects of geothermal development on water quality in Imperial Valley, Lawrence Livermore National Lab, Livermore, CA. report UCRL-81177 Conf-780708-16 (1978).

References (continued)

[20] Sammel, Edward A., Hydrogeologic appraisal of the Klamath Falls geothermal area, Oregon. U. S. Geol. Surv. Prof. Pap. 1044-G (1980).

[21] Stauffer, R. E., Jenne, E. A. Ball, J. N., Chemical studies of selected trace elements in hot-spring drainages of Yellowstone National Park, U. S. Geol. Surv. Prof. Pap. 1044-F (1980).

[22] Breckenridge, Roy M. Hinckley, Bern S., Thermal Springs of Wyoming, Geol. Surv. Wyo., Bull. 60, 1-104 (1978).

[23] Bamford, R. W., Christensen, O. D., Multielement geochemical exploration data for the Cove Fort-sulphurdale known geothermal resource area, Beaver and Millard counties, Utah, DOE/ET/28392-28 from NTIS (1980).

[24] Beeland, Mrs. Gene V., Survey of environmental regulations applying to geothermal exploration, development and use, EPA - 6001 7-78-014 NTIS PB-281 023 (February 1978).

[25] Hussey, Elaine T., Coming to Grips with Geothermal, pp. 321-323 Geothermal Resources Council, Transactions 3 Davis, California, (1979).

[26] Thompson, John Michael, Arsenic and fluoride in the upper Madison River System:, Environ. Geol. 3 [1] 13-21 (1979).

[27] Aggett, J., Aspell, A. C., Release of Arsenic from geothermal sources, Report #NP-23412 (1978) in Chemical Abstracts 91: 112211e, (1979).

[28] Nishiyama, Eisuke et al, Arsenic removal from geothermal water, Patent Japan 79 113, p. 952 in Chemical Abstracts 91: 198609a (1979).

[29] Chiba, Shigehilco et al, Silicate and Arsenic removal from geothermal water, Patent Japan 80 31,437 in Chemical Abstract 93: 101313e (1980).

[30] Fernelius, Wayne A., Fulcher, Martin K., East Mesa geothermal test site, J. Environ. Eng. Div. (Am. Soc. Civ. Eng.) 105 [EE1] 13-32 (1979).

[31] Clark, Patrick J. et al, Arsenic and Selenium in Texas Lignite, Int. J. Environ. Anal. Chem., 7 [4] 295-314 (1980).

[32] Chadwick, Robert A. et al, Sulfur and trace elements in the Rosebud and McKay coal seams, Colstrip field Montana, Mont. Geol. Soc. Annu. Field Conf. Guides. 22 167-175 (1975).

References (continued)

[33] Fiene, F. L. Kuhn, J. K., Gluskoter, H. J., Mineralogic Affinities of Trace Elements in Coal, pp. 29-58 Proceedings: Symposium on Coal Cleaning to Achieve Energy and Environmental Goals held at Hollywood, Florida in September 1978 Volume 1, NTIS YPB 299383 (1978).

[34] Moore, Richard T., Chemical analysis of coal samples from the Black Mesa field, Arizona, Circ.-Ariz. Bur. Mines 18 pp. 14 (1977).

[35] Roehler, Henry W., Geology and energy resources of the Sand Butle Rim NW quadrangle, Sweetwater County, Wyoming, US Geol. Surv. Prof. Pap. 1065-A pp. 54 (1979).

[36] Swaine, D. J., Trace elements in coal, Trace Subst. Environ. Health 11 107-116 (1977).

[37] Beranova, Eva, Reduction of the content of sulfur and arsenic of ferritic brown coal by biological leaching, translated abstract in Chemical Abstracts 93: 12486w (1980).

[38] Wildeman, Thomas R., Meglen, Robert R., Analysis of oil shale materials for element balance studies, Chapter 14 Analytical Chemistry Of Liquid Fuel Sources, Advances in Chemistry Series 170, Edited by Uden, Peter C., Siggid, Sidney and Jensen, Howard B., pp. 195-212, (American Chemical Society, Washington, D. C. 1978).

[39] Ford, Charles T., Boyer, James F., Effects of coal cleaning on elemental distributions, pp. 59-90 Proceedings; Symposium on Coal Cleaning to Achieve Energy and Environmental Goals held at Hollywood, Florida in September 1978, Volume 1, NTIS PB 299383 (1978).

[40] Cannon, Helen L., Swanson, Vernon E., Contributions of major and minor elements to soils and vegetation by the coal fired Four Corners Power Plant, San Juan County, N.M. USGS prof. paper 1129-B in Shorter Contributions to Geochemistry 1979 (U.S. Gov. Printing Office, Washington D. C. 1980).

[41] Coles, David G. et al, Chemical studies of stack fly ash from a coal-fired power plant, Environ. Sci. Technol. 13 [4] 455-459 (1979).

[42] Smith, Richard D., Campbell, James A., Nielson, Kirk K., Concentration dependence upon particle size of volatilized elements in fly ash, Environmental Sci. Technol. 13 [5] 553-558 (1979).

References (continued)

[43] Fisher, Gerald L. et al, Filtration studies with Neutron-activated coal fly ash, Environ. Sci. Technol. 13 [6] 689-693 (1979).

[44] Smith, Richard D., Campbell, James A., Nielson, Kirk K., Volatility of fly ash and coal, Fuel 59 [9] 661-665 (1980).

[45] Moore, George T., Elig, Victor J., Evaluation of arsenic and selenium emissions from a coal fired power plant, Proc. 71st Annu. Meet. Air Pollut. Control Assoc. paper 78-34.2 pp. 15 3 (1978).

[46] Arvesen, J., Filtration of heavy metals from flue gas: literature studies and experiments, in Chemical Abstracts 92 :115580f (1980).

[47] Ondov, John M., Ragaini, Richard C., Biermann, Arthur H., Emissions and particle-size distributions of minor trace elements at two western coal-fired power plants equipped with cold-side electrostatic precipitators, Environ. Sci. Techol. 13 [8] 946-953 (1979).

[48] Ondov, John M., Ragaini, Richard C., Biermann, Arthur H., Elemental emissions from a coal-fired power plant, Comparison of a venturi wet scrubber system with a cold-side electrostatic precipitator, Environ. Sci. Technol. 13 [5] 598-607 (1979).

[49] Ondov, John M., Ragaini, Richard C., Biermann, Arthur H., Evaluation of two particulate collection alternatives for trace element removal at coal-fired power plants. Proc. 71st Annual Meet. - Air Pollution Control Assoc. Paper 78-34.6 pp. 24 3 (1978).

[50] Mann, Robert M. et al, Trace elements of fly ash: emissions from coal-fired steam plants equipped with hot-side and cold side electrostatic precipitators for particulate control, Report EPA/908/4-78/008 RAD-78-216-137-09 Radian Corp., Austin, Texas Available NTIS PB-295040, (1978).

[51] Goetz, L. et al, Heavy Metals: Environmental research related to energy production, Manage, Control heavy Met. environ. Int. Conf. (1979) in Chemical Abstracts 92: 1681606 (1980).

[52] Churey, Dorothy et al, Element concentrations in aqueous equilibrates of coal and lignite fly ashes, J. Agric, Food Chem. 27 [4] 910-911 (1979).

[53] Theis, T. L. et al, Field investigation of trace metals in groundwater from fly ash disposal, J. water pollut. Control Fed. 50 [11] 2457-2469 (1978).

References (continued)

[54] Wood, K. G., Environmental impact of coal ash on tributary streams and near shore water of Lake Erie, Final Report, State Univ. Coll. Fredonia N. Y. Avail. NTIS #COO-2726-5 pp.73 (1978).

[55] Evans, David W., Wiener, James G., Horton, John H., Trace element inputs from coal burning power plant to adjacent terrestrial and aquatic environments, J. Air Pollut. Control Assoc. 30 [5] 567-573 (1980).

[56] Shan, Henry, Pollution control and energy conversion efficiency of advanced power technologies, AICHE Symp. Ser. 76 [196] pp. 1-16 (1980).

[57] Filby, Royston H., Khalil, Samir R., Trace elements in the solvent refined coal processes SRC I and SRC II, Proc. Symp. Potential Health Environ. Eff. Synth. Fuel Technol. (1978).

[58] Fruchter, J. S. et al, High-precision trace element and organic constituent analysis of oil shale and solvent-refined coal materials, Chapter 18, Analytical Chemistry of Liquid Fuel Sources, Advances in Chemistry Series 170, Edited by Uden, Peter C., Siggia, Sidney, Jensen, Howard B., (American Chemical Society, Washington, D. C. 1978).

[59] Koppenaal, D. N., Trace element studies on coal gasification process streams, U. of Missouri, Columbia Report TID-290-22 pp. 219 available from NTIS, (1978).

[60] Forney, A. J. et al, Trace elements and major component balances around the Synthane PDU gasifier, U. S. Env. Prot. Agency Report EPA-60012-76-149, Symp. Proc. Env. Aspects Fuel conversion tech. II PB-257 182 (1975).

[61] Allen, John M., Duke, Kenneth M., Multimedia emissions from pressured fluidized-bed combustion of coal, Proc. 71st Annu. Meet. - Air Pollut. Control Assoc. paper 78/69.8 5 (1978).

[62] Fisher, Dennis D. Monitoring emissions from an in situ coal gasification experiment Proc. 2nd annu. UCG Symp. Morgantown, W. Va. pp. 242-248 MERC/SP-7613 (1976).

[63] LaRue, D. M., Reismann, G. A., Materials for in situ processing systems, U. S. Dept. Energy Rep. No. Conf-791014, pp. II-63-64, (1979).

[64] Youngberg, D., Santoro, Hydrogeologic Evaluation of Hanna Coal gasification site, Proc. 7th UCC Symposium, (Fallen leaf lake, Ca. Sept. 8-11, 1981).

References (continued)

[65] Schubert, Jeffery P., Groundwater contamination problems resulting from coal refuse disposal, Mine Drains Proc. Int. Min Drain. Symp. 1st (1979).

[66] Skogerboe et al, Environmental effects of western coal surface mining Part III: The water quality of Trout Creek Co., Colorado State U. Ft. Collins Co., Report EPA160013-79/008 order No. PB-292701 (1979).

[67] Parkinson, Gerald, Shale oil get the final nod, Chemical Engineering, 88 [18] 47-51, (Sept. 7, 1981).

[68] Fox, J. P., McLaughin, R. D., Thomas, J. F., Poulson, R. F., The partitioning of As, Cd, Cu, Hg, Pb, and Zn during simulated in-situ oil shale retorting, 10th Oil Shale Symposium Proceedings, pp. 223-237, Colorado School of Mines Press, Golden, Colorado (July 1977).

[69] Curtin, D. J., Dearth, J. D., Everett, G. L., Grosball, M. P. and Myers, G. A., Arsenic and nitrogen removal during shale oil upgrading, American Chemical Society Div. Fuel Chem. 23 [4] 18-29 (1978).

[70] Burger, E. D., Curtin, D. J., Myers, G. A., Wunderlich, D. K., Prerefining of Shale Oil, American Chem. Soc. Div. Pet. Chem. 170th National meeting Chicago Aug. 24-29 (1975).

[71] Myers, Gary A., Wunderlich, D. K., Shale Oil Treatment, U.S. Patent 3,804,750 assigned to The Atlantic Richfield Co. (April 16, 1974).

[72] Myers, Gary A., Guard bed system for removing contaminant from Synthetic oil, U. S. Patent 3,876,533 assigned to Atlantic Richfield Co. (April 8, 1975).

[73] Myers, Gary A., Slurry system for removal of contaminant from synthetic oil, U. S. Patent 3,933,624 assigned to Atlantic Richfield Co. (Jan. 20, 1976).

[74] Curtin, Daniel J., Method of removing contaminant from hydrocarbonaceous fluids, U. S. Patent 3,954,603 assigned to Atlantic Richfield Co., (May 4, 1976).

[75] Burger, Edward D., Curtain, Daniel J., Edison, Robert R., Method of removing contaminant from a hydrocarbonaceous fluid, U. S. Patent 4,003,829 assigned to Atlantic Richfield Co. (Jan. 18, 1977).

References (continued)

[76] Curtin, D. J., Method of removing Contaminant from hydrocarbonaceous fluid, U.S. Patent 4,029,571 assigned to Atlantic Richfield Co. (June 14, 1977).

[77] Myers, Gary A., Wunderlich, Donald K., Synthetic Oil treatment, U.S. Patent 4,051,022 assigned to Atlantic Richfield Co., (September 27, 1977).

[78] Wunderlich, Donald K., Removing contaminant from hydrocarbonaceous fluid, U. S. Patent 4,069,140 assigned to Atlantic Richfield Co., (January 17, 1978).

[79] Styring, Ralph E., Method of regenerating used contaminant-removing material, U. S. Patent 4,083,924 assigned to Atlantic Richfield Co., (April 11, 1978).

[80] Wunderlich, Donald K., Processing Shale Oil Cuts by hydrotreating and removal of arsenic and/or selenium, U. S. Patent 4,113,745 assigned to Atlantic Richfield Co. (January 9, 1979).

[81] Sullivan, Richard F., Process for upgrading arsenic-containing oils, U. S. Patent 4,141,820 assigned to Chevron Research Co., (Feb. 27, 1979).

[82] Jensen, Harbo P., Hydroprocessed shale oil including thermally treating and coking steps, U. S. Patent 4,142,961 assigned to Chevron Research Co., (March 6, 1979).

[83] Jensen, Harbo P., Process for Treating hot Shale Oil effluent from a retort, U.S. Patent 4,181,596 assigned to Chevron Research Co., (Jan. 1, 1980).

[84] Jensen, Harbo P., Method for removing arsenic from shale oil, U. S. patent 4,188,280 assigned to Chevron Research Co., (Feb. 12, 1980).

[85] Chappell, W. R., Runnells, D. D., Toxic trace elements and oil shale production, 11th Annu. Trace Substances Environ. health Conf. Columbia, Mo., 617-9/77 in Petroleum Abstracts No. 266,001 (1977).

[86] Stollennerk, K.G., Runnells, D. D., Leachability of arsenic, selenium, molybenum, boron and fluoride from retorted oil shale, Proc. of 2nd inter-amer confed Chem. Eng. and Asian pacific Chem. Eng. 2 pp. 1022-1036 (1977).

[87] Nakand, Tonolcuni, Industrial waste solidification, Japanese Patent 79 143,772 (09 Nov 1979) in Chemical Abstracts 92 :134797y (1980).

References (continued)

[88] Amano, Ryuzo, Heavy metal fixation in municipal incinerator flue dust, Japanese patent 79 23909 (17 Aug 1979) in Chemical Abstracts 92 :28192n (1980).

[89] Nakane, Mitsun, Solidification of powdery industrial wastes, Japanese patent 80 1,830 (09 Jan 1980) in Chemical Abstracts 92 :203099m (1980).

[90] Hounslow, Arthur, N., Ground Water geochemistry: arsenic in land fills, Groundwater 18 [4] 331-333 (1980).

[91] Sandesara, Mahendra D., Fixation of heavy metals for secure land fill disposal, Toxic Hazard, Waste Disposal, 4 127-133 (1980).

[92] Mehta, Anil, Twidwell, Larry, Fixing arsenic heaving materials U. S. Patent Appl. 15,079 21 Dec 1980 in Chemical Abstracts 93 :53236h (1980).

[93] Schofield, John T., Sealosafe, Toxic Hazard, Waste Disposal 1 297-319 (1979).

[94] Personal Communication, Environmental Protection Agency, Region VI, Dallas, Texas, October 1981.

DISCUSSION

A. Furst: Do the grasses that grow in the processed shale concentrate any form of arsenic?

R. A. Schraufnagel: No, the grasses that grow on processed shale do not concentrate the arsenic. This has been confirmed by Wilard L. Lindsay and Robert E. McFadden in studies at the Department of Agronomy at Colorado State University. The uptake of arsenic, selenium, flourine, boron, and molybdenum in western wheat grass, four wing salt bush and alfalfa was studied for three different processed shales in the greenhouse and on field plots. The maximum arsenic detected in the plants was 1.5 ppm even though the soil contained up to 80 ppm arsenic.

F. E. Brinckman: There have been a number of reports and studies conducted regarding bioleaching of kerogen or oil shale for the recovery of oil. In your survey, did you see any evidence of work regarding bioleaching of processed or spent shale for recovery of any critical elements, especially arsenic?

R. A. Schraufnagel: That sounds like an intriguing idea. Arsenic tends to be concentrated in oil shale initially when compared to other geologic deposits. Shales were originally deposited by the accumulations of microorganisms in a huge fresh water lake. The organisms probably accounted for at least some concentration of the trace metals present. However, I saw no reference to using microorganisms to further concentrate the trace metals in shale for possible recovery.

F. E. Brinckman: This is an active industry. In beneficiation of low-grade ore such as copper, perhaps 20% of copper production in the United States is by microbial leaching. This seems to be a prime possibility for commercial arsenic recovery.

R. A. Schraufnagel: I know that people are looking at novel means for recovering arsenic in the processed shale. Likely, a novel process will be required to economically recover the metals. If you want to use processed shale as a source of arsenic, there is almost an unlimited supply of feedstock.

SESSION II - INDUSTRIAL USES

Session Chairman: G. Donald Munger
Diamond Shamrock Corporation
1100 Superior Avenue
Cleveland, Ohio 44114

INTRODUCTION

I think you'll find the subject matter presented in this session diverse and interesting. As you will note, the uses that will be covered include glass, agricultural, industrial chemicals, feed additives, and the use of arsenic in wood preservatives. In addition, there will be a presentation on the standard reference materials certified for measurement of arsenic.

4

ARSENIC: GLASS INDUSTRY REQUIREMENTS

Richard J. Bauer
Corning Glass Works
HP-ME2-(E-6)
Corning, NY 14830

The specialty glass industry has a diversified, critical need for arsenic as oxides or high quality arsenic acid. This paper presents the extent of use and the various purposes for which arsenic is uniquely useful in these applications. The major functions include: (1) fining, (2) oxidation-reduction buffering, (3) interacting to affect control of surface and interfacial tension during crystal growth in opals and glass ceramics. A brief outline of the efforts to eliminate and reduce arsenic use is presented. The alternatives and broad economic implications involved, if it was not available for use, are discussed briefly. The results of environmental and energy constraints over the past decade as well as the thrust of melting process changes occurring in the industry are presented, as they impact arsenic utilization.

1. Introduction

This paper will describe the specialty glass industry needs for arsenic. This industry is a highly specialized, diversified small segment of the "glass industry." It does not include flat glass, containers, or fibers. It comprises less than ten percent of the tonnage of the glass industry as a whole.

Corning Glass Works, a leader in these specialized businesses produces over 60,000 products by utilizing over 1,000 compositions and over 300 different raw materials each year. Our processes run 24 hours per day, 7 days per week and many operate two years without stopping. We make batch quantities in pots and crucibles for special low volume requirements. We employ a wide range of forming processes including sophisticated high speed automatic pressing and blowing, continuous draws, and handshop operations which have little changed over centuries.

Products containing critical amounts of arsenic are used in a great many applications: television and electronics; communications and information processing; food preparation, storing, and serving; aerospace and defense; transportation; illumination; science; medical; optical.

2. Extent of Use

Corning Glass uses nearly 1,000 tons of liquid arsenic acid per year in its domestic operations. Currently we use bulk tanker truck quantities at nine locations and drums at four intermittent use sites.

Arsenic trioxide had for years been the traditional form of arsenic used in the glass industry. This powder form was always handled with great care in the mixing and batch transport areas because of its obvious toxicity. In the 1970's, however, arsenic in the trivalent state was declared to be an inhalation carcinogenic hazard. Although we knew of no cases related to our operations, Corning elected to eliminate all powdered forms of arsenic from use in our plants.

By the end of the 1970's, we had converted our domestic operations to the liquid form: 75% H_3AsO_4, arsenic acid. This move converted the input form to the essential pentavalent oxidation state. The liquid form helped achieve improved dispersion of arsenic in the batch and simultaneously aided in control of fugitive dust.

Analysis of the many glasses which contain arsenic would yield ranges of total retained or chemically bound arsenic, expressed as weight percent As_2O_3, from as low as 0.05% to a few low volume types of as much as 5%. The majority of these glasses and glass ceramics would show a range of 0.1 to 0.9%.

3. Use of Arsenic Compounds in Glass Manufacture

3.1. Fining

Fining is the term used today to describe the clarification of the molten glass. Various minerals and chemicals are mixed carefully to facilitate melting and insure homogeneity. After being delivered to the melting furnaces, usually as a continuous process, it goes through several stages of progressive refinement before being delivered to the forming process. In the early stages of melting, fusion and other chemical and thermal processes occur, leaving gaseous inclusions. These are defects generally detrimental to the properties of the desired product. The process of removing these gas bubbles is known as fining.

The gas bubbles are generally caused by one of three basic sources: (1) air entrapped within the batch, (2) gases evolved from complex reactions within the refractories, (3) gases produced during the melting process.

Chemical fining agents are materials which can release gases, at the elevated temperatures created relatively late in the fusion processes of melting. When this occurs, new bubbles are formed and gas originating in the glass melt diffuses into existing bubbles, with the result that they rise more quickly out of the melt.

If chemical, rheological, and thermal conditions are correct, another phenomenon known as resorption or bubble collapse can occur to clarify very small bubbles known as seeds.[1,2][1]

This process is often enhanced if oxygen is the fining gas released. The chemical solubilities and partial pressures of the various gases present, and the localized imbalance created when a diffusion enhanced bubble leaves the vicinity; can cause small volumes of gases to be quickly resorbed into the glass network. The bubble vanishes. [3,4]

Arsenic in the pentavalent state is an effective fining agent at high temperatures. Depending critically upon the compositional family of the glass being melted, at temperatures above 1400°C, it is by far the most effective oxygen generator available. [5]

Typical products fined by arsenic include television and other cathode ray tube display faceplates; Vycor™ products and the flexible, chemically strengthenable flat glass used in aircraft and space craft windows; many lead and lead crystal glasses; high reliability resistors.

3.2 Redox and color control

The second extremely important need for arsenic in the glass melting process is as an oxidation-reduction buffer. The most common example is the oxidizing action upon the iron contamination introduced with the batch materials.

Iron is easily reduced by the thermal equilibriums and progressive building of the glassy network which requires oxygen in most matrixes. In the divalent state iron typically imparts a blue or greenish color to the product. Arsenic (V) oxidizes it to the trivalent state yielding a yellow color. The presence of precisely selected third elements, such as oxides of nickel, cobalt, or rare earths, create complementary colors and the final product is rendered neutral or colorless. [3,6]

Television face plates, tableware and fine lead crystal and stemware are examples of this use.

Analogous reactions occur with many elements in the melting process. Most transition elements and rare earths can exhibit color in at least one oxidation state which can be either a help or a hindrance to the final product requirements.

In this specialty glass area, some compositions often require very high temperatures as compared to the majority of glasses manufactured in other segments of the industry. Melt zones in our furnaces frequently

---

[1]Figures in brackets indicate literature references at the end of this paper.

reach temperatures above 1600°C. Only arsenic, of the oxygen releasing fining agents, is effective at these temperatures.

3.3. Surface tension aid in crystal growth

The third and probably most complex use of arsenic, which is growing in importance and not well understood, is related to surface and interfacial tension. In opal glasses and glass ceramics it can be critically important to the final product properties.

Spontaneous opals which depend upon calcium fluoride crystals for their opacity are often found to be directly dependent upon the presence of arsenic for their properties. In addition to the fining and the redox actions described earlier, arsenic here plays a critical role in creating the proper conditions to allow an adequate nucleation and crystal growth to occur in a single, very rapid cooling cycle during the forming process [7]. The size of the phase separated crystals of CaF_2 largely determines the opacity attained. According to light scattering theory, and especially the Rayleigh equation, effectiveness increases as the radius of particle raised to the 6th power [8].

Perhaps more complex; but even more critically important, is a whole class of high technology glass ceramics emerging since the late 1950's. These are polycrystalline materials produced by the controlled devitrification of glass. They depend upon controlled nucleation and the growth of fine grained, randomly oriented crystals with some residual glass. They must be free of gaseous inclusions and usually exhibit high strength. They require controlled color, usually whiteness, in many product applications. Proper nucleation and more critically, crystal growth rate which, with other factors, determines crystal size are dependent upon the presence of arsenic. [9]

Typical products included in the glass ceramic class include the missile nose cones known as radomes, MACOR™, The Counter that Cooks™, Rangetoppers™, Corning Ware™, Centura™, VISIONS™ and other new transparent products such as windows for wood burning stoves and fireplaces.

3.4. Significant glass constituent

A fourth general area of arsenic use in glass is the small volume, unique glasses in which it becomes an important part of the glass network itself.

There are a few glasses in the optical lens market where its special dispersions are useful: long wave length blue regions, and where applications require a high infrared permeability.[6]

A very common application involves the medical thermometer: the buried white stripe is usually a very dense opal whose major phase is an arsenate.

4. Melting Process Changes

The broad efforts aimed at the problems of the availability and costs of energy have had a great impact on the melting of glass. Natural gas has long been the favorite fuel for specialty glass production. Oil is often a severe threat to quality . . . due to higher nominal amounts of sulfur and other potentially coloring contaiminants in these specialized products.

One way to counter the fossil fuel concerns of cost and availability is to move toward furnaces which utilize electricity to introduce the required heat.

4.1. Conventional melting

The traditional glass furnace is a horizontal refractory lined container in the shape of a rectangle. Raw materials, and recycled glass of the proper composition, are fed in at one end and molten glass is conditioned and withdrawn from the other. All heat is supplied by fossil fuel burning above the pool of molten glass. For many of the glasses in the specialty field, this becomes analogous to brewing a cup of coffee in a large sugar cube. Refractories capable of withstanding the chemical and thermal attack during the melting process are difficult to develop and produce.

4.2. All electric melting

Corning has developed over the past twenty years, a process in which cold raw materials and cullet, the term we use to describe the recycled glass, is distributed over the top of the molten glass and is maintained in a steady state dynamic process we call cold crown vertical melting. In these melting furnaces, all energy is introduced as electricity below the surface and there is much more efficient melting. It allows controlled temperature zones, and essentially no air pollution since there is no stack or extensive regeneration or recuperation of heat from the melting process. Higher melting temperatures are possible and the glass experiences high average thermal history.

The melting process is much more sensitive, however, to both batch material homogeneity and oxidation-reduction phenomena than conventional melting. Gases, once released, must either escape through the unmelted batch blanket or be trapped in the molten glass and be potentially swept into the conditioned glass and into the product.

Fining takes on a completely new set of ground rules and problems. Redox and the phenomenon of seed collapse and resorption become paramount in the selection of the agents used and the precise levels allowed. There is only limited free surface available in the conditioning zones immediately prior to forming.

4.3. Boosting

Between the two extremes of continuous melting furnace types is, of course, the widely used, growing technology, of boosting. Fossil fuels are used in conjunction with electrodes submerged in the molten glass either through the sides or bottom of the modified horizontal melting furnaces.

The advantages of this hybrid melting units are several: (1) utilization of existing capital - the melting unit can be repaired many times over a 20-40 year interval before that stage is reached when you must rebuild from the true foundations; (2) costs and availability of energy can be optimized as the local economics and constraints come to bear; (3) the electrification usually means an additional factor now exists in evaluating the type and need for stack devices. Colder, less violent melting conditions in the melter above the raw batch usually means less physical and chemical carryover to be dealt with as potential air pollution. And, finally, (4) there remains in the hybrid melter a quiet zone for high temperature fining to be accomplished in the more conventional sense.

Corning is committed to move toward all electric melting. The process will require additional decades to complete. Each type of glass requires specially tailored melting processes to allow its unique salable properties to be successfully produced. Technologies are not fully in place and the costs of conversion at a rapid pace are probably prohibitive. Certainly this is so far the industry as a whole. Adequate power is often not yet available; fossil fuels will apparently continue to be viable, though costly.

4.4. Electrode concerns

Electrodes for the electric melting furnaces are defined by a compromise determined by the requirements of the glass matrix being produced, melting furnace configuration, costs, and life. Typical electrode materials are carbon (graphite), molybdenum, tin oxide, and platinum (or its alloys).

Arsenic is compatible with the two latter materials; but with carbon and moly it is easily reduced to the elemental vapor state at elevated temperatures. This greatly reduces electrode life and prevents their use beyond relatively early stages of the melting process. Current densities, dictated by electrode shapes and molten glass resistivity, coupled with a desire to minimize electrode-glass contact area, force difficult compromises if arsenic has no effective substitute in a given matrix.

5. Reduction of Use Levels

Corning has actively pursued a multifaceted program of reexamining arsenic use for the past decade. The goals of energy cost programs, particulate stack emission controls, and the implied problems of electric melting have been major carriers of the justification burden to reevaluate each constituent in each glass and its melting process. It must also be stated that in some cases, specifications have been renegotiated with our customers to allow higher seed counts and wider color variation in the product.

As electric boosting became a proven reality in many of our furnaces, with more controlled higher average melt zones and less violent early melt zone conitions, we have been able to reduce the level of arsenic used in the matrix.

Several glasses have been greatly modified to eliminate arsenic. Others have been, through melting process changes, able to successfully compete with use reductions of up to 30-50%. We have been totally successful in preventing use increases on the basis of weight percent in the matrix.

Devices for stack pollution control have been installed on some melting furnaces and, in some of these, dust can be recycled to the melting process, with very careful control, without detrimental effects.

The only growth we foresee in the use of arsenic is with new, very dense spontaneous opals and especially glass ceramics both as opaque and clear products.

6. Conversion to Arsenic Acid

By the end of the 1970's we had converted all operations to the liquid arsenic acid. While several of our glass ceramic production locations had used the liquid form since the very early 1960's, this was not an easy or cost effective conversion: (1) Extensive engineering efforts and modifications in our batch plants were required. The previous technology in place was largely based on dry state weighing mixing, and transport of the batch. (2) The liquid form was not readily available, indeed our two original sources both abandoned the business due largely to environmental pressures and costs. When available, our purity requirements largely aimed at avoiding coloring oxide contaminations have proved difficult to attain. (3) Our competitors have not generally followed this approach. The liquid form, delivered, costs nearly twice as much as the powder form. Semi-processed mixed powders containing both arsenic and antimony are even more economical as delivered.

7. Alternatives - Chemical

The most commonly named potential substitute is the one often used in conjunction with arsenic: antimony. Used in some glasses as antimony trioxide, and in most commercial high volume production as the pentavalent form available as sodium antimonate; it is a most effective fining agent at temperatures below about 1400°C as an oxygen generator.[5,6] It does find wide use in many lead containing glasses. Indeed the presence of both arsenic and antimony is more effective than either when used alone. It, unfortunately, is not effective, in many of the opals and glass ceramics either alone or in combination with other agents used to control redox, color, or crystal growth. It can even be used, occasionally, as a reducing agent.

The halogens, especially chloride and fluoride are used in many glass matrixes. Chloride fining, usually as sodium and occasionally as potassium salts, is used in the hard borosilicates. It however does not control redox and can, with water and alkali-boron balances in the process, create particulate air pollution problems. Fluoride, if present, mostly affects lowered viscosity and often later enters into crystalline phases. It almost always is an air pollution concern, and is relatively ineffective at high temperatures.

The various oxidation states of sulfur make this the most commonly used agent for fining in the lime glasses which dominate the container and flat glass segments of the industry. While limited applications do exist in the specialty glass field, it is almost an anathema when involved in lead, borosilicate, and glass ceramic applications.

Other metal oxides, especially tin when used with antimony oxides have shown some remarkable attributes in fining and redox control with some matrixes. In no case studied to date, however, can even this combination replace arsenic to produce an equivalent product.

8. Alternatives - Physical

Theoretically there are many physical means to affect fining action. Most are as yet far too costly or unreliable in the melting environment to be called successful. [2,6]

The method most discussed is created by thermal shock: a very rapid increase in average glass batch temperature followed by a relatively rapid cooling cycle or soaking zone. In practice, this is an extreme acceleration of the normal melting process and is close to another phenomena known as reboil. It creates almost an instant foaming activity far enough back in the fining zone to allow clarification to occur by Stokes Law fining and later bubble collapse.

Vacuum fining, if it can be created in the process, speeds bubble rise and removes gases from the melt. To date it has proved difficult and expensive.

Mechanical agitation in several different approaches can be as simple as stirring and as subtle as flowing past a sharp edge. Ultrasonics have yet to be proven commercially.

Diffusion bubbling, a fairly theoretical approach consisting of controlled introduction of selected gases as tiny bubbles late in the melting process, has had little practical success. It has the promise of redox control.

The most common practical solutions to the problem have been simply to slow down the throughput in the furnace or extend the melt zone path length through which the fining process must reach proper equilibrium. Both are costly. The first in high production costs on a unit basis and the second adding capital and energy requirements.

9. Utilization Efficiency

In a theoretical sense arsenic oxide as a fining agent must be present in the pentavalent state late in the highest temperature zones of the melting process, release an oxygen molecule, and revert to the trivalent state in the fused glass. [2,4,5]

In practice, the specialty glass processes produce a significant amount of cullet or scrap and trim glass from the forming processes. Nitrates and other excess oxygen materials are added to the furnace blended in the raw material mix. The intent is to help ensure that the arsenic in the cullet is also again in the pentavalent state and the oxygen is available at the proper time.

In addition to the redox control, the raw batch must contain fresh arsenic acid in sufficient amount to compensate for that material lost in the melting process.

Before Corning began to reduce arsenic use levels, measurements indicated percent retention ranges from 70 to 99%. The weighted averages for domestic operations was approximately 91%.[10] The increased use of electrical energy and the general 20-30% use level reduction have significantly raised that early figure.

Of the lost material, much is trapped in the slag and dusts of the regenerative system. Data taken by ourselves, consultants from EPA contracts, and constant industry monitoring indicate that over 95% of it is captured before it exits the tank stacks and enters the atmosphere.[6,10,11,12] Here the efforts in use reduction, electric boosting, electrostatic and fabric filter devices have certainly been effective. The stack design requirements are checked via dispersion models using a variety of programs and cross checked with stationary field equipment when applicable. Ambient air data are nominally at less than microgram per cubic meter levels around our plants. All of Corning's operations are well within local, state and federal requirements for work place and air pollution requirements.

10. Economic Implications

We exist in a worldwide competitive environment. This competition comes both from within the glass industry and from other technologies outside our industry. Constraints in this country are often not yet imposed by other nations with different priorities. The specialty glass industry has unique and critical needs for arsenic, many of which do not have technological substitutes, most of which do not have economical substitutes.

With the pressure placed upon the use of arsenic trioxide, and arsenic in general, during the past decade; those who can avoid it have done so, those who cannot must endure.

The controls and limits currently in force for the workplace and the environment have been met with significant technological effort, capital expense, and competitive sacrifice. To further tighten any part of them will be counterproductive: many technologies will be forced to grow in more favorable locations.

References

[1] Grey, W. T., Internal Communications, file L 525.5, Corning Glass Works, Corning, N.Y.

[2] Rindone, G. E., Fining (Part II), The Glass Industry, pp. 561-566, (Oct. 1957).

[3] Weyl, W. A., The use of arsenic in the glass industry, The Glass Industry 23 [7] pp. 253-258, (July 1942).

[4] Cable, M., Clarke, A. R., Haroon, M. A., The affect of arsenic on the composition of the gas in seed during the refining of glass, Glass Technology 10 [1], pp. 15-21, (Feb. 1969).

[5] Cameron, R. A., Kinetics of arsenic--antimony fining, presented at symposium on gases in glass, 67th annual meeting of American Ceramic Society, Philadelphia, PA, May 1-6, 1965.

[6] Peters, A., Zur situation des arsenverbrauchs unter besonderer Borucksichtigung der Glasindustria (The use of arsenic with particular reference to the glass industry) Glastechn. Ber., 50 [12] pp. 328-335, 1977.

[7] Dumbaugh, Flannery, Hares, U.S. patent No. 3,681,098, opal glass compositions comprising calcium fluoride, Aug. 1, 1972.

[8] Blau, H. H., Diffusing glasses for illumination, Ind. and Engr. Chem., 25 [8] pp. 848-852, (Aug. 1933).

[9] Beall, G. H. and Doman, R. C., Processing, properties, and uses of glass ceramics, (Proceedings of 10th International Conference: Science of Ceramics, Technische Universitat Berlin, Sept. 1979) Book: Vol. 10, Hausner, H., ed, pp. 25-36, Deutsche Keramische Gesellschaft, (1980).

[10] Mosely, G. H., Comments and data on arsenic in glass and arsenic emissions from glass melting units. Letter to J. R. O'Connor, U.S. EPA, Air Quality Planning Agency, Aug. 28, 1978, (Corning Glass Works memo to EPA).

[11] Thalman, Meyer and White, Arsenic glass manufacturing emission test report, U.S. EPA, EBM Report 78-GLS-3 (Feb. 1979).

[12] Thalman, Meyer, White, Arsenic glass manufacturing emission test report, U.S. EPA, EBM Report 78-GLS-4 (Feb. 1979).

DISCUSSION

E. A. Woolson: In 1978, the Environmental Protection Agency's Cancer Assessment Group estimated that there were three cancer deaths per year in the glass industry. You indicated that you did not see any adverse effects. Do you care to comment on that document?

R. J. Bauer: We put out a statement dated August 28, 1978: "Comments and Data on Arsenic in the Glass and Glass Emissions in the Melting Units." This contained a great amount of information which is on file with the EPA. To paraphrase our position, the EPA statement generally concerned an assumption that over 50% of the arsenic was lost in the glass melting process. We violently dispute that, and have "acres of papers" of proof to demonstrate otherwise. The assumption also was made that of the lost material, most went up the stack; and there are all sorts of arguments still raging today on this issue. I can't comment on this one way or the other. You can apply an infinite number of dispersion and incidence models. In summary, however, we think that number was quite high.

E. A. Woolson: You did say that the arsenic levels now meet OSHA air standards.

R. J. Bauer: Yes.

E. A. Woolson: You also said that antimony trioxide was used frequently. Do you have any idea of antimony levels outside the plant or in the workers' quarters?

R. J. Bauer: Not specifically, but I'd like to make a few comments. The antimony, in general, is used as sodium antimonate and in very low temperature, non-violent melting conditions. We feel that if you can find arsenic in the glass, the antimony is usually at a lower level. However, in the case of television glasses, for instance in the face plates, the antimony level is almost twice as high. We've almost completely eliminated the arsenic from those same applications. In most cases, antimony is found at lower quantities, with that one major exception.

5

ROLE OF ARSENICAL CHEMICALS IN AGRICULTURE

John R. Abernathy
Associate Professor - Weed Research
Texas Agricultural Experiment Station
Route 3
Lubbock, Texas 79401

Arsenical chemicals currently used in cotton production as herbicides and harvest aids are MSMA (monosodium methanearsonate), DSMA (disodium methanearsonate), arsenic acid (orthoarsenic acid), and cacodylic acid (hydroxydimethylarsine oxide). Herbicidal control of grassy and broadleaf weeds is obtained with MSMA, DSMA, and cacodylic acid, while arsenic acid and cacodylic acid are used as desiccants and defoliants in cotton. Estimates of acres treated and total pounds used in the U.S. annually are DSMA and MSMA (10-12 million acres, 23 million pounds), arsenic acid (2.1 million acres, 7.3 million pounds), and cacodylic acid (200,000 acres, 574,000 pounds). Major geographic areas of use of MSMA and DSMA correspond to the heaviest weed pressure areas of the South and Southeast cotton producing areas. The major use area for arsenic acid as a desiccant is the High Plains and Central Texas cotton production areas. Most of the cacodylic acid is used as a cotton defoliant in Arizona and California.

Economic and alternative chemical possibilities have been evaluated. With the advances in selective herbicide application equipment, glyphosate [N-(phosphonomethyl) glycine] may become a viable alternative to approximately 46% of the MSMA and DSMA use. Alternatives for the remaining 54% of MSMA and DSMA usage do not currently exist. Only one potential alternative harvest aid chemical exists. Many studies elucidate the plant and soil residue aspects of the arsenical chemicals used in agriculture.

Today I would like to discuss the role that arsenical chemicals have in agricultural crop production, how they are used, the amounts used, their importance, and perhaps other alternative materials that could be used. In past history there have been a number of arsenical materials that have found roles in agriculture. However, today there are only primarily four materials. They are MSMA, DSMA, arsenic acid and cacodylic acid. Looking at these four materials in terms of their phytotoxicity to rats, their acute oral LD-50 values are 100 for arsenic acid, 1800 for MSMA, and 2600 for cacodylic acid. These values compare to aspirin at 1750 and 120 for paraquat (1,1′-dimethyl-4, 4′ bipyridinium ion).

The major use of the arsenical chemicals is in cotton production; therefore, I would like to briefly review U.S. cotton production. Projected cotton yields for the 1981 season are approximately 15 million bales, which have been grown on approximately 13.5 million acres of land. We consider a bale of cotton to be approximately 500 pounds of lint. Assuming that growers might get $.50 per pound of lint across the U.S. for this year's crop, that places the value of the U.S. cotton crop at about $3.8-$4.0 million.

For those of you who may not be familiar with cotton, it is produced in the southern states of the U.S., southern areas of California, Arizona, New Mexico and the High Plains area of Texas as well as other parts of Texas, southwest Oklahoma, Arkansas, Louisiana, Mississippi, Georgia, Alabama and some in the Carolinas, with a little bit in Florida. Of course, being from Texas, it is the most significant cotton state, with about 7 million acres of cotton production and anticipated yields being about 5.8 million bales, so roughly, a little over one-third of the cotton is produced in Texas.

In looking at the state map of Texas, I show you specific geographic areas. The High Plains area is where cotton is the most concentrated with about 4.5 million acres. The geographic areas of cotton production are important as we move on through this presentation in terms of use patterns of the chemicals and importance of their availability.

Over the past 10 years, a definite shift has occurred in the areas where cotton is grown. Traditionally we think of the southeast and the Delta as the cotton growing parts of the country, and indeed, they originally were. But today, the majority of the cotton is produced in the Southwestern U.S.

Turning our attention to MSMA and DSMA, these are two materials used specifically as herbicides or in combination with other herbicides in cotton. They are applied either as a post-emergence directed spray, which means that they are directed underneath the crop for the control of weeds that are already up and growing; or they are applied as a post-emergence topical, or above the crop, at stages prior to the time that the bole begins to develop on cotton. These materials are specifically grass herbicides. They have very good activity on Johnsongrass, but they also increase the activity of other herbicides for control of broadleaf weed species. They have been used extensively in the Mississippi Delta and the southeast, as well as other parts of the country for many years with very high satisfaction and at economical costs. Between 10 and 12 million acres of land are treated with MSMA and DSMA each year which is approximately 23 million pounds of these products being utilized. A herbicide that has come into the picture in recent years that may be a potential alternative for some MSMA and DSMA use is glyphosate.

With the advent of selective application equipment such as the rope application systems which allow herbicides to be wiped onto Johnsongrass or other weeds above the crop, glyphosate becomes an economical consideration at \$3 - \$5 per acre. The use of glyphosate would be an alternative only for the post-emergence topical application of MSMA and DSMA (46% of use). Alternatives for the remaining 54% of MSMA and DSMA usage do not currently exist.

At this point, I need to discuss more about cotton production and the importance of harvest aid chemicals. There are several ways of harvesting cotton - the first was by hand, actually hand picking the individual boles and putting them into a sack and into the trailer. Today we use mechanical harvesting, the first of which is a mechanical picker which spins fiber out of the boles of the cotton and collects it. The picker is utilized in the Delta, southeast states, southwest Oklahoma, South Texas, California and Arizona. With a picker, the harvesting operation may occur two or three times as the boles become open. Also this system utilizes different varieties of cotton, which take a longer season to mature.

A mechanical stripper is used in areas such as western and central Texas and southwestern parts of Oklahoma. Cotton stripping means that a machine takes all of the cotton in one pass. Therefore, all of the leaves must be off the cotton in order for this piece of equipment to operate.

Stripper harvesting is becoming more and more important because of economic considerations. A tractor-mounted stripper costs \$10,000 - \$15,000 compared to \$80,000 for a picker. Traditionally, cotton has been put into trailers and hauled to the cotton gin. Today, cotton is dumped into a piece of equipment called a module builder. The module builder is capable of compressing the cotton into a module which contains about 10 bales of cotton as compared to a trailer that held two or three bales. The module is then loaded onto the truck, transported to the gin in that fashion, and is fed directly into the gin. The use of the module system has dramatically changed our harvesting operation. We can harvest cotton much more rapidly because we are not having to wait on the turn-around time of trailers coming out of the gin. But in doing so, it is even more imperative that clean, dry cotton is harvested. If green leaves or damp cotton is packed into a module, the cotton heats up thereby decreasing seed and lint quality. Thus harvest aid chemicals are essential for preparing cotton for harvest.

Harvest aid chemicals can be divided into two groups. Defoliants are materials which cause the cotton leaves to drop from the plant. Defoliants are used in conjunction with picker harvested cotton. Current defoliants do not work well enough for stripper harvest. The other harvest aid group are desiccants which completely kill the plant and prepares the cotton plant for stripper harvest.

All of this is background information leading up to the use of arsenic acid. Arsenic acid is used as a desiccant on about 2.1 million acres of cotton annually, primarily on the Texas High Plains, Black Lands of Texas, and parts of southwest Oklahoma. Currently about 7.3 million pounds per year are being utilized. There is one alternative product, paraquat, which is also labelled for use on cotton. Comparing the economics of these two, arsenic acid usually runs about \$4-\$6 per acre, compared to paraquat at \$5-\$9 per acre. We get very concerned because these are the only two products that are labelled as desiccants on cotton and it is imperative that we have the use of these products in the Texas High Plains as well as other parts of Texas.

The remaining product that we have not discussed is cacodylic acid. It is used as a foliar herbicide which can be used prior to planting crops for the burn-down of weeds, but its primary use today is as a cotton defoliant, mainly in Arizona and California. Its use is on about .2 million acres, or about .57 million pounds being sold annually. In considering soil residues and plant residues, much work has been done in that specific area with these materials and this subject will be addressed later on in the program.

DISCUSSION

W. Peters: I wanted to ask about the use of monosodium methylarsonate (MSMA) and disodium methylarsonate (DSMA) when applied to the cotton plants. Does anyone have a sense of what they may break down into after a period of time? Are they stable compounds? Is there any way that this can become an inorganic form of arsenic and become an atmospheric pollutant?

E. A. Woolson: Nobody really knows what happens to MSMA in a cotton plant or above a cotton field. If it behaves as it does when sprayed on grass, dimethyl or trimethylarsine will be formed either on the plant leg or by microorganism on the plant or in the soil. You will get some volatile alkylarsines released into the air. It hasn't been measured above cotton, but is probably formed and released. In terms of incorporation into the plant, a little bit of MSMA will probably stay as the methanearsonate. However, it will also form more complex organic compounds. MSMA that is post-directed and falls on the soil will be metabolized or incorporated. That is, it will be taken up by the plant and probably incorporated as an organic complex, not as inorganic arsenate.

W. Peters: Thank you. I have one other question. Has there been any ambient sampling done for atmospheric levels of arsenic around these cotton fields. I know Cyril Durenberger has done a study, but is there anything in addition to that?

J. R. Abernathy: There was work in our area conducted several years ago, which involved sampling in all directions from cotton gin operations for primarily arsenic-type material.

E. A. Woolson: In terms of the cotton field itself, to the best of my knowledge no sampling has ever been done to determine particulates or vapor phase arsenic levels. It's something that needs to be done and there are some plans to do that.

J. R. Abernathy: One thing I might add right here is that when we talk about the use of arsenic acid in cotton harvesting, some of the gin trash goes into the livestock industry. On the cotton that has arsenic acid applied to it, it specifically states on the label, "DO NOT FEED THIS GIN TRASH TO LIVESTOCK." As far as feeding studies with the other materials, I'm sure that has been done.

R. D. Wauchope: These compounds have been under continuous review by the EPA for some years. The real potential problem that I see is misuse, especially with the herbicides. We've got to educate the user because I don't think the farmers are aware of the close scrutiny these chemicals are receiving. In the cases where they are being misused, this misuse is leading to residues in food crops that should not be there. My experience, of course, is in the Mississippi Delta area and I wonder if you can tell me if this is a problem in the Texas High Plains area.

J. R. Abernathy: I would have to say we don't use that much MSMA/DSMA on the Texas High Plains. If we look at Central Texas/South Texas where we have greater weed pressures and greater Johnson grass pressures, there is a considerable amount used. There are definitely some times when it is used and it shouldn't be. Primarily, it is put on the cotton crop too late. When it is put on after the flowering stage, it not only hurts the yield of the crop, but it concentrates the residues. And then there are some problems with spray drift on some adjacent crops. But from our standpoint, in Texas overall, I don't think we're in that bad a situation. I think there are four states that are cleared for overtop applications. Up until a couple of years ago we did not have a clearance for overtop use in Texas. It was always a directed spray from an airplane.

W. Peters: What percent of the cotton panhandle would be stripper versus picker?

J. R. Abernathy: A hundred percent of the cotton on the Texas High Plains is stripper harvested.

W. Peters: As I understand Ed Woolson, did you indicate that you were not aware of air sampling on MSMA or DSMA at the time of application or was that a general statement?

E. A. Woolson: There has been no publication in the literature dealing with air sampling of the cotton crops be it MSMA, DSMA or arsenic acid.

W. Peters: As far as I know that is true. I want to point out that there are data in the EPA files where air sampling has been done from the point of application to air pilots to ground crews and to ground spray applicators. I think this gives some indication of exposure.

J. R. Abernathy: Yes, there was considerable amount of work done several years ago on the residue levels of arsenic acid on clothing.

E. A. Woolson: But, John, those deal specifically with exposure to applicators and I believe that's what you're saying also for MSMA. In terms of environmental air sampling after the initial spray, I don't think anyone could speciate or detect the low levels. Correct me if I'm wrong.

J. R. Abernathy: I think that's right. Thank you very much.

6

THE CONTINUING NEED FOR INORGANIC ARSENICAL PESTICIDES

John C. Alden
Woolfolk Chemical Works, Inc.
P. O. Box 938
Fort Valley, Georgia 31030

This paper deals with the continuing need for inorganic arsenical pesticides in special use situations where no alternative product is available or where registered alternatives are not effective or dependable. A brief history and the present status of inorganic arsenical pesticides in the United States are given. Some of the uses which should be continued are discussed. These include use as a herbicide on turf, as a growth regulator for grapefruit, as an insecticide for turf and fly control in poultry houses, and as a fungicide for grapes and certain cotton fabrics. Products covered are Lead Arsenate, Calcium Arsenate, Sodium Arsenite, and 10,10'-oxybisphenoxarsine. Worker exposure to arsenical pesticides in manufacturing and during application is discussed along with suggested methods of reducing exposure.

1. Introduction

It was with some reluctance that I accepted this assignment. The subject has already been well investigated and reported in the responses to the Federal Register Arsenic and Lead Notice of 1971 made by industry, research, user and other groups. It was again updated in the responses to the Rebuttable Presumption Against Registration (R-PAR) issued by EPA in November 1978. It was very well covered by the report prepared for EPA by the 1972 Special Pesticide Review Group and amended in 1975. Much of what follows comes from these reports. The rest comes from my own knowledge, and I hope will bring you up-to-date on the present status of these products and their usefulness.

2. History

Our company was one of the earlier producers of arsenical pesticides which at one time consisted of approximately ten companies. Four produced both Lead Arsenate and Calcium Arsenate, five produced Lead Arsenate only, and one produced Calcium Arsenate only.

Calcium Arsenate and Lead Arsenate were the mainstay of the insecticide industry from the early 1900's until the advent of the organic pesticides following World War II. Because they were less phytotoxic they largely replaced the earlier arsenicals, Paris Green and London Purple, which had been in use since about 1865. During the peak use period from the 1930's through the 1940's, an average of approximately 50,000,000 pounds of each was used annually. The peak year for use of Lead Arsenate was 1944 when 90,000,000 pounds [1] was used. As much as 80,000,000 pounds [2] per year of Calcium Arsenate has been used.

Lead Arsenate was used primarily on fruit and vegetable crops and Calcium Arsenate on cotton. At their peak, Lead Arsenate was registered for use on forty-one feed and food crops and Calcium Arsenate on eighty-three. Both were also registered for non-food use such as turf, ornamentals and shade trees. Presently there are twenty-seven registered crops or sites for Lead Arsenate use and thirty-four for Calcium Arsenate [5].

These are somewhat moot statistics, however, since there are no domestic commercial producers of Calcium Arsenate at the present and only one producer of Lead Arsenate, which is used for the limited grapefruit market in Florida. The last production of the old line companies was probably in 1974 for Calcium Arsenate and 1975 for Lead Arsenate. Since that time, the remaining markets have been supplied by scattered inventories and in the case of Lead Arsenate at least by imports.

Most of the old arsenical producers were out of the business by the time the FR Notice on Lead and Arsenic was issued in 1971, but there were still at least two producing Lead Arsenate and two producing Calcium Arsenate at that time. The coup de grace was administered to the industry by release in 1974 of the Allied and Dow studies and the OSHA decision to drastically lower the arsenic air standards a few years later.

As can be seen from table 1, the air tolerance for arsenic was finally established at 1/50 the exposure limit formerly allowed.

Table 1. OSHA arsenic air standards

WORKER EXPOSURE LIMIT As mg/m^3			
REGULATIONS 1968	NIOSH RECOMMENDATION 1973	PROPOSED OSHA REGULATIONS 1975	FINAL OSHA REGULATIONS 1978
0.5	.05	.004	.01

Since all of the arsenical pesticide plants had been built in the early part of the century, it would have been prohibitively costly to modernize them enough to meet these standards. The Woolfolk plant was completed in 1925. So with a decreasing market, the few remaining plants had no choice but to suspend operations. The last arsenical production at our plant was in 1975. However, as you can see from the

[1]Figures in brackets indicate the references at the end of this paper.

following graphs, our company still had a sizable market for Lead Arsenate and a stabilizing market for Calcium Arsenate which also gave some indication of increasing.

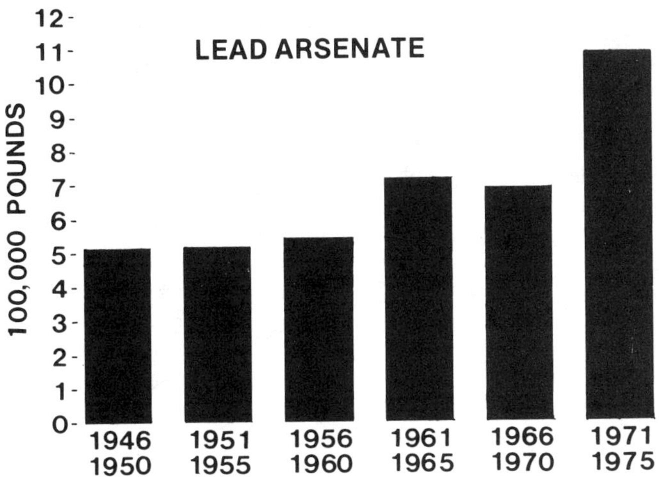

Figure 1. Woolfolk average annual production Lead Arsenate.

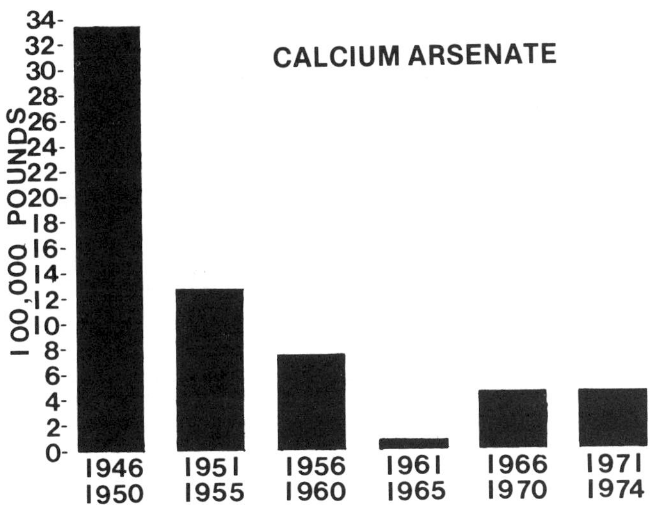

Figure 2. Woolfolk average annual production Calcium Arsenate.

As can be seen from these graphs, it is obvious that there was a demand and there was a need for these products at the time manufacturing was voluntarily suspended.

3. Important Uses

The following five tables summarize the uses that still remain important for the inorganic arsenical pesticides and also for the organic arsenical, 10,10'-oxybisphenoxarsine, which I was also asked to cover.

Table 2. Sodium Arsenite uses

CROP/SITE	PURPOSE	AREA	ESTIMATED ANNUAL USE
Grapes	Measles	Calif. 5-10% Acreage Table Grapes Only	
	Dead Arm	Calif. 30,000 Acres	90,000 lbs. [3]
USDA [1] Livestock Quarantine	Texas Cattle Fever Tick	Mexican Border	---

There is no other control for measles; potential loss has been estimated at $7,000,000 [3]. It is essential for guarding against introduction of Texas cattle fever tick into the United States from Mexico [3].

Table 3. 10,10'-oxybisphenoxarsine uses

CROP/SITE	PURPOSE	AREA	ESTIMATED ANNUAL USE
Cotton Fabric for Boats [3]	Fungi Control	Tropical & Subtropical Areas	---
Vinyl Films [3]	Fungi Control	Shower Curtains, Wall Coverings	

This compound is considered more effective than available substitutes [3]. In 1970, 90% of all flexible polyvinyl chloride films were treated with this material. The value of these films ran to the tens of millions of dollars [3].

Table 4. Lead Arsenate uses

CROP/SITE	PURPOSE	AREA	ESTIMATED ANNUAL USE
Grapefruit	Control Acidity	Florida 40,000 Acres	300,000 lbs. [4]

There is no substitute for an arsenic compound in obtaining the growth regulatory effect of acid reduction in grapefruit [4]. However, Calcium Arsenate could replace Lead Arsenate. Absence of arsenical growth regulators could severely affect the growers of 100,000 acres of grapefruit in the state of Florida.

Table 5. Calcium Arsenate uses

CROP/SITE	PURPOSE	AREA	ESTIMATED ANNUAL USE
44 Crops Flowers/Ornamentals	Snail Bait	---	---
Grapefruit (if registered)	Control Acidity	Florida 40,000 Acres	230,000 lbs.
Poultry Houses	Fly Control	Southeast	300,000 lbs.
Turf	*Poa annua*	Midwest, Mid-Atlantic NE, 35,000 Acres	1,400,000 lbs.
	Soil Insects	Above Plus Turf Nurseries	---

Snail bait use of Calcium Arsenate, usually in combination with metaldehyde, is important because it has a broad range of registered uses. Other potential controls have very limited registrations [4].

Use of Calcium Arsenate on grapefruit as a substitute for Lead Arsenate is a real possibility. It is slightly more effective and does not contain lead. An IR-4 petition for a tolerance for Calcium Arsenate, based on arsenic, was submitted to EPA in 1976 but was not considered at that time because of the R-PAR of inorganic arsenicals [4].

While there are other chemicals, mostly organophosphates, which are used for control of fly larvae in poultry manure, the ability of flies to develop resistance demonstrates the importance of having a number of different classes of insecticides available for their control. Calcium

Arsenate could be used in a schedule with other effective products to reduce the risk of resistance.

We have recently conducted a survey of golf courses in the Midwest and Northeast and find a continuing need for Calcium Arsenate for control of Poa annua. Other products have not proven satisfactory, and in the absence of Calcium Arsenate, some courses have tried to manage and live with this grass. Golfers who have played on Poa annua infested turf know what a poor playing turf and putting surface it presents. Table 6 shows a consensus rating of potential control materials.

Table 6. Herbicidal effectiveness for Poa annua

USE AREA	DCPA	BETASAN	BENFLURALIN	CALCIUM ARSENATE
Golf Greens	Not Recommended	Fair Not Adequate	Not Recommended	Good, Best On Market
Fairways	Good	Fair Not Adequate	Fair To Good	Good, Best On Market

An additional benefit from the use of Calcium Arsenate on turf is control of soil insects which is becoming more difficult and expensive after the loss of the chlorinated hydrocarbon insecticides.

Though to the best of my knowledge Paris Green is not now being produced, I should call your attention to its possible need as a mosquito larvicide in case of resistance to organic insecticides. Oil is not effective when vegetation in summer and fall prevents adequate penetration and granular Paris Green may need to be used [3].

4. Availability and Future of Inorganic Arsenicals

Shortly after closing our arsenical plant in 1975, we also discontinued production of Sodium Arsenite.

We do not intend to resume production of Sodium Arsenite or Lead Arsenate, but present plans are to resume manufacture of Calcium Arsenate in a new facility and in a different form. The new flowable product which we plan to produce should greatly reduce worker exposure to airborne arsenic. Since it will be supplied as a liquid and will be applied as a coarse spray, applicator exposure will also be reduced.

This product has been tested for efficacy by Dr. W. H. Daniel, Turf Specialist at Purdue University. Preliminary tests indicate that it is more efficacious than formulations previously used, but certainly it is just as effective on Poa annua. Just recently tests with this product were also run, under his supervision, to measure worker exposure during application to turf.

We also have had this formulation in tests at the Lake Alfred, Florida, Citrus Experiment Station and have every reason to believe that it can be used as a substitute for Lead Arsenate on grapefruit, if registration is granted.

Tests have also been scheduled for control of housefly and soldier fly larvae in poultry manure. Though results are not in at this time, we expect efficacy to be just as good as with other forms of Calcium Arsenate.

While we no longer produce Sodium Arsenite, there is at least one producer of this product in California, and I think at least one additional one in the United States.

10,10'-oxybisphenoxarsine is produced by a single producer and to the best of my knowledge, this compound continues to be available [3].

The future of all these arsenical compounds depends, of course, on the ability of the manufacturers to meet OSHA standards, the future regulatory actions of EPA, the economics of maintaining registration, and the ultimate cost to the user.

References

[1] Broom, R. A., Niagara Chemical Division, FMC Corporation, Brief on Lead Arsenate in response to "F. R. Arsenic and Lead Notice" on behalf of the Industry Task Force for Agricultural Arsenical Pesticides.

[2] Metcalf, Flint and Metcalf, Destructive And Useful Insects, Fourth Edition (McGraw-Hill, Inc., 1962).

[3] United States Environmental Protection Agency, Special Pesticides Review Group, Office of Pesticide Programs, Arsenical Pesticides, Man And The Environment, 1972, Rev. 1975.

[4] United States Department of Agriculture Response to the Rebuttable Presumption Against Reregistration of Inorganic Arsenicals.

[5] Woolfolk Chemical Works, Inc., Response to the Rebuttable Presumption Against Reregistration of Inorganic Arsenicals.

DISCUSSION

E. A. Woolson: What specific geographic area and environmental conditions are required for Poa Annua (Annual Bluegrass) to be a problem?

J. C. Alden: It's a cool season weed and is found in the mid-west, the northeast, up into the Atlantic seaboard around Maryland and up the coast. The other area is in the south, but calcium arsenate is not necessary for use in the south. We estimate that about 35,000 acres, primarily golf courses, are treated. It would be used by professional people who are experienced and have knowledge of the proper means of application.

E. A. Woolson: They probably have to be registered pesticide applicators anyway for other compounds.

J. C. Alden: Yes, that is correct.

E. A. Woolson: To the best of your knowledge, have any insects ever developed a resistance to the arsenicals? You said that house flies would develop resistance to organic materials. I know plants seemed to have adopted their biochemical systems to deal with high levels of arsenic.

J. C. Alden: I think the coddling moth in the northwest did develop resistance to lead arsenate, but that's the only one. Flies, to the best of my knowledge, have never developed any resistance.

7

ARSENIC IN ELECTRONICS

Robert K. Willardson
Cominco American Incorporated
Electronic Materials Division
E. 15128 Euclid Avenue
Spokane, WA 99216

The use of arsenic in the electronics industry covers a wide spectrum involving consumer, communications, industrial instrumentation, government/military, and data processing applications. Of major importance as a group V element it is used to impart n-type conductivity in the group IV semiconductors germanium and silicon, to alloy with group IV metals, and to form important semiconductor compounds with group III and group V metals. One of the most important compound semiconductors is gallium arsenide and its alloys. Products include room temperature lasers, discrete microwave devices, solar cells, and photoemissive surfaces as well as light emitting diodes and displays. Gallium arsenide diodes and lasers are used in couplers and fiber optic communications. Possibly its most important use will be monolithic microwave integrated circuits for communication and ultra high speed data processing.

1. Introduction

The semiconductor industry is the youngest market for metallic arsenic and its compounds, being barely 30 years old. In terms of product quality it is the most demanding market because the performance of microelectronic circuits is so strongly affected by certain critical impurities that these must be controlled in the parts per billion range. The purity of arsenic required, the form in which it is used and the specific applications are shown in Table I.

The production of high purity arsenic has become a logical business extension for some of the lead and copper smelters which process arsenic bearing feed materials and, therefore, historically have had to pay attention to the factors pertinent to dealing with this element. Today, practically all of the metallic arsenic used by the semiconductor industry is produced by a few integrated smelters in North America, Europe, and Japan. The same, however, does not apply to high purity arsenic compounds, and in particular, to arsine gas. The production and distribution of this compound is handled by specialty companies which offer an array of cylinder packaged high purity gases to the semiconductor industry.

TABLE I - ARSENIC IN ELECTRONICS

As Grade	Form	Use	Major Market Sector
69+	GaAs	Discrete Microwave Devices Monolithic Microwave Integrated Circuits	Communications Government/Military Industrial/Instrumentation Electronic Data Processing
69	GaAs GaAlAs	Solar Cells	Energy
69	GaAs GaAlAs InGaAs	Lasers Optoelectronic Couplers Photocathodes	Communications Industrial/Instrumentation Government/Military
69	GaAsP	Light Emitting Diodes Displays	Consumer Industrial/Instrumentation
59	Se-As Alloys	Dry Copiers	Data Processing
49 and lower grades	Pb-Sb-As Alloys	Batteries	Industrial Energy

2. Production

Historically, arsenic oxide and metallic arsenic have been produced as by-products in copper smelters. Lead concentrates also contain arsenic. Cominco Ltd. recovers metallic arsenic from its lead smelting and refining operations {1}. Anode slimes from the Betts electrolytic refining process are the starting material. After drying, the slimes are melted in an oxidizing atmosphere. The volatilized arsenic and antimony oxides are collected and reduced in the presence of suitable lead-bearing materials to yield a lead-arsenic-antimony alloy. Crude metallic arsenic is recovered from this alloy by vacuum distillation. After a chloride treatment to remove traces of cadmium, a second stage of distillation results in commercial grade metallic arsenic suitable for the production of various alloys with controlled composition {1}.

Several additional refining operations are necessary to produce the 69+ arsenic required for the manufacture of high purity III-V semiconductor compounds. Table II shows a typical analysis of high purity arsenic as well as that of a lower grade. Note that zinc, cadmium, sulphur, and silicon which must be reduced to a few parts per billion for gallium arsenide microwave devices are relatively high in the lower grade of arsenic and higher than desired for microwave devices in the 69+ grade material.

TABLE II - TYPICAL ANALYSES HIGH PURITY ARSENIC

	PPB by Weight			PPB by Weight	
	49+	69+		49+	69+
Al	800	7	Mn	15	(10)
Bi	(25)	(25)	Ni	30	(30)
Cd	200	(30)	Pb	250	30
Ca	50	3	Si	5000	30
Cu	20	(10)	Na	200	2
Cr	40	(10)	Ag	300	(10)
In	(15)	(15)	Sb	1600	60
Fe	800	4	S	400	9
K	50	2	Sn	(15)	(15)
Mg	70	1	Ti	200	(6)
			Zn	30	(10)

( ) Not detected with detection limits given in parentheses.

Arsenic exists in at least three allotropic forms. The major product sold commercially is the crystalline alpha form. Initially, it has a bright shiny metallic luster, but it oxidizes readily in air and is soon covered with a thick black coating that is soft and easily abraded. The alpha form has a low electrical resistivity and is very brittle.

The beta form is amorphous, black, vitreous and glassy, much like obsidian. It is a non-conductor, and is very brittle. It has a very high resistance to oxidation.

The gamma form is amorphous with a dull grey columnar structure that is soft and friable. It is stable towards oxidation at room temperature and may be stored or shipped in air with no change in appearance. It alloys with lead more rapidly than do the alpha or beta forms, probably because of its columnar amorphous structure. The columns are 0.1 to 0.5 mm in diameter, which easily break into successive fragments until an aspect ratio of about one is reached. As a result, the sandy fines do not contain a large fraction of small particles which can become and remain airborne. For uses such as arsenic diffusion sources for silicon wafers, the gamma form has an advantage over both the alpha and beta forms because crushing and the resulting dust and possible contamination are avoided.

Both the beta and gamma forms are unstable at higher temperatures and convert to the alpha form at 360°C.

3. Batteries

Battery grids operate in a corrosive environment and require significant mechanical strength to bear the load of their own weight as well as the stresses arising from changes in temperature and changes in volume of the active materials on charge and discharge. The solid solubility of antimony in lead is about 3.5 weight percent (w/o) at the eutectic temperature and decreases to 0.1 w/o at room temperature. Precipitation hardening gives a substantial improvement in the tensile strength and creep properties of the alloy, e.g., the tensile strength can be increased as much as a factor of three {2}.

Arsenic serves an important function in the antimonial lead alloys used in storage battery grids for industrial batteries. In amounts as low as 0.01 w/o and as high as 0.5 w/o, arsenic improves the castability and refines the microstructure of lead antimony alloys. Concentration of 0.2 - 0.4 w/o gives the resulting grid increased resistance against positive plate electrolytic corrosion during the life of the battery {3}.

In recent years lead/acid batteries with lower maintenance requirements have been introduced. This has reduced the requirements for arsenic in two ways: first, a change to lead-calcium and lead-calcium-tin alloys and second, a reduction in the amount of antimony and arsenic in some of the lead-antimony-arsenic grids. By lowering the antimony content to 3 w/o or less, the electrochemical activity of the positive plate is lowered. This lowers the overcharge current and reduces the water losses and maintenance substantially in lead-antimony-arsenic batteries {2}.

Current annual usage of arsenic in lead acid storage batteries is estimated to be 1000 tons.

4. Electrophotography

Amorphous selenium-arsenic alloys have been important materials for electrophotography (plain-paper copying and duplicating machines) for more than two decades. Amorphous selenium is an excellent photoconductor for this application. It is an insulator, which in the dark will hold a charge with applied fields of up to 1.5 x 10^5 V/cm. It has an adequate hole mobility of 0.13 cm/V sec and a relatively long deep trapping lifetime.

The transition temperature from the excellent photoconductive amorphous to a high-dark-conductivity crystalline form is a relatively low 35°C. Should an area crystallize it would prove useless as a photoconductor since it could not hold a charge. The addition of arsenic increases this transition temperature to as high as 180°C at 40 w/o arsenic. The rate of crystallization of selenium with 1 w/o arsenic at 35°C is ten times slower than that in pure selenium {4}.

The addition of arsenic provides other important improvements, including an increase in the electron range, a decrease in the dark-decay rate, an increase in acceptance potential for negative charging and an extension of the spectral response to the red end of the spectrum. Selenium films containing 25 w/o arsenic have essentially panchromatic response in the visible region.

Amorphous arsenic triselenide not only has the advantage of a broader spectral sensitivity than pure selenium but superior resistance to abrasion as well as crystallization.

The primary configuration of selenium-arsenide alloys for application in electrophotography is in films about 60 microns in thickness. These are prepared by vapor deposition. Adhesion to the substrate is important. A match of thermal coefficient of expansion of the substrate and the photoconductor is vital to prevent bonding failures during large excursions in temperatures. The thermal-expansion coefficient of selenium-arsenic alloys is a strong function of the arsenic content. A glass with 40 w/o arsenic and 60 w/o selenium has the same coefficient of expansion as aluminum {4}.

The amount of arsenic used annually in electrophotography is in the range of 10 to 20 tons.

5. Elemental Semiconductors

Arsenic and phosphorus are used to impart n-type conductivity to silicon and germanium. High-purity 69 alpha arsenic is used with high-purity silicon to grow single crystals of the desired resistivity. Wafers of boron doped p-type silicon are diffused or ion implanted with arsenic in very small regions to make integrated circuits. Figure 1 shows a cross-section of an HMOS gate or device, which utilizes an arsenic ion-implanted region for a source and drain.

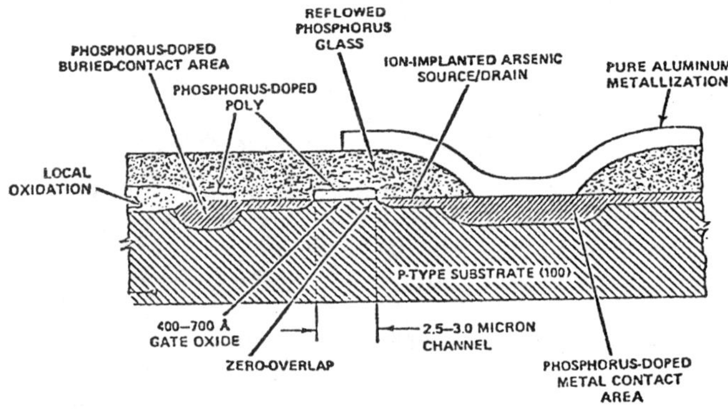

Figure 1. Cross-section of an HMOS circuit element.

Four, sixteen, or sixty-four thousand of these regions may be implanted into a small silicon chip only 5 or 6 mm on a side. The implant depth is only about one micron. The amount of arsenic in this small chip is about the same as can be found in one drop of ordinary drinking water, yet it is essential to the working of each of those thousands of semiconductor devices.

Metal-oxide-semiconductor integrated circuits (MOS IC) have grown from medium scale integration (MSI) in 1965, (10^1 to 10^2 gates per chip) to very large scale integration (VLSI) manufacturable today, (10^4 to 10^5 gates per chip). Figure 2 shows this growth in numbers of gates per chip and also the growth in the size of the chips being used.

The worldwide annual usage of arsenic in silicon and germanium devices is only a few dozen kilograms.

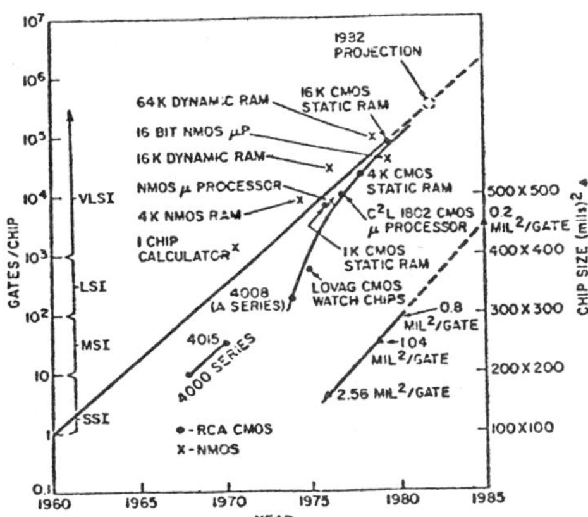

Figure 2. Size and density of integrated circuit chips as a function of time (after Douglas {4}).

6. Gallium Arsenide

Complementary to the group IV elemental semiconductors are the semiconducting compounds made from the group III and group V elements. These III-V compounds are GaAs, GaP, InP, InSb, InAs, GaSb, AlAs, AlSb, and AlP in order of commercial importance. Alloys of these compounds make possible a wide variety of properties that can be applied in an array of applications.

The crystal structure for these compounds is the same as for silicon and germanium, i.e., a diamond lattice. However, for a compound with two different atoms it is called the zinc blende lattice.

In a GaAs crystal, the Ga and As atoms occupy adjacent positions in the lattice, and the bonding between the atoms is tetrahedral, as for column IV elements. Each Ga atom is surrounded by four As atoms and each As atom is surrounded by four Ga atoms as is shown in Figure 3.

The arsenic and phosphorus are very reactive with most materials and have high vapor pressures at the melting point of most of these compounds. Therefore, in the 1950s and 1960, they were produced in sealed quartz vessels. To pull crystals, complicated magnetic coupling was used. In the 1970s the use of boric oxide glass and an over pressure of argon came into wide usage. In high pressure systems, where the argon pressure could be as high as one hundred atmospheres, the compound GaAs was both formed in situ by direct reaction of the elements and then the crystal pulled. This system eliminated silicon contamination from the all quartz system and made 75 and 100 mm diameter crystals of up to 5 kilogram size practical.

The current usage of arsenic in GaAs based devices is between 5 and 10 tons/year. By 1990 this will increase to between 30 and 50 tons/year.

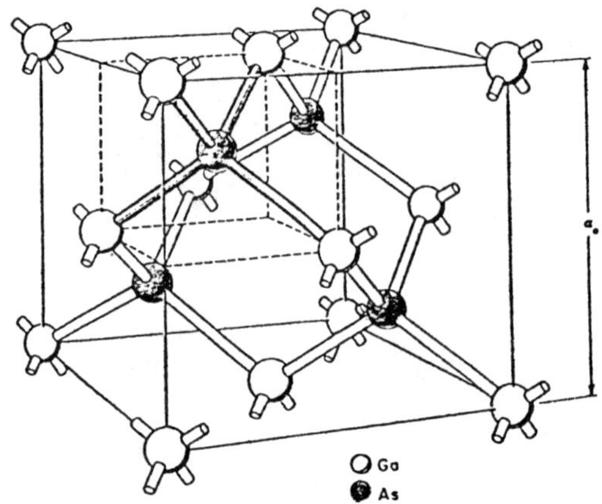

Figure 3. Crystal structure of GaAs

6.1. Light emitting diodes

Light emitting diodes or LEDS were discovered nearly two decades ago and during the past decade have become quite commonplace as the ubiquitous red panel lights and numerical displays on everything from hand calculators and clocks to instrument panels and television sets.

The red light is produced in an alloy of 60 w/o GaAs - 40 w/o GaP{6}. This alloy is grown on a 400 micron thick wafer of GaAs or gallium phosphide and is about 10 microns thick. It is epitaxially deposited in vapor phase growth from the thermal decomposition of arsine and phosphine. The gallium is transported as a subchloride by passing HCl gas over the molten metal. Figure 4 shows a schematic representation of the equipment and gases used in the epitaxial growth. The Se and Zn are used to dope the layers with the desired concentrations of n-and p-type impurities, respectively.

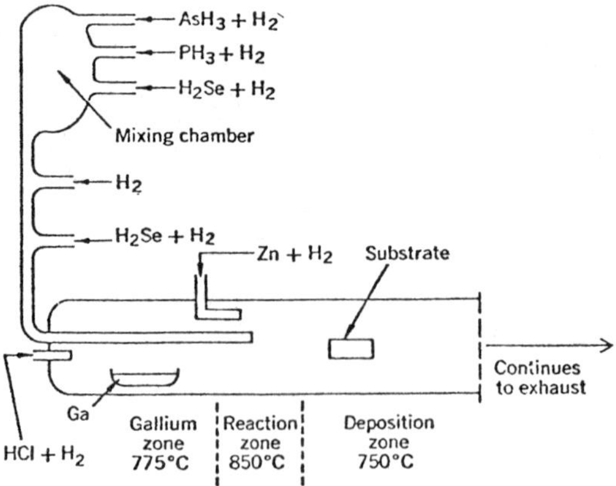

Figure 4. Schematic of equipment used to prepare vapor epitaxial GaAsP (after Nuese, Kressel and Tadany {6}).

Increasing the amount of GaP can produce yellow light and pure gallium phosphide the green light.

6.2 Solar Cells

Solar cells are photovoltaic devices which convert sunlight directly into direct current electricity. They have been an important part of the space program since it began and are capable of making a significant impact on terrestrial energy needs {7 }.

Most silicon solar cells have efficiencies of 10% to 14% at AM1. The AM1 spectrum represents the sunlight at the earth's surface under optimum conditions and leads to a total incident power of 1.0 KW/m^2. Under similar conditions the best silicon, GaAs, and GaAlAs solar cells have efficiencies of 15%, 19% and 23%, respectively. A cross-section of a GaAlAs heterojunction solar cell is shown in Figure 5.

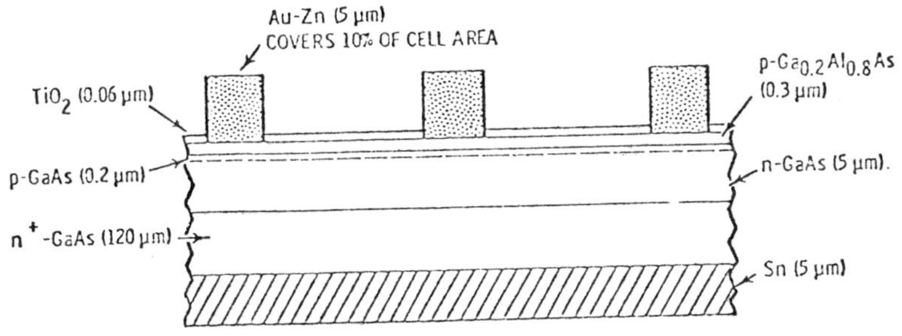

Figure 5. Cross-section of a GaAlAs solar cell

The solar cell is constructed by first growing an epitaxial layer of n-type GaAs on a heavily doped n-type GaAs substrate. Next deposited is a $Ga_{0.2}Al_{0.8}As$ layer doped with a p-type dopant such as Zn which diffuses about 0.2 microns into the n-type GaAs layer during the $Ga_{0.2}Al_{0.8}As$ growth. This produces a p/n junction in the GaAs across which the photovoltage appears when the surface of the cell is illuminated. A coating of TiO_2 about 600 A in thickness is used to reduce reflection losses from the $Ga_{0.2}Al_{0.8}As$ surface.

It is important to note other advantages that GaAs and GaAlAs solar cells have over those made from silicon. The efficiency of GaAs solar cells increases slightly up to 80°C. and then decreases at only one-half the rate of that of silicon cells. They can be operated at temperatures up to 350°C. as compared to 200°C. for silicon. Light absorption by GaAs is much more effective than by silicon. This makes it possible to use cells with less thickness and design more efficient cooling. Concentrations of up to 5000 suns have been used successfully on GaAS. By using inexpensive 1000 sun concentrators, one cell could generate as much power as a thousand cells without concentrators. The resistance to nuclear radiation is also superior in GaAs.

Because of production costs and the limited amount of gallium available, GaAs and GaAlAs solar cells will probably never become a major source of terrestrial power. The world's annual supply of gallium is currently about 40 tons and it would be difficult to increase this by more than a factor of 10. However, gallium arsenide has important advantages over silicon and will be used in many applications where efficiency and operating temperature are of prime importance. For example, in the 1980s GaAlAs solar cells will find practical application in space shuttle missions.

6.3 Optoelectronic Devices

Pulsed GaAs laser diodes emitting peak power in excess of 5 W are useful in ranging and infrared illumination. These usually operate at duty cycles of about 0.1% with pulse lengths of 50-200 ns. Operating current densities can be 40,000 A/cm^2.

Visible homojunction lasers made using Ga(AsP) alloys can efficiently produce red light with a wavelength of 6750 angstroms.

The development of optical fibers with low (<10 db/Km) absorption "windows" in the vicinity of 0.8-0.9, 1.0-1.1, and 1.5-1.6 microns has given impetus to the development of optical communication fiber systems using diode and injection laser sources. Figure 6 shows the elements of a fiber optic system - the diode or laser transmitter, the fiber cable and the photodiode detector. Fiber optics provides complete electrical isolation, no short-circuit possibilities, immunity to radio-frequency interference, transmission security, wide signal bandwidth, no crosstalk, no spark or fire hazard, wide-operating temperature, light weight and low cost.

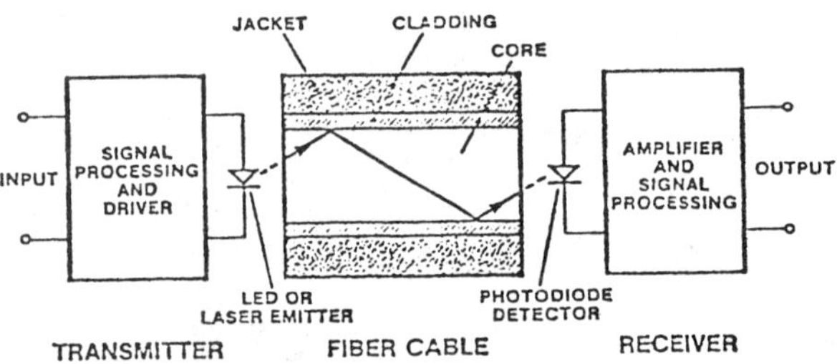

Figure 6. Elements of a fiber optic system

Telecommunications is the most notable of the current users of fiber optic technology (with telephone links within and between major U.S. cities already installed and many more scheduled for installation). Other industries-business machines, industrial controls, medical equipment and CATV-are also beginning to incorporate fiber optics technology into their end products. The system requirements vary depending on the desired bandwith and transmission distance. In general, it is desirable that the diodes be capable of modulation at very high frequencies (100-1000 MHz range), that the spectral line width be small to reduce pulse dispersion, and that the beam width be small in order to improve the radiation coupling into the fibers. Power in the milliwatt range from the CW laser diode is usually adequate.

The AlGaAs system provides lasers in the 0.8-0.9 micron wavelength range. Figure 7 shows a cross-section of a AlGaAs laser. The laser light is generated in the very thin 100 to 3000 angstrom thick $Al_{0.1}$, $Ga_{0.9}As$ expitaxial layer.

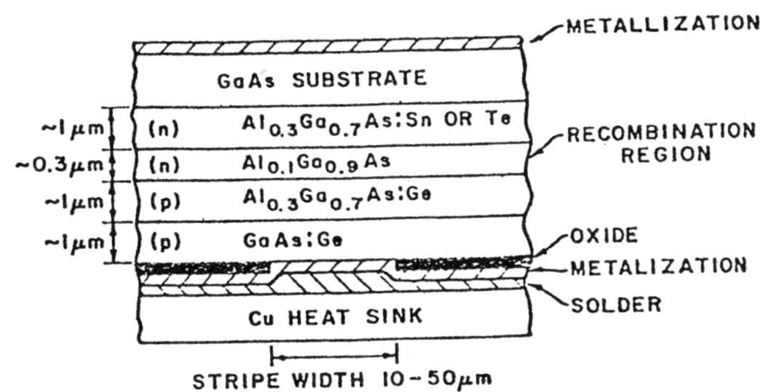

Figure 7. Cross-section of GaAlAs laser diode designed for CW operation in the 0.80-0.83 micron spectral range. After Kressel and Butler {9}.

The $Al_{0.3}Ga_{0.7}As$ layers are optical walls which confine the light and help to produce a more efficient laser. Confinement in the horizontal direction is provided by the special oxide-isolated metal-stripe base contact {9}. The various layers of AlGaAs are grown by a liquid expitaxy technique on a highly perfect GaAs substrate {9}.

Other layer compositions provide lasers for different wavelength requirements. For example: $In_{0.16}Ga_{0.84}As$ gives lasers at the 1.075 micron wavelength. For long distance communication, it is desirable to use a GeO_2SiO_2 core glass composition which has very low 0.2 db/km loss in the 1.5 to 1.6 micron wavelength range. For this wavelength region the quaternary compound InGaAsP is used. Detectors for this wavelength region are also made from InGaAsP {8}.

6.4 Monolithic Microwave Integrated Circuits

Gallium arsenide devices can have higher operating frequencies, lower power consumption and noise as well as superior resistance to nuclear radiation than comparable silicon devices. One should also note that in small volume, gallium arsenide substrates are now almost 50 times as expensive as silicon substrates and will probably always be at least 5 to 10 times more expensive. This is because 1) gallium arsenide has two and one-half times the density of silicon, 2) gallium is relatively expensive, and 3) gallium arsenide crystal growth is more complicated and hence much more expensive.

The GaAs field-effect transistor (FET) is well suited for monolithic microwave circuits. Progress has been made in crystal growth and large slices are now available in 50, 60, 75 and 100 mm diameters and in large quantities for processing. GaAs material uniformity has improved markedly and ion-implantation technology has been successfully applied to monolithic microwave circuits.

Interest in gallium arsenide centers on four major applications: (1) low noise FETs for front-end circuits in microwave receivers, (2) high-power FETs for microwave transmitters, (3) integrated circuits (ICs) for communication net works, and (4) ICs for signal processors. These are applications where silicon cannot do the job {10}.

Probably the most complex GaAS microwave circuit produced to date is a receiver with an oscillator (2.1 to 2.5 GHz), balanced mixer, drive circuitry, and IF preamplifier on a single chip. This unit was made at Hewlett-Packard. Mixers operating at 31 and 110-GHz as well as oscillators working at 12 to 20-GHz have been integrated on GaAs.

Developments in digital GaAs integrated circuits are also very impressive. Rockwell has already developed a 1008-gate, 8x8 bit multiplier with an execution time of 5 nanoseconds compared with 45 nanoseconds for an equivalent commercial silicon part. A 16x16-bit multiplier IC with about 10,000 gates that will produce a 32-bit product in less than two nanoseconds is under development {10}.

Fujitsu has developed a gallium arsenide logic device with switching times of 30 picoseconds at room temperature and only 17 picoseconds with one-tenth the power dissipation of conventional gallium arsenide circuits when operated at-196°C {11}.Fujitsu officials predict that this exotic circuit will be used to make the high-speed, low-power micro-micro circuits of the coming generation of super computers.

Gallium arsenide has been used for almost two decades to make discrete microwave devices, but in the future all of our lives will be influenced by gallium arsenide monolithic microwave integrated circuits for communication and new exotic integrated circuits for ultra high-speed data processing.

6.5 Photoemissive Surfaces

Negative-electron-affinity photocathodes represent a significant new class of seminconductor electron-emission device. Their original conception and subsequent development have derived from a rather detailed knowledge of the surface physics {12}.

Devices utilizing GaP, GaAsP, GaAs, InGaAs, InGaAsP, and Si are commercially available at the present time. Figure 8 shows yield curves for conventional S-1 and S-20 cathodes. The S-type cathodes are GaAsP, the M-type are GaAs, and the A1, A2, and A3 are different inGaAsP compositions.

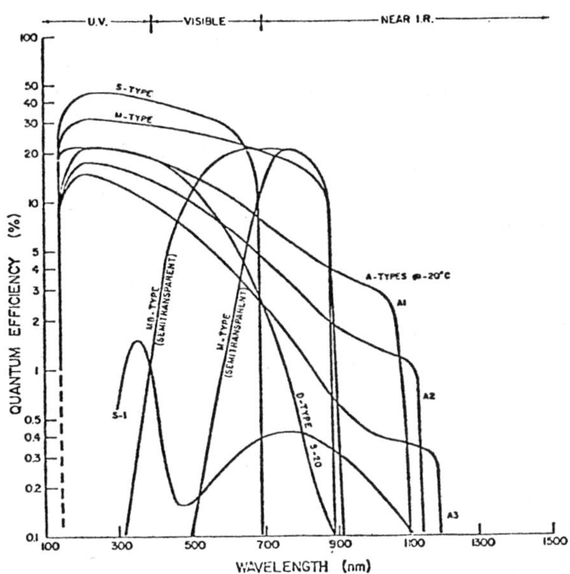

Figure 8. Quantum Yields from Commercial Photomultiplier Cathodes
S-Type are GaAsP
M-Type are GaAs
A-Type are InGaAs
S-1 and S-20 are conventional cathodes (after Escher {12}).

The GaAs photomultiplier tube was the first negative-electron-affinity photomultiplier tube commercially introduced and is the most commonly available. Advantages of the GaAs photomultiplier tube over conventional tubes are the high quantum efficiency (25%) nearly flat to wavelength of 0.855 microns and low cathode dark-current emission at $300^{\circ}K$. Other compound semiconductor cathodes which are commercially available in photomultiplier tubes include GaAsP alloys for the highest cathode quantum efficiency in the visible region and low dark current, and InGaAsP or InGaAs alloy cathodes for broad response out to wavelengths of approximately 1.1 microns {12}.

The superior performance often comes, however, at a higher cost and more restricted operating range when compared with conventional cathodes. For these reasons, the negative-electron-affinity cathode will likely remain a "premium" specialized device for some years to come. Nevertheless, the proven potential coupled with a diversity of applications assures a firm future for this semiconductor photocathode, which usually utilizes an arsenic compound.

7. Environmental Considerations

The use of arsenic in electronics is potentially hazardous when it is used in the form of arsine or arsenic trichloride in the semiconductor production area and during chemical etching of arsenic-compound wafers or scrap materials because arsine can be produced. In general, the major risk occurs in the handling of "empty" arsine tanks, GaAs scrap and epitaxial reactor residues, particularly if attempted by personnel who are not adequately trained and cautious.

Routine and special urinalyses have provided detailed information on the water supplies for and dietary habits of both GaAs workers and control individuals, but little evidence of arsenic exposure has been found.

However, while tests of urine and blood provide excellent indications of acute exposure (after a period of twelve to twenty-four hours), they are not applicable for monitoring chronic levels of toxic materials. Chronic arsenic poisoning may result from continued exposure to small amounts of arsenical compounds since the uptake is cumulative. It is known that arsenic has a special affinity for the keratinous structure of hair and nails. These materials are commonly analyzed to monitor the arsenic. Silver diethyldithiocarbamate (SDDC) colorimetric and atomic absorption methods are used for these analyses but usually require an excessive amount of sample (several hundred milligrams). Nails grow at the rate of about 0.1 mm per day and the cumulative growth of all fingernails for one to three months is required for a single analysis. Spark source or secondary ion mass spectroscopy are highly sensitive and require less than one hundred micrograms of sample for an analysis. A single sliver of nail is a sufficient sample {13}.

Care must be taken to separate arsenic surface contamination from that actually accumulated in the nail. This can be done by cleaning the sample. The nail to be analyzed can be degreased by successive two minute immersions in boiling reagent grade trichloroethylene, acetone and methanol, followed by copious rinsing in deionized water. The nail is then etched in aqua regia (1:1 reagent grade HNO_3:HCl) for five minutes. After copious rinsing in deionized water, it is immersed in methanol until it is mounted for analysis. Typical analyses for arsenic before and after etching which removed the surface of the nail to a depth of one micron are shown in Table II, {13}.

TABLE II. Arsenic Content (PPMW)

	As-Received	Cleaned
Employee A	5.6	1.3
Employee B	1.6	1.9
Employee C	2.0	2.2
Employee D	17.4	1.4

It is recommended that fingernails both as received and etched be analyzed. The as-received sample provides information on the work environment and habits of employees, either of which may require improvement or corrective action. Samples from control individuals never exhibited the high as-received arsenic analyses shown by GaAs workers A & D. The analysis of an etched sample provides the desired data on the actual arsenic content of the nail. The use of hair samples is not recommended because of possible surface contamination. The nail from the little finger or toe is preferred because it is shorter and therefore would make it possible to detect an increase in arsenic level at an earlier stage. Monthly sampling is recommended, but weekly sampling is feasible with mass spectrometry. Typical data taken over a nine-month period for several workers is shown in Table III {13}.

TABLE III

Comparison of Analyses by Spark Source Mass Spectrometry and Silver Diethyldithiocarbamate* Methods (PPMW As)

Employee	Aug	Sept	Oct	Nov	Dec	Jan	Feb	Mar	Apr
E	7.2	1.3	0.6	1.4	-	1.6 0.4 *	2.3	-	1.5
F	2.7 3.3 *	-	1.0 1.6 *	-	-	0.9 2.0 *	-	1.8	1.5
G	4.5 7.1 *	-	1.1	3.7 4.4 *	-	- 2.8 *	5.7	5.3	-
H	3.7 9.0 *	1.6	1.4	3.2	-	- 11.6 *	-	2.8	0.6
I	2.7 230 *	2.0	1.0	2.4	-	- 2.8 *	-	1.2	-

* Samples not etched and analysis by silver diethyldithiocarbamate. All other samples were analyzed by mass spectrometry and were etched with the exeption of those analyzed in August.

None of the SDDC colorimetric samples were etched, but all of the mass spectrometry samples were etched except for those analyzed in August. Increasing amounts of arsenic surface contamination are obvious in the August SDDC analysis for employees G, H and I, respectively. Surface contamination is also evident in the August mass spectrometry analysis of the nails of employee E and the January SDDC analysis for employee H. No analyses were made in December for these particular employees. All of the sample material in January from employees G, H and I was used for the SDDC analyses. The other omissions in the mass spectrometry data were due to vacations.

Three groups of workers were studied: (1) The control group of which employee F, a secretary was representative. (2) Workers in GaAs production areas whose arsenic dust sampling was very low. Employees G, H and I are representative of this group. (3) Workers involved in the clean-up of residues in GaAsP epitaxial reactors and GaAs crystal growing equipment, who are represented by employee E. This latter group wore special protective breathing apparatus during their clean-up operations.

Within the scatter of the data on etched samples from over twenty workers, the three groups could not be differentiated in terms of arsenic uptake. The workers in the GaAs production areas did not have anymore arsenic in their nails than the control group. Mass spectrometry is a viable technique for monitoring the physiological uptake of arsenic. Its primary advantage lies in a small sample size that facilitates a higher sampling rate. This higher sampling rate provides a more immediate warning of the onset of toxic levels of arsenic.

8. Summary

There are many and varied applications for semiconducting compounds. One of the most important compound semiconductors is gallium arsenide. Discovered in the early 1950s, extensive studies were made on the basic properties of GaAs as well as the discovery of laser action and microwave device characteristics in the 1960s. Practical new products including room temperature lasers, discrete microwave devices, photoemissive surfaces and in particular light emitting diodes and displays gained wide acceptance in the 1970s. By 1990 new products utilizing arsenic in the form of GaAs and its alloys will be a billion dollar industry.

Gallium arsenide phosphide is used to make red and yellow light emitting lamps, displays, and lasers. Layers of GaAs and GaAlAs make efficient heterojunction lasers and operate continuously at room temperature, as well as being the most efficient solar cells made to date. Indium arsenide, indium gallium arsenide, and indium gallium arsenide phosphide are used commercially as photoemitters, secondary-electron emitters, and cold-cathode emitters. Gallium arsenide diodes and lasers are used in couplers and fiber optic communications, but the new low-loss optical-fiber cables for long distance telephone lines use InGaAs and InGaAsP infrared light emitters and detectors.

Gallium arsenide has been used for almost two decades to make discrete microwave devices, but in the future all of our lives will be influenced by gallium arsenide monolithic microwave integrated circuits for communication and ultra high speed data processing.

Although the usage of arsenic in compound semiconductors is growing rapidly, the total quantity used will probably never exceed 100 tons/year.

References

1. Hirsch, H.E., The Recovery of Metallic Arsenic in Cominco's Lead Operations, Lead-Zinc-Tin '80 (Proceedings of a World Symposium on Metallurgy and Environmental Control sponsored by the TMS-AIME Lead, Zinc, and Tin Committee at the 109th AIME Annual Meeting, Las Vegas, Nevada, February 24-28, 1980), J. M. Cigan, T.S. Mackey, and T.J. O'Keefe, eds., pp. 360-372 The Metallurgical Society of AIME, Warrendale, PA, 1980.

2. May, G.J., Lead Alloys for Low-Maintenance Batteries, Metallurgical and Materials Technologist 12, {10}, 546-548 (1980).

3. Borchers, H., and Reuleaux, M., Effect of Small Arsenic Additions on the Age-Hardening Properties of Pure, Low-Antimony Lead Alloys, Metal 34, {9}, 811-820 (1980)

4. Berkes, John S., Materials Properties of Selenium-Based Amorphous Alloys, (Proceedings of the Second International Conference on Electrophotography Washington D.C. October 24, 1973), Book, pp. 137-141 (Society of Photographic Scientists and Engineers, Washington, D.C., 1974).

5. Douglas, E.C., Advanced Process Technology for VLSI Circuits, RCA Engineer September/October 1980.

6. Nuese, C.J., Kressel, H., and Ladany, I., The Future of LEDs, IEEE Spectrum 9, 28-38, 1972

7. Hovel, Harold J., Solar Cells, Book, "Semiconductors and Semimetals", Book, R. K. Willardson and A.C. Beer, eds., Vol. 11 Academic Press, New York, 1975.

8. Montgomery, J.D., and Dixon, F.W., State of the Art in Long Wavelength Fiber Optic Sources and Detectors, Proceedings of the 31st Electronic Components Conference, pp. 364-370, (Institute of Electrical and Electronics Engineers, New York, 1981).

9. Kressel, H., and Butler, J.K., Hetrojunction Laser Diodes, Book, "Semiconductors and Semimetals", R.K. Willardson and A.C. Beer, eds., (Academic Press, New York, 1979).

10. Gallium Arsenide Integrated Circuits, Circuits Manufacturing 21, 32-42, 1981.

11. Abe, M., Mimura T., Yokoyama N., and Ishikawa, H., New Technology towards GaAs LSI/VLSI for computer Applications, (IEEE Gallium Arsenide Integrated Circuit Symposium, San Diego, California, October 27-29, 1981, Book, Research Abstracts GaAsIC Symposium, paper No. 7, (Institute of Electrical and Electronic Engineers, New York, 1981).

12. Escher, John S., NEA Semiconductor Photoemitters, Book, "Semiconductors and Semimetals," R.K. Willardson and A.C. Beer, eds., Vol. 15, pp. 195-300, Academic Press, New York, 1981).

13. Janus, A., Brumbaugh, G., and White, W., Monitoring Physiological Levels of Arsenic by Spark Source Mass Spectrometry, unpublished.

8

ARSENICALS AS FEED ADDITIVES FOR POULTRY AND SWINE

C. E. Anderson
Vice President-Domestic Marketing
Salsbury Laboratories, Inc.
2000 Rockford Road
Charles City, IA 50616

A great deal of information exists, both published and unpublished, about organic arsenicals and their use in poultry and swine. My purpose then is to bring together the highlights of the pertinent knowledge about these valuable chemicals and their role in poultry and swine nutrition and disease prevention.

1. Introduction

The use of arsenicals as animal feed additives has its roots in human medicine. Out of these early investigations came Salvarsan, discovered by Dr. Paul Erlich, a German scientist, for the control of syphilis; tryparsamid, for combating sleeping sickness; and mapharsen, another drug for syphilis. Another anti-syphilis drug, atoxyl, was used for chickens as an antibacterial as early as 1907, but tests were inadequate to show conclusive results.

During the 1930's, extensive testing was done to determine the ingredient, or ingredients, which were present in a poultry drinking water medication used to help control coccidiosis, a costly debilitating disease of chickens. Dr. Neal Morehouse, *et al*., determined that copper arsenite was that ingredient. Many other inorganic compounds of arsenic were then quickly tested, but all were found too toxic for practical use. These researchers then turned their attention to organic arsenicals, some of which had been used for treatment in man, and had been found reasonably safe at therapeutic levels. Dozens of organic arsenicals were investigated before suitable compounds were found which could be fed successfully and safely. Of these, 3-Nitro-4-hydroxyphenylarsonic acid, which I shall henceforth call 3-Nitro, or roxarsone, had the most activity against cecal coccidiosis.

The present widespread use of the leading arsenical 3-Nitro is almost a paradox. Its use as a growth stimulant in poultry and swine stems from what researchers call serendipity--a discovery by chance. In other words, the scientists who discovered the growth stimulating property of 3-Nitro were not searching for such a benefit, but for a practical drug to control coccidiosis in chickens. In fact, the drug is no longer used primarily for the control of coccidiosis, but for growth stimulation, improved feed conversion, better feathering, increased egg production, and pigmentation. The USDA Beltsville Station confirmed these findings.

2. How arsenicals are used

Although 3-Nitro was the initial organic arsenical to be used as a feed additive, three other organic arsenicals were introduced over the next few years and all four are in use today. These arsenicals and their FDA-approved claims are:

Table I

ANIMAL	DRUG	USE LEVEL	INDICATIONS FOR USE
Chickens	Arsanilic Acid	0.005-0.01% (45-90 g/ton)	For increased rate of gain, improved feed efficiency, and improved pigmentation. Increased egg production and feed efficiency in chickens (breeders and layers).
Swine	Arsanilic Acid	0.005-0.01% (45-90 g/ton)	For increased rate of gain and improved feed efficiency in growing swine.
Swine	Arsanilic Acid	0.025-0.04% (225-360 g/ton) (Feed for 5-6 days)	For control of swine dysentery.
Chickens	Roxarsone--3-Nitro	0.0025-0.005% (22.5-45 g/ton)	For increased rate of gain, improved feed efficiency, and improved pigmentation. Increased egg production and feed efficiency in chickens (breeders and layers).
Swine	Roxarsone--3-Nitro	0.0025-0.005% (22.5-45 g/ton)	For increased rate of gain and improved feed efficiency in growing swine.
Swine	Roxarsone--3-Nitro	0.02% (180 g/ton)(feed 5-6 days)	For control of swine dysentery
Turkeys	Carbarsone	0.025-0.0375% (225-337.5 g/ton)	As an aid in the prevention of blackhead.
Turkeys	Nitarsone	0.01875% (168.75 g/ton)	As an aid in the prevention of blackhead.

3. Improved performance and economics of use

No report would be of any value without data to substantiate the economic feasibility of drugs as they are used in livestock production.

"The Hays Report," by Dr. Virgil W. Hays, University of Kentucky, developed for the Office of Technology Assessment, United States Congress, compares the use of arsenicals and widely used antibiotics for the improvement in broiler chicken performance.

Table II

SUMMARY OF CHICK PERFORMANCE TO ABOUT 8 WEEKS

	Chick Weight				Feed/Gain			
Antibiotic	n[a]	-	+	% Imp.[b]	n[a]	-	+	% Imp.[b]
Tetracycline	88	1192.1[c]	1236.1[c]	3.69	53	2.60[c]	2.54[c]	2.31
Penicillin	54	1299.2	1337.3	2.93	38	2.54	2.47	2.76
Bacitracin	39	1425.2	1438.7	0.95	33	2.27	2.22	2.20
Arsenicals	59	1285.5	1329.7	3.44	50	2.54	2.46	3.15
Bambermycins	32	1483.3	1517.9	2.35	32	2.06	2.02	1.94
Lincomycin	4	1734.5	1812.2	4.48	4	2.12	2.05	3.30
Nitrofurans	3	1414.3	1442.3	1.98	2	2.05	2.02	1.47
Oleandomycin	7	1483.0	1549.4	4.48	7	2.25	2.21	1.78
Total number	286				219			
Weighted avg.		1313.0	1351.6	2.94		2.42	2.36	2.48

a-number of comparisons (experiments).
b-percentage improvement in performance due to antibiotic.
c-average weight (g) per bird and feed/gain for birds fed diets without (-) or with (+) antibiotics.

The Hays Report, 1979.

This compilation of data shows the value of antibiotics and arsenicals in improving broiler performance in birds fed to eight weeks of age. This can be noted in the percent improvement for both chick weights and feed per pound of gain. Note that the data for arsenicals exceeds by 17 to 27% the average improvement for all medications in both categories.

Because of the combination of traditional low prices to the consumer, and increased costs of producing broilers, improvements in broiler performance--improved weights and feed conversion--are more significant today than ever before to the highly sophisticated integrated broiler operations working with shrinking margins.

Now let's take a look at how 3-Nitro benefits influence the return on 50,000 broilers.

Table III

SUMMARY OF ADDITIONAL RETURN ON A 50,000 BIRD BROILER FARM WHEN USING AN ARSENICAL

Broiler gain based on table II showing 3.44% improvement.

With an arsenical

50,000 birds x 4.3445 lbs.	= 217,225 lbs. live weight

Without an arsenical

50,000 birds x 4.20 lbs.	= 210,000 lbs. live weight
	7,225 lbs. extra weight

Broiler feed/gain based on table II showing 3.15% improvement

With an arsenical: 217,225 lbs. x 1.937 lbs. feed/gain	= 420,765 lbs. feed
Without an arsenical: 210,000 lbs. x 2.00 lbs. feed/gain	= 420,000 lbs. feed
	765 lbs. extra feed

7,225 lbs. extra weight x $0.40	=	$ 2,890.00
Less 765 lbs. extra feed x $0.08	=	61.20
Less arsenical: 108.6 lbs. x $0.75	=	81.45
Additional return		$ 2,747.00

The 3.44% improvement in gain and the 3.15% improvement in feed conversion noted in table II provide an additional return of $2,747.00 from 50,000 broilers, compared to broilers not receiving an arsenical in their feed.

The following summary (table IV) covers the use of antibiotics and arsenicals in laying hens regarding egg production, feed per dozen eggs, and hatchability. Although the use of arsenicals in these tests does not match the average improvement achieved in egg production and feed per dozen eggs, they did show a significant improvement in hatchability.

Table IV

SUMMARY OF EGG PRODUCTION,
FEED PER DOZEN EGGS AND HATCHABILITY

	Egg Production, %				Feed/doz. Eggs, lb.				Hatchability, %			
Antibiotic	n[a]	-	+	% Imp.	n	-	+	% Imp.	n	-	+	% Imp.
Tetracycline	39	52.9	59.2	11.91	22	5.50	5.01	8.91	16	74.9	76.0	1.47
Arsenicals	33	59.8	61.2	2.34	30	5.43	5.36	1.29	17	72.3	76.5	5.81
Penicillin	27	54.3	57.3	5.52	19	5.75	5.46	5.04	13	75.6	78.6	3.97
Bacitracin	12	63.2	63.8	0.95	5	4.38	4.28	2.28	2	74.6	79.8	6.97
Bambermycins	5	47.8	52.0	8.79	5	7.50	6.62	11.73	1	84.2	86.3	2.49
Erythromycin	37	66.4	67.3	1.36	13	4.42	4.36	1.36	7	84.6	84.9	0.35
Others and Combination	91	62.1	63.9	2.90	28	4.89	4.64	5.11	13	78.6	81.7	3.94
Weighted avg.	244	59.9	62.3		122	5.30	5.05		69	76.2	78.8	
Avg. Improvement,%				4.01				4.72				3.41

a-number of comparisons of hens fed diets without (-) or with (+) antibiotics.

The Hays Report, 1979.

These data (table V) from three Midwest swine operations demonstrate that 3-Nitro does not show the significant improvements as seen with broilers. But in consideration of the relatively long growing period for swine and the much heavier weights involved, arsenicals do contribute to additional profits in performance and to generally better health.

Table V

SUMMARY OF SWINE PERFORMANCE USING ARSENICALS
DURING FINISHING PERIOD

Test Company	No. of Pigs	Days on Test	Avg. Gain			Feed/Gain		
			Controls Pounds	3-Nitro Pounds	% Imp.	Controls Pounds	3-Nitro Pounds	% Imp.
Golden Sun Feeds, IA	48	106	170.77	173.05	+1.34	3.02	3.02	0.00
Hastings Pork, NE	60	91	132.60	137.57	+3.75	3.72	3.65	+1.88
S.B. Penick, MO	48	98	162.29	165.29	+1.85	3.16	3.08	+2.53
Av.			155.22	158.64	+2.20	3.30	3.25	+1.52

Using the data you have just seen for swine, let's look at a farm producing 1,000 market hogs to determine the additional return that can be attributed to an arsenical versus no arsenical.

Table VI

SUMMARY OF ADDITIONAL RETURN ON 1,000 MARKET HOGS WHEN USING AN ARSENICAL

Increase in arsenical gain based on table V showing 2.20% improvement

With an arsenical

1,000 pigs x 204.4 lbs.	= 204,400 lbs. live weight

Without an arsenical

1,000 pigs x 200 lbs.	= 200,000 lbs. live weight
Extra weight	4,400 lbs.

Feed/gain ratio based on table V showing 1.52% improvement

With an arsenical: 204,400 lbs. x 3.2006 lbs. feed/gain	= 654,202.64 lbs. feed
Without an arsenical: 200,000 lbs. x 3.25 lbs. feed/gain	= 650,000.00 lbs. feed
Extra feed	4,202.64 lbs.

4,400 lbs. extra weight x $0.50	= $ 2,200.00
Less 4,203 lbs. extra feed x $0.07	= 336.24
Less arsenical: 327 lbs. x $0.75	= 245.25
Additional return	$ 1,618.51

Projection of the data in table VI--the 2.20% improvement in gain and 1.52% improvement in feed use--result in an additional return of $1,618.51 for 1,000 market hogs over hogs not receiving the arsenical.

4. Arsenic-An essential element

Walter Mertz, Director of the Beltsville Human Nutrition Research Center, Science and Education Administration, USDA, writing in Science, Volume 213, 18 September 1981, states that research during the past three decades has added arsenic to the list of essential elements. A broad definition of an essential element, widely accepted, is this:

"An element is essential when a deficient intake consistently results in an impairment of a function from optimal to suboptimal and when supplementation with physiological levels of this element, but not of others, prevents or cures this impairment."

5. Tissue residues

It has been determined that nearly all livestock and poultry carry detectable amounts of arsenic in their body tissues. This is understandable since we know that arsenic is a constituent of many feedstuffs and is especially high in products of marine origin, such as fish meal and crab meal.

Analysis of 18 non-medicated commercial poultry rations from eight different states, by the late Dr. J. W. Cavett, averaged 0.39 ppm of arsenic. Arsenic analysis of 210 different tissues from animals receiving non-medicated feeds indicate that arsenic may frequently be present at a level of 0.1 to 0.2 ppm.

When initiating an arsenical into feeds, the arsenic content of tissues rises quite rapidly, but in six to ten days reaches a plateau and, on continued use of the feed additive, tends to decrease as indicated in table VII.

Table VII

TYPICAL DEPLETION PATTERN OF 3-NITRO FROM CHICKEN TISSUES UNDER CONTROLLED EXPERIMENT

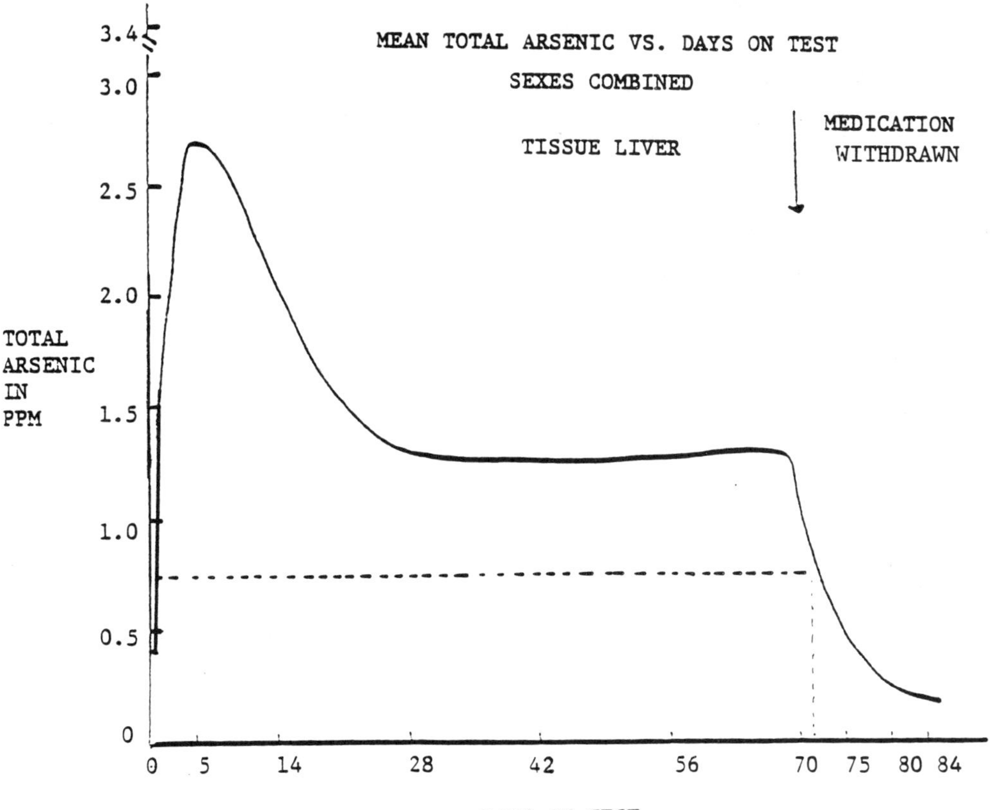

In a series of studies where 3-Nitro was included in the feed, the arsenic level increased to an average of 2.7 ppm in the liver tissue by the fourth day but had dropped to less than 1.5 ppm by the third week. The flocks represented in these tests received 3-Nitro in the feed for 70 days when it was withdrawn. Two days later, the arsenic level of the tissue was about 0.7 ppm, or below the tolerance level of 2 ppm established by the Food and Drug Administration (FDA).

6. Tolerance level increased

The FDA increased the tolerance level of 3-Nitro (roxarsone) from 1 ppm to 2 ppm in uncooked, edible by-products of chickens and turkeys in 1973. The reasons for this action are summarized in the Federal Register, Vol. 38, No. 124--June 28, 1973 (in part):

"Data from a 3-generation rat reproductive study; 2-year feeding studies in the rat, mouse and dog; and a 2-year skin painting study, concludes that a tolerance of 2 ppm is consistent with the public health and will not compromise human safety.

"Since there is no change in the required 5-day withdrawal period nor in the established use level of roxarsone, there will be no increase in levels of actual muscle tissue residue above the established safe tolerance of 0.5 ppm. Thus there is no increase in human exposure to arsenic residues.

"Certain segments of the human diet contain arsenic as a natural component at levels higher than the new tolerance of 2 ppm. For example, the National Marine Fisheries Service reports arsenic in flounder at 2.1 to 9.4 ppm, in domestic cod at 1.2 to 7.4 ppm, and in haddock at 3.5 to 9.0 ppm. Oysters can have arsenic concentrations as high as 25 ppm."

7. But what about toxicity

One can summarize the mass of toxicity data of roxarsone accumulated at Salsbury Laboratories and at independent laboratories with these conclusions:

Tests confirm that roxarsone fed at recommended levels does not produce microscopic changes in tissue which could give rise to disease.

Safety of roxarsone was further demonstrated at London University where tests showed a lack of harmful metabolites and proved that roxarsone and its lone metabolite pass almost entirely from the bodies of treated animals within a short period of time.

By way of overall summary:

The success of organic arsenicals used today in animal feeds is based on early use in human medicine. Continued research has refined organic arsenicals to a role of vital economic importance to operations feeding poultry and livestock in the worldwide food chain.

The national presence of arsenic in animals and feedstuffs does not peril the use of the element as a feed additive. Toxicological studies demonstrate low toxicity and rapid depletion of organic arsenic from animal tissue poses no threat to animal or human safety.

The recent recognition of and acceptance of arsenic as an essential trace element places it on the threshold of new discovery to an even greater role in aiding food production through improved efficiencies and disease control.

DISCUSSION

E. A. Woolson: There are a couple of things that you said which may have confused people. The levels that are found in fish and other marine fauna are of organic arsenical compounds which apparently are not metabolized in humans. The feed additive organic compounds that are fed to chickens and to swine likewise are not metabolized in the animals. The essential trace element that you were talking about is arsenate. The ingestion of arsenate is not the same as ingestion of organic arsenic compounds. That point may be missed but it is an important distinction.

C. E. Anderson: Thank you.

9

THE USE OF ARSENIC AS A WOOD PRESERVATIVE

William J. Baldwin
Forest Products Group
Koppers Company, Inc.
Pittsburgh, PA 15219

Inorganic pentavalent arsenical compounds of varying formulation have been used as wood preservatives in substantial quantities for more than 40 years. There are essentially two types of arsenical wood preservatives used in commercial practice, which are: Chromated Copper Arsenate [CCA] and Ammoniacal Copper Arsenate [ACA]. The manufacture of most of these products is regulated by OSHA; however EPA governs the impregnation of the wood. For the most part, CCA preservatives are supplied to the treating plant as EPA-registered products in a 50% concentrate. Both preservatives, as well as their water-diluted treating solutions, are handled in closed-system operations. Arsenic exposure levels fall well below the OSHA Workplace Standard action level for inorganic arsenic. Several health studies of workers at CCA treating plants have been performed. Neither signs of acute nor chronic toxicity associated with the work environment were observed. Wood treated with arsenical preservatives is non-staining, odorless, clean, paintable, lightweight, extremely durable, and safe. This versatility is reflected in its widespread usage. For example, in 1979 approximately one hundred and fifty million cubic feet of wood were treated with these preservatives, versus fifteen million cubic feet in 1969. Pressure treatment with CCA and ACA preservatives not only extends the life of our forest resources, but also reduces this country's dependence on foreign oil. Studies have shown that the arsenic in CCA and ACA solutions forms insoluble arsenate compounds fixed within the wood structure which chemically are different than inorganic arsenic per se.

Introduction

Although arsenic is an alleged carcinogen, no animal system has yet been found to verify this contention.[1] [1] In fact, strong evidence is mounting that arsenicals have value against cancer.[2,3] Several studies have also indicated that arsenic in small amounts is essential and beneficial to cell metabolism.[4,5,6]

[1]Figures in brackets indicate the literature references at the end of this paper.

Arsenic has two valence states, arsenite (+3) and arsenate (+5), which exhibit markedly different biochemical effects. The arsenic used in wood preservative formulations is the less toxic or pentavalent form which has not directly been associated with carcinogenesis.[7] In the treatment process, these chemicals are pressure-impregnated and chemically fixed in the wood cells via reaction with the wood sugars to form highly insoluble arsenate precipitates. These preservatives have been in use for decades without significant history of harmful effects to workers in the producing or treating industries, the public at large, wildlife, livestock, or the environment in general when used properly. This paper will briefly review the use of pentavalent arsenical compounds as wood preservatives and summarize recently generated data that has been submitted to the Environmental Protection Agency (EPA), but had not previously been available to the scientific community.

Arsenical Wood Preservatives

Untreated wood generally incurs rot and/or insect damage within three to five years. Since Biblical times innumerable preservative treatments have been applied to wood in order to enhance its durability and longevity, especially in soil contact and weather-exposed applications. To satisfy this need, the waterborne arsenical wood preservatives have evolved from the known biocidal activity of various metallic salts. Among the oldest is Fluor Chrome Arsenic Phenol (FCAP) which was used as early as 1918 in the preservative treatment of wood in the United States. Although modifications of this patented system are on EPA's list of registered pesticides, it has nearly been phased out for pressure treatment uses and replaced with the more leach resistant arsenicals described below. In fact, imports are the only source of supply for the current, limited demand.[8]

Two additional arsenical preservatives have also been in use for several decades. The first, Chromated Copper Arsenate (CCA), was initially patented in the United States by Kamesam in 1938. Historical information on the development of this preservative is summarized by Hartford.[9] Today CCA exists as three separate formulations in this country designated Types A, B, and C. Although some quantities of Types A and B are used in wood preservation, Type C has gained rapid acceptance since its introduction in 1968, and today is the dominant formulation due to its optimum cost and minimum leachability.

Ammoniacal Copper Arsenate (ACA) is the second arsenical preservative system which combined with CCA represent over 99 percent of the arsenical wood preservatives used in the United States.[10] ACA, origially patented in 1939, is an ammoniacal solution of copper and arsenic. The nominal preservative composition and raw materials which can be used to formulate both these arsenical wood preservatives are based on the American Wood-Preservers' Association (AWPA) standards and provided in Table 1.[11]

TABLE 1

Nominal Composition for the Arsenical Preservatives

	CCA*			ACA**
	A	B	C	
Hexavalent chromium, as CrO_3	65.5%	35.3%	47.5%	....
Copper, as CuO	18.1%	19.6%	18.5%	49.8%
Arsenic, as As_2O_5	16.4%	45.1%	34.0%	50.2%

*The preservative shall be made up of water soluble compounds selected from the following groups each in excess of 95 percent purity on an anhydrous basis:

Hexavalent chromium - e.g., potassium or sodium dichromate, chromium trioxide.
Bivalent copper - e.g., copper sulfate, basic copper carbonate, cupric oxide or hydroxide.
Pentavalent arsenic - e.g., arsenic pentoxide, arsenic acid, sodium arsenate or pyroarsenate.

**The weight of the ammonia in a treating solution shall be from 1.5 to 2.0 times the weight of the copper oxide.

Table 1 provides the approximate composition of the arsenical preservatives; however, the standard allows a narrow range of compositions about these percentages. More importantly, it should be noted that the less toxic, pentavalent arsenic is required in these preservatives.

The production of all arsenical preservatives is regulated by the Occupational Safety and Health Administration (OSHA) in accordance with the Inorganic Arsenic Standard.[12] CCA preservatives, for example, are primarily produced as 50 percent liquid concentrates and supplied to the treating plants by tank truck. CCA is also formulated as a 70 and 72 percent concentrate which is shipped in drums. These EPA-registered products are primarily sold for export where a more concentrated product is desired due to significant transportation charges.

Pressure Treating Process

Both CCA and ACA concentrates must be diluted to working strength of 0.75 - 6.0% prior to use. The degree of dilution will vary depending upon the wood species, type of wood product, and the anticipated exposure or end use. The AWPA has issued standards which set forth the minimum chemical retentions deemed necessary for a given product based on the foregoing variables. Preservative retentions are described as pounds of total active chemicals, on a comparative oxide basis, per cubic foot of wood (pcf). Since the arsenic content of the three types of CCA and ACA varies, wood treated with different preservatives to identical total chemical retentions will not contain the same amount of arsenate. However, AWPA standards establish quality control and inspection procedures to insure product compliance with the applicable specification requirements.

EPA regulates the application or pressure treatment of wood with arsenical preservatives. This treatment utilizes a closed system process. Moreover, the use of liquid concentrates essentially eliminates worker exposure. The untreated wood is loaded onto small rail or tram cars which are pushed into the treating cylinder using locomotives, forklifts or similar equipment. The cylinder door is sealed via a pressure-tight door, either manually with bolts or hydraulically, and a vacuum applied to remove most of the air from the cylinder and the wood cells. Treating solution is then pumped into the cylinder and the pressure raised to about 150 psi. The total treating time will vary, depending on the species of wood, the commodity being treated, and the desired chemical retentions, but in all instances, the treating process remains a closed system. At the end of the process, the excess treating solution is pumped out of the treating cylinder and back to storage for re-use. The cylinder door is opened, and the trams loaded with treated wood are pulled from the cylinder.

Fixation of Arsenical Preservatives

CCA is unique among wood preservatives in that the chemicals actually react with the wood substrate. This reaction is termed "fixation" because the preservative compounds in the treating solution are fixed in the treated wood in a highly insoluble state. Fixation accounts for permanency of the preservative in the treated wood, which in turn, explains the leach resistance and durability of the product. Although the leach resistance of the arsenical formulations varies slightly due to their respective composition, reports by Fahlstrom [13] and Henry and Jeroski [14] indicate the relative leach resistance of the arsenical preservatives to be CCA-A $>$ CCA-C $>$ CCA-B $>$ ACA. The fixation mechanism is complex and the reactions involved are primarily dependent upon the fixative used, wood species, preservative formulation and concentration, and temperature. In all instances, however, highly insoluble arsenate preservatives are precipitated in the wood. [15]

In CCA, the fixative is chromic acid or alkali metal dichromate. A number of investigators have studied this complex process as noted by Arsenault [16], but the most thorough study has been reported by Dahlgren and Hartford. [17] Fixation involves the reduction of the hexavalent chromium to trivalent chromium with formation of a complex mixture of insoluble chromates. Insoluble arsenates of copper and chromium are also precipitated in the treated wood. Although some of the initial reaction products are not completely stable, they are very insoluble and do convert into stable compounds.

The length of the fixation period is temperature sensitive and theoretically can last from several hours at 45°C to two months at 5°C. Studies measuring the relative efficiency of the fixation mechanism have shown that drying at 21°C will fix 70 percent of the arsenic within the first few hours, 95 percent within four days and 99 percent within five days. [18,19] However, movement of the arsenate compounds from the treated wood into the environment can only be accurately shown by actual service tests. Such testing has in fact shown very little difference between above-ground and below-ground retentions in CCA treated telephone poles after 32 years of service.[16]

Air Monitoring and Health Studies Performed at Various Treating Plants

Over 95 percent of the CCA and ACA preservatives supplied to the treating plants are in the liquid form. Liquid concentrates present less exposure and are handled in a closed system operation as discussed previously. Bulk liquid concentrate is the easiest and safest product to handle. Most plants utilize a completely closed system operation in which the liquid concentrate and dilution water are meter-mixed. In these plants, exposure to the concentrate and the diluted preservative is essentially eliminated as evidenced by the air-monitoring data provided in Table 2. Such data clearly demonstrate that, under normal conditions, workers in ACA and CCA plants are exposed to arsenic levels well below the OSHA workplace standard action level of 0.005 mg/M^3.

TABLE 2

Air Monitoring Data at Wood Preserving Plants that Use CCA or ACA [20]

Type of Sample and Location	Concentration of Arsenic mg/M^3	Monitored by
Area Sample - near CCA cylinder door	<0.008	Insurance Company
Area Sample - loading and unloading of three CCA cylinder charges (202 min.)	0.0006	NIOSH Contractor
Personnel Monitoring - treatment helper during loading and unloading of three CCA cylinder charges (80 min.)	0.0045	NIOSH Contractor
Personnel Monitoring - mean of 27 full-shift measurements from various personnel at a CCA plant (plant 1)	0.0021	Koppers Co., Inc.
Personnel Monitoring - mean of 20 full-shift measurements from various personnel at a CCA plant (plant 2)	0.0007	Koppers Co., Inc.
Personnel Monitoring - treatment operator during loading and unloading of an ACA cylinder (29 min.)	<0.0052	NIOSH Contractor
Personnel Monitoring - treatment helper during loading and unloading of an ACA cylinder (27 min.)	<0.0058	NIOSH Contractor

Cross-sectional health studies of treating plant workers exposed to CCA-C preservative and the treated wood have been performed at two different locations.[20] The areas of investigation were chosen on the basis of the known and alleged toxicology of the components of CCA, i.e. arsenic, chromium, and copper, to which the workers were potentially exposed. The tests performed on the 109 participants included: (1) medical and work history questionnaires, (2) special Chemex questionnaire, (3) chest x-ray, PA view, (4) pulmonary function, (5) clinical chemistries, (6) hematology, (7) urinalysis, (8) sputum cytology, (9) urinary arsenic, and (10) urine cytology.

To our knowledge those are the first studies of this type to be performed at a treating plant utilizing arsenical wood preservatives and will undoubtedly serve as the basis for future studies. Concurrent full-shift personnel monitoring[2] at the plants indicate minimal exposure to arsenic. This data was substantiated in a summation of the medical teams' findings which stated in part that the general health of the plant population was good, and did not reflect the presence of adverse health effects from the work environment. Moreover, evidence of occupationally related diseases of the lung, liver, kidneys, skin and cancer were sought but not found.

The Treated Product

Wood treated with arsenical preservatives is non-staining, odorless, clean, paintable, lightweight, and safe. Extensive reviews have been written on the use of CCA preservatives in Europe and their relation to public health.[21] More recently, independent studies performed by the Occupational Health and Safety Division of Alberta, Canada [22] and Versar, Inc., in a study commissioned by U. S. EPA, [23] also strongly indicate that exposure to the chemicals in arsenically treated wood is minimal due to the highly insoluble characteristic of the preservative.

Other pertinent studies include a retrospective analysis of cancer mortality among carpenters in Hawaii who worked with arsenically treated wood.[24] The authors of this study conclude that exposure to dust from such wood is not associated with increased risk of total cancer, lung cancer, or lymphatic cancer. This data, which constitutes the best epidemiological evidence available on treated wood, clearly shows that excess respiratory cancer mortality was not observed in the carpenters exposed to arsenically-treated sawdust. In fact, OSHA has acknowledged the difference between inorganic arsenic and the arsenate compounds found in the treated wood. OSHA notes,

> In the Budy-Rashad study, the carpenters were not exposed to pentavalent arsenic, but rather to a stable arsenic-wood complex. Thus, carpenter exposure to this arsenic-wood complex cannot be considered *a priori*, equivalent to exposure to unbound pentavalent arsenic.[3]

---

[2]This data is supplied in Table 2 for the plants designated as 1 and 2, respectively.

[3]43 Federal Register 19,599 (May 5, 1978)

Studies by Hood [25] and Peoples [26] support the fact that CCA and ACA wood preservatives are fixed in the wood as relatively innocuous compounds. In the first study, a paste of CCA treated sawdust (26 grams treated to 0.66 pcf) was applied to a shaved area of pregnant rabbits during days 7 through 20 of gestation. The patches were changed every four days and a control group was treated in the same manner using untreated wood sawdust. This test method represents 6600 mg sawdust/Kg body weight and is equivalent to four dermal exposures of 45 mg As_2O_5/Kg body weight. Neither teratogenic or fetotoxic effects were observed. Furthermore, the arsenic levels in the blood samples taken from the sacrificed animals (treated and controls) were the same. Finally, no skin irritation was observed in the group exposed to CCA treated sawdust.

In a second study, Hood fed pregnant mice a diet of commercially prepared laboratory food mixed with ten percent (w/w) CCA treated sawdust (0.66 pcf). Two control groups were used. The first was fed a diet without sawdust and the second control group was fed a diet consisting of ten percent untreated sawdust. The experimental diets were fed continuously from day 1 of gestation through day of sacrifice (gestation day 18). Again, no teratogenic effects were observed in any of the three groups. Additional supporting evidence is provided in a study by Peoples in which 1500 mg of finely ground ACA or CCA-C treated sawdust was applied to the closely clipped skin of dogs. The excreted urine was monitored for arsenic compounds both before and during the three days in which the patches were applied, but there was no evidence of dermal absorption from either preservative.

The safe use of CCA and ACA treated wood is also documented by its service history in the agricultural market. Treated grape stakes have long been used with no reported phytotoxic effects on the grape plants and no migration of the arsenate compounds into the grapes.[27] The Food and Drug Administration's (FDA) "Market Basket Survey" has consistently shown that arsenic levels in tomatoes are below the analytical level of detection despite the increased usage of arsenical treated wood for tomato stakes.[28] Moreover, even though CCA and ACA treated wood has been increasingly used in applications such as cattle feed bunks, stalls, and poultry brooders for the last ten years, the FDA survey has shown a decrease in the arsenic content of dairy, meat and poultry products, again approaching the level of detection. Arsenical treatment extends the life of a mushroom tray 3-4 times as compared to an untreated tray, but chemical analysis shows that there is no difference in the arsenic content for mushrooms grown on either arsenical treated or untreated trays.[29]

Based on the previously discussed characteristics, it is evident that arsenically treated materials are suitable for practically all uses without restriction. As recently noted by the United States Department of Agriculture (USDA), "arsenicals are the only preservatives suitable for materials to be used in critical, structural applications in enclosed, habitable space in residential and other buildings."[30] Such is the case in the "All Weather Wood Foundation" (AWWF), a federally approved system gaining in popularity throughout the United States due to the cost savings to both the builder and homeowner.

The versatility of arsenically treated wood is reflected in its increased usage as shown in figure 1.

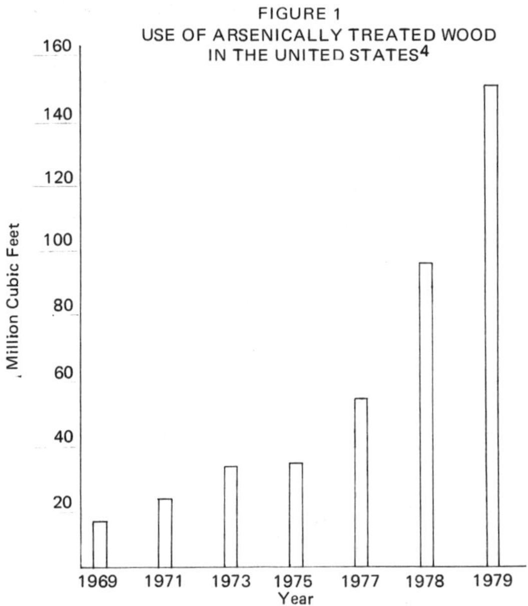

FIGURE 1
USE OF ARSENICALLY TREATED WOOD
IN THE UNITED STATES[4]

[4]1969–1977 AWPA Proceedings, 76, pp. 343–345 (1980).
1978–USDA, Biological Assessment of Pentachlorophenol, Inorganic Arsenicals, and Creosote, p. 1–17 (1980).
1979–Koppers Company, Inc.

For example, in 1969 about fifteen million cubic feet of wood were treated with arsenicals; however, in 1979 almost 150 million cubic feet were treated with CCA and ACA preservatives. This ten-fold increase is something we can all understand, but the significance of 150 million cubic feet of treated wood is not as obvious. Therefore by using USDA conversion factors [31] and a price of $300 per thousand board feet, one can show that the 1979 production of arsenically treated wood has a value of over $700 million or more than the net sales of many of the Fortune 500 Companies.

Treatment with CCA and ACA preservatives benefits this country in other ways. For example, one piece of arsenically treated wood will provide the service of six or more untreated pieces. This extends the life of our forest resources, not only for the timber products but also for recreational and other competing non-timber uses. The use of these waterborne arsenical preservatives, also reduces this country's dependence on foreign oil. Studies by the Economic Analysis Branch of EPA estimate that if all pentachlorophenol treated material and all creosoted lumber were treated with arsenicals, the net reduction in No. 2 fuel oil usage would be more than 32 million gallons annually.[32]

Summary

There is increasing evidence that at most arsenite acts as a co-carcinogen, and that arsenate has no carcinogenic effects. Only pentavalent arsenic (arsenate) compounds are used in the formulation of CCA and ACA preservatives. Health studies at CCA treating plants and air monitoring at both CCA and ACA plants document the minimal exposure at such facilities. Wood treated with CCA or ACA preservatives is fixed within the wood and forms a stable arsenic-wood complex. An earlier epidemiological study on carpenters exposed to arsenically treated sawdust combined with recent animal studies demonstrate that CCA and ACA treated wood is relatively innocuous.

References

[1] Sunderman, Jr., F. W., Mechanisms of metal carcinogenesis, Biolog. Trace Element Research 1, 63-86 (1979).

[2] Schrauzer, G. N., White, D. A., and Schneider, C. I., Inhibition of the genesis of spontaneous mammary tumors in C_3H mice: effects of selenium and of selenium-antagonistic elements and their possible role in human breast cancer, Bioinorg. Chem., 6, 265-270 (1976).

[3] Frost, D. V., In defense of arsenic, C&EN, 59 [40], 4 (Oct. 5, 1981).

[4] Muth, O. H., Whanger, P. D., Weswig, P. H., and Oldfield, J. E., Occurrence of myopathy in lambs of ewes fed added arsenic in a selenium-deficient ration. Amer. J. Vet. Res., 32 [10], 1621-1623 (1971).

[5] Nielsen, F. H., Givand, S. H., and Myron, D. R., Evidence of a possible requirement for arsenic by the rat. (Abstract.), Fed Proc. 34, 923 (1975).

[6] Anke, M., Grun, M., and Partschefeld, M., The essentiality of arsenic for animals, in Trace Substances in Environmental Health, D. D. Hemphill, ed., Vol X, pp. 403-409 (Univ. of Missouri, Columbia, Mo., 1976).

[7] American Wood Preservers Institute, Response to EPA PD 1, OPP-30000/29, Inorganic Arsenic, pp. 19-225 (Feb., 1979).

[8] USDA, Biological Assessment of Pentachlorophenol, Inorganic Arsenicals, and Creosote, p. 1-8, (1980).

[9] Hartford, W. H., Chemical and physical properties of wood preservatives and wood-preservative systems, in, Wood Deterioration and Its Prevention by Preservative Treatments, Vol. II Preservatives and Preservative Systems, D. D. Nicholas, ed., pp. 34-36 (Syracuse University Press, Syracuse, N.Y., 1973).

[10] Parker, C. L., US EPA Contract No.: 68-01-5965, The MITRE Corporation, p.1, 1981.

[11] American Wood-Preservers' Association Standard P5-81.

[12] 29 CFR Section 1910.1018, Inorganic arsenic.

[13] Fahlstrom, G. B., Gunning, P. E., and Carlson, J.A., Copper-chrome-arsenate wood preservatives: a study of the influence of composition on leachability, Forest Products J., 17, [7], 17-22 (1967).

[14] Henry, W. T., and Jeroski, E. B., Relationship of arsenic concentration to the leachability of chromated copper arsenate formulations, American Wood-Preservers' Assoc., 63, 187-192 (1967).

[15] Hartford, pp. 99-103.

[16] Arsenault, R. D., CCA-treated wood foundations, a study of permanence, effectiveness, durability, and environmental considerations, Proc. American Wood-Preservers' Assoc., 71, 126-149 (1975).

[17] Dahlgren, S. E., and Hartford, W. H., Kinetics and mechanism of fixation of Cu-Cr-As wood preservatives, Holzforschung, 26 [2], 62-69; 26 [3], 105-113; 26 [4], 142-149 (1972). Dahlgren, S. E., Kinetics and mechanism of fixation of Cu-Cr-As wood preservatives, Holzforschung, 26 [4], 58-61 (1974).

[18] McMahon, W., Hill, C. M., and Koch, F. C., Greensalt--a new preservative for wood, Bell Telephone Monograph B-1342, (1942).

[19] Jain, J. C., and Lagus, A., Studies on the fixation of wood preservatives in timber, Journal of the T. D. and P. A., 6 [4], 10-17.

[20] American Wood Preservers Institute, Response to EPA PD 2/3, OPP-30000/28C, Inorganic Arsenicals, pp. 97-107 (May, 1981).

[21] British Wood Preserving Association Safety Liaison Panel, Arsenic: its use in wood preservative formulations in relation to public health (Sept., 1968).

[22] Bowhay, D. C., and Horsern, H. R., Community health aspects of arsenical preserved woods, Occupational Health and Safety Division, Alberta, Canada (1976).

[23] Versar, Inc., Technical and microeconomic analysis, Task III--Arsenic and its compounds, EPA 560/6-76-016, p. 77, 1976.

[24] Budy, A. M., and Rashad, M. N., Cancer mortality among carpenters in Hawaii, presented at the Third International Symposium on Detection and Prevention of Cancer, New York, April 29, 1976.

[25] Unpublished 1979, American Wood Preservers Institute (Feb., 1979), pp. 223-225.

[26] Unpublished 1979, American Wood Preservers Institute (May, 1981), pp. 87-88.

[27] Huxter, Jr. W. T., Treated wood and the grape grower, N. C. State Univ. Agric. Extension Service (1974); Levi, M. P., Huisingh, D., and Nesbitt, W. B., Uptake by grape plants of preservatives from pressure-treated post not detected, Forest Products Journal, 24 [9], 97-98 (1974).

[28] Jelinek, C. F., and Corneliussen, P. E., Levels of arsenic in the United States food supply, Environ. Health Perspectives, 19, 83-87 (1977).

[29] American Wood Preservers Institute, Response to EPA PD 2/3, OPP-30000/28C, Vol. 3, Exhibit 20 (May, 1981).

[30] USDA, p. 8-103.

[31] USDA, pp. 1-15, 1-19.

[32] EPA, Energy implications of improved pest management, Economic Analysis Branch, Benefits and Field Studies Division, OPP, pp. 3-5 (March, 1980).

DISCUSSION

E. A. Woolson: In the human exposure studies, what was the mean arsenic level of the urine? Do you have those figures?

W. J. Baldwin: The urinary arsenic level for all workers was below the limit of detection which was 3 and 1 micrograms in the two Koppers' studies.

10

NBS STANDARD REFERENCE MATERIALS CERTIFIED FOR ARSENIC

Robert Alvarez
Office of Standard Reference Materials
National Bureau of Standards
Washington, D.C. 20234

The accurate determination of arsenic concentrations is difficult but necessary to establish sound scientific data bases for industrial, environmental, epidemiological, and nutritional investigations. Interlaboratory studies, in which identical samples have been analyzed, frequently show unacceptably large variations among laboratories. This is true not only when different methods are used but even when the same method is used. One method of establishing the reliability of analytical data for arsenic is through the use of chemical composition Standard Reference Materials issued by NBS. SRM's are used worldwide to calibrate instrumentation, validate experimental data, develop methods of known accuracy, and to refer experimental data to a common base. The first such materials, issued by NBS in 1906, were cast irons certified for chemical composition. Today there are approximately 1000 different SRM's. Examples of SRM's certified for arsenic in the industrial category are: four glasses (two certified for As_2O_3 and As_2O_5), cast irons, low-alloy steels, and unalloyed copper...; in the environmental category are: two coals, a coal fly ash, urban particulate matter collected from the atmosphere, sediments, plant materials, and water; and in the nutritional category are: oyster tissue, bovine liver, and rice flour. In addition, a high-purity arsenic trioxide SRM is being renewed. These SRM's have been developed as a result of demonstrated needs by individuals, industry, professional societies, trade associations, government agencies, and academia.

1. Introduction

The analysis of materials for arsenic is necessary to define its role in industrial, environmental, and biological processes. But to formulate valid conclusions, an investigator must be assured of the accuracy of the analytical data.

Interlaboratory studies in which identical samples have been analyzed frequently show unacceptably large variations among laboratories. This is true not only when different methods are used but even when the same method is used. In one of these studies, the concentrations of arsenic reported in the same silicate material ranged from 25 μg/g (parts-per-million) by atomic absorption spectroscopy to 570 μg/g by x-ray fluorescence while the result obtained by optical emission spectroscopy was 330 μg/g [1]. Laboratories using the same method also obtained discordant results.

The state of development of analytical methodology for inorganic nutrients in foods has been described as either sufficient, substantial, conflicting, or fragmentary [2]. In this classification, methodology for arsenic is considered conflicting; i.e., the probability of obtaining the correct concentration is considered only "fair".

What can investigators do to increase the accuracy of the experimental data? The use of Standard Reference Materials (SRM's) issued by the National Bureau of Standards (NBS) for this purpose is discussed by Wolf [3]. In an analytical procedure, SRM's are analyzed at the same time as the unknowns and the elemental concentrations obtained are compared with the certified values listed in the NBS Certificate of Analysis. Results within the estimated uncertainty limits of the certified values would validate the concentrations obtained for the unknowns. In his paper, Horwitz also suggests the use of SRM's "to verify the correctness of a series of analyses" [4].

There are approximately 1000 different SRM's certified for chemical composition or physical properties [5]. The largest category comprises those certified for chemical composition. They are issued with Certificates of Analysis which list certified and uncertified values. Certified values are based either on the results of a definitive method or on the concordant results of two or more independent methods.

The preferred method of certification is a definitive method. This high precision method is based on a sound theoretical foundation and has been thoroughly tested and found to provide results essentially free of systematic error. One such method is isotope dilution, thermal source mass spectrometry, which can be used to determine elements that have two or more stable isotopes. Unfortunately, arsenic is a mononuclidic element and cannot be determined by this high accuracy method. Therefore, certification of the arsenic content of a proposed SRM is based on analytical determinations by two less accurate but independent methods, such as atomic absorption spectroscopy and neutron activation.

Representative SRM's certified for arsenic will be discussed under three categories: industrial, environmental, and biological. But, before considering these matrix SRM's, a high-purity arsenic trioxide SRM will be considered because of its use to prepare standard solutions of known arsenic concentration.

A new lot of high-purity arsenic trioxide is being analyzed to replace SRM 83c, which is out-of-stock. The purity of SRM 83c was certified as 99.99 percent based on its effective reducing power. The assay was based on a direct comparison with SRM 136b, potassium dichromate. The estimated experimental error is approximately 0.02 percent. The purity of SRM 83d, the renewal, is expected to be comparable to that of the previous issue.

[1]Figures in brackets indicate the literature references at the end of this paper.

1. Industrial SRM's

A number of chemical composition SRM's that have been developed for industrial use include certified concentrations of arsenic. Table 1 lists SRM's issued for use by the glass industry. When SRM's 89 and 91 were issued in 1932, estimated uncertainties were not provided for the certified concentrations.

Table 1. NBS glass SRM's certified for arsenic

SRM No.	Name	Concentration, weight percent	Number of certified elements
89	Lead-Barium Glass	0.03 As_2O_3	17
		0.36 As_2O_5	
91	Opal Glass	0.091 As_2O_3	15
		0.102 As_2O_5	
620	Soda Lime Glass (Flat)	0.06 $\pm$ 0.01 As_2O_3	10
621	Soda Lime Glass (Container)	0.03 $\pm$ 0.005 As_2O_3	12

For use in quality control of metals, a number of metal SRM's have been issued which are certified for chemical composition. These are listed in table 2. The low-alloy-steel and unalloyed-copper SRM's provide a range of certified concentrations of arsenic. Each SRM is also certified for other important elements that affect the quality and use of the metal.

Table 2. NBS metal SRM's certified for arsenic

SRM No.	Name	Concentration[a], weight percent
5L	Cast Iron	<0.005
365	Electrolytic Iron	(0.0002)
1261a-1265a	Low Alloy Steels	0.0002 to 0.09
494-500,1251-1253	Unalloyed Copper	<0.2 to 244 μg/g
53e	Bearing Metal	0.057
127b	Solder	0.01

[a]Values are certified except when appearing in parentheses.

2. Environmental SRM's

Arsenic is one of thirteen elements classified as priority pollutants by the U. S. Environmental Protection Agency. Because of this designation, studies are being conducted to determine its concentration in fossil fuels and other materials from which the amounts released into the environment can be calculated. In addition, other studies focus on the concentration of arsenic in air, water, and sediments in order that the environmental pathways can be traced and the efficiency of clean up procedures evaluated.

For these studies, environmental SRM's have been developed to assist investigators in validating their experimental data. The concentrations of arsenic and the estimated uncertainties are listed in table 3. In addition to arsenic, the concentrations of other elements of environmental significance have also been determined and certified. in these SRM's.

Table 3. NBS environmental SRM's certified for arsenic

SRM No.	Name	Concentration,[a] μg/g		Number of elements certified
1643a	Trace Elements in Water	0.076	± 0.007	Be, Cd, Cr, Pb, Se, and 12 others
1648	Urban Particulate Matter	115	± 10	Cd, Cr, Pb, and 6 others
1632a	Trace Elements in Coal (Bituminous)	9.3	± 1	Cd, Cr, Pb, Hg, U, and 9 others
1634a	Trace Elements in Fuel Oil	(0.11)[b]		Pb, V, and others[b]
1635	Trace Elements in Coal (Subbituminous)	0.42	± 0.15	Cd, Cr, Pb, Th, U, and 9 others
1633a	Trace Elements in Coal Fly Ash	145	±15	Cd, Cr, Hg, Pb, Th, U, and 14 others
1645	River Sediment	(66)		Cd, Pb, and 11 others
1646	Estuarine Sediment	11.5[b]		Cd, Hg, Pb, and others[b]

[a] Values are certified except when appearing in parentheses.

[b] Tentative (Certificate of Analysis for the SRM to be issued in early 1982).

3. Biological SRM's

The Food and Drug Administration has monitored toxic elements in foods for many years. Mercury, lead, cadmium, arsenic, selenium, and zinc are the elements of highest priority [6]. On the other hand, three of these elements, arsenic, selenium, and zinc are also classified as essential elements; i.e., indispensable for life [7].

Consequently the biochemical role of arsenic is dependent on exposure. The interactive effect of other elements is also important in these studies. This evaluation requires accurate trace chemical composition data on a variety of substances. Table 4 shows the biological SRM's which have concentration values for arsenic and other elements of nutritional and environmental significance. As an example the Certificate of Analysis for the Oyster Tissue SRM appears in the Appendix.

Table 4. NBS biological SRM's certified for arsenic

SRM No.	Name	Concentration,[a] μg/g	Number of elements certified
1566	Oyster Tissue	13.4 ± 1.9	Cr, Cu, Pb, Se, and 15 others
1567	Wheat Flour	(0.006)	Cu, Na, Se, and 7 others
1568	Rice Flour	0.41 ± 0.05	Cu, Na, Se, and 9 others
1573	Tomato Leaves	0.27 ± 0.05	Cu, Fe, Mn, and 11 others
1575	Pine Needles	0.21 ± 0.04	Cu, Fe, Pb, and 12 others
1572	Citrus Leaves	3.1[b]	
1577a	Bovine Liver	(0.047)[b]	

[a]Values are certified except when appearing in parentheses.

[b]Tentative (Certificate of Analysis for the SRM to be issued in early 1982.)

4. Summary and Discussion

Arsenic is a difficult element to determine accurately especially when it is present at trace levels. In addition to the previous references, which present discordant results for arsenic determinations, another interlaboratory study reenforces the difficulty of obtaining reliable data for arsenic [8]. In this study, the concentrations of 2 selected trace elements in a homogeneous oyster tissue material were determined. The values reported by 23 laboratories for arsenic ranged from 0.016 to 163 μg/g--the largest reported for any of the 12 elements.

SRM's are developed in response to requests from professional societies, standards organizations, trade associations, companies, academia, Federal, State, and local government agencies, and individuals. The use of SRM's with certified concentrations of constituents has been found to improve the accuracy of these determinations in samples. The NBS Office of Standard Reference Materials would be pleased to consider requests for SRM's from the Chemical Manufacturer's Association.

References

[1] Gladney, E. S., Perrin, D. R., Owens, J. W., and Knab, D., Elemental Concentrations in the United States Geological Survey's geochemical exploration reference samples - A Review, Anal. Chem. 51 [9], 1557-1569 (1979).

[2] Stewart, K. K., Nutrient analysis of foods: state of the art for routine analysis (Proc. Nutrient Analysis Symposium, Assoc. of Official Analytical Chemists, Arlington, Va., Oct. 15-18, 1979) pp 1-19.

[3] Wolf., W. R. Inorganic nutrient analysis of foods, ibid, pp 64-84.

[4] Horwitz, W., Kamps, L. R., and Boyer, K. W., Quality assurance in the analysis of foods for trace constituents. J. Assoc. Off. Anal. Chem. 63 [6], 1344-1354 (1980).

[5] NBS Standard Reference Materials Catalog 1982-1983 Edition, NBS Special Publ. 260, Nat'l Bur. of Stds., Washington, D.C.

[6] Jelinek, C. F., and Corneliussen, P. E., Levels of arsenic in the United States food supply, Environ. Health Perspectives 19, 83-87 (1977).

[7] Mertz, W., The essential trace elements, Science 213, 1332-1338 (1981).

[8] Fukai, R., Oregioni B., and Vas, D. Interlaboratory comparability of measurements in trace elements in marine organism: results of intercalibration exercise on oyster homogenate., Oceanologica Acta, 1, [3], 392-396 (1978).

APPENDIX

National Bureau of Standards

Certificate of Analysis

Standard Reference Material 1566

Oyster Tissue

This Standard Reference Material is intended primarily for use in calibrating instrumentation and validating methodology for the chemical analysis of marine animal tissue.

Certified Values of Constituent Elements: The certified values for the constituent elements are shown in Table 1. Certified values are based on results obtained by reference methods of known accuracy; or alternatively, from results obtained by two or more independent and reliable analytical methods. Non-certified values are given for information only in Table 2. All values are based on a minimum sample size of 250 mg of the dried material.

NOTICE AND WARNINGS TO USERS

Expiration of Certification: This certification is invalid after 5 years from the date of shipping. Should it become invalid before then, purchasers will be notified by NBS.

Storage: The material should be kept tightly closed in its original bottle and stored in a desiccator at temperatures between 10-30 °C. It should not be exposed to intense sources of radiation, including ultraviolet lamps or sunlight.

Use: A minimum sample weight of 250 mg of the *dried* material (see Instructions for Drying) is necessary for any certified value in Table 1 to be valid within the stated uncertainty. The bottle should be shaken well before each use, and closed tightly immediately after use.

The statistical analysis of the data was performed by K. R. Eberhardt and H. H. Ku of the Statistical Engineering Division.

The overall direction and coordination of the analytical chemistry measurements leading to this certificate were performed in the NBS Center for Analytical Chemistry by P. D. LaFleur.

The technical and support aspects involved in the preparation, certification, and issuance of this Standard Reference Material were coordinated through the Office of Standard Reference Materials by R. Alvarez.

Washington, D.C. 20234
December 12, 1979

George A. Uriano, Chief
Office of Standard Reference Materials

(over)

Instructions for Drying: Before weighing, samples of SRM 1566 should be dried to constant weight by one of the following procedures:

1. Reduced-pressure drying at room temperature for 48 hours over $Mg(ClO_4)_2$ in a vacuum desiccator at approximately 1.3×10^4 Pa (100 mm Hg).
2. Vacuum drying at room temperature for 24 hours at a pressure of approximately 30 Pa (0.2 mm Hg) using a cold trap.
3. Freeze drying for 20 hours at a pressure of approximately 3 Pa (0.02 mm Hg).

Source and Preparation of Material: The oysters for this reference material were obtained by the FDA Bureau of Shellfish Sanitation from a commercial source. They had been shucked, frozen, and packaged in sealed plastic bags. The oyster material was ground, freeze-dried, and powdered at the U.S. Army Natick Research and Development Command, Natick, Mass., under the direction of L. Hinnegardt and G. C. Walker. At NBS, preliminary analyses of the material homogeneity indicated that an improvement in homogeneity would be required to establish more reliable certified values for a minimum sample size of 250 mg. Accordingly, the material was cryogenically ground by J. R. Moody and J. Matwey. It was then blended and bottled at NBS, after which it was again freeze-dried at the Natick, Mass., laboratory.

Homogeneity Assessment: Randomly selected bottles of SRM 1566 were sampled and tested for homogeneity by neutron activation and atomic absorption spectrometry. No inhomogeneity was observed for the following elements determined by neutron activation: Na, Cl, V, and Mn. The values for Mg, K, Cu, Zn, and Cd determined by atomic absorption spectrometry were within the imprecision of the method; however, Ca does exhibit some inhomogeneity--approximately 4% relative standard deviation.

Analysts:

Center for Analytical Chemistry, National Bureau of Standards:

1. J. V. Bailey
2. C. Blundell
3. T. J. Brady
4. M. Diaz
5. L. P. Dunstan
6. M. S. Epstein
7. M. Gallorini
8. E. L. Garner
9. T. E. Gills
10. J. W. Gramlich
11. R. R. Greenberg
12. S. Hanamura
13. S. Harrison
14. E. F. Heald
15. G. M. Hyde
16. W. F. Koch
17. W. R. Kelly
18. R. M. Lindstrom
19. G. J. Lutz
20. L. A. Machlan
21. W. A. MacCrehan
22. E. J. Maienthal
23. J. Maples
24. O. Menis
25. J. D. Messman
26. J. R. Moody
27. L. J. Moore
28. T. J. Murphy
29. P. J. Paulsen
30. T. C. Rains
31. H. L. Rook

Cooperating Analysts:

32. University of Tokyo, Tokyo, Japan; Y. Dokiya (NBS Guest Worker).
33. Division of Chemistry, National Research Council of Canada, Ottawa, Canada; S. Berman, A. Desaulniers, J. McLaren, A. Mykytiuk, D. Russell, and S. Willie.
34. Ibaraki Electrical Communication Laboratory, Nippon Telegraph and Telephone Public Corporation, Tokai, Ibaraki, Japan; K. Kudo and K. Kobayashi.
35. Food Research Division, Health Protection Branch, Tunney's Pasture, Ottawa, Ontario, Canada; R. W. Dabeka, A. D. McKenzie, and H. B. S. Conacher.

Table 1. Certified Values of Constituent Elements

Element[1]	Content[2], Wt. Percent	Element[1]	Content[2], Wt. Percent
Calcium[b,d]	0.15 ± 0.02	Potassium[d]	0.969 ± 0.005
Magnesium[a,d]	0.128 ± 0.009	Sodium[b,f]	0.51 ± 0.03

Element[1]	Content[2], μg/g	Element[1]	Content[2], μg/g
Arsenic[a,f,g,h]	13.4 ± 1.9	Nickel[a,e,h]	1.03 ± 0.19
Cadmium[a,d,e,f,h]	3.5 ± 0.4	Rubidium[d,f]	4.45 ± 0.09
Chromium[d,e,f]	0.69 ± 0.27	Selenium[a,e,f]	2.1 ± 0.5
Copper[a,c,e,f]	63.0 ± 3.5	Silver[a,f]	0.89 ± 0.09
Iron[b,c,e,f]	195 ± 34	Strontium[b,d]	10.36 ± 0.56
Lead[a,d,e,h]	0.48 ± 0.04	Uranium[d]	0.116 ± 0.006
Manganese[a,c,f]	17.5 ± 1.2	Zinc[a,c,d,e,f,h]	852 ± 14
Mercury[a,f]	0.057 ± 0.015		

1. Analytical Methods:
   [a]Atomic absorption spectroscopy
   [b]Atomic emission spectroscopy, flame
   [c]Atomic emission spectroscopy, inductively coupled plasma
   [d]Isotope dilution mass spectrometry, thermal ionization
   [e]Isotope dilution mass spectrometry, spark source
   [f]Neutron activation
   [g]Photon activation
   [h]Polarography

2. Based on dry weight. (For drying instructions, see the section of this certificate on Instructions for Drying.) The estimated uncertainty is given as 95 percent tolerance limits for coverage of at least 95 percent of the measured values of all bottles of SRM 1566. For a given element, the following statement can be made at a confidence limit of 95 percent. "If the concentrations were measured for all bottles, at least 95 percent of these measured values should fall within the indicated limits." The concept of tolerance limits is discussed in Chapter 2, Experimental Statistics, NBS Handbook 91, 1966, and page 14, The Role of Standard Reference Materials in Measurement Systems, NBS Monograph 148, 1975.

Table 2. Non-certified Values of Constituent Elements

Element	Content[1] (Wt. Percent)
Chlorine	(1.0)
Sulfur	(0.76)
Phosphorous	(0.81)
	(μg/g)
Bromine	(55)
Cobalt	(0.4)
Fluorine	(5.2)
Iodine	(2.8)
Molybdenum	(≤0.2)
Thallium	(≤0.005)
Thorium	(0.1)
Vanadium	(2.8)

[1]Based on dry weight. (For drying instructions, see the section of this certificate on Instructions for Drying.)

SESSION III - BIOMEDICAL PERSPECTIVES

Session Chairman: William H. Lederer
Koppers Company, Inc.
1201 Koppers Building
Pittsburgh, PA 15219

INTRODUCTION

The toxicology of arsenic is a complex subject. Not only does arsenic exist in pentavalent (+5) arsenate and trivalent (+3) arsenite states but also in organic forms. Arsenic is a toxic chemical but is considered by some investigators to be an essential trace element and has been used as a medicinal. Medicinal uses of arsenic have included treatment of psoriasis, parasitic diseases and use as a tonic. Arsenic can be toxic to animals, plants and bacteria yet its actions are selective.

The acute toxicology of inorganic arsenic has been well known for a long time with trivalent arsenites generally considered more acutely toxic than pentavalent arsenates. The chronic effects of inorganic arsenic are not well clarified and even less is known about the acute and chronic effects of organic arsenicals.

Inorganic arsenic compounds are irritants of the skin, mucous membranes and the eyes. Additionally, it has been suggested on the basis of epidemiological studies in man, that arsenic is either a carcinogen, co-carcinogen or promotor. However, no animal model for arsenic carcinogenicity has been found. Arsenic is the only chemical considered to be a carcinogen, other than perhaps benzene, for which no animal (rodent) model for carcinogenicity is known.

It has been proposed that only trivalent arsenite compounds are the suspected carcinogenic agents with the pentavalent forms lacking carcinogenic activity. Trivalent inorganic arsenic is known to be biologically converted to pentavalent arsenic but the reverse has not been established. Thus an understanding of the metabolism of arsenic in man and animals is of essential importance in evaluating the chronic effects of arsenic.

Arsenic has also been suggested to be fetotoxic and/or teratogenic but the concentrations at which these effects occur and the form of arsenic are important considerations. Finally, the genotoxic activity of arsenic has been the subject of considerable concern. The following presentations will review these areas.

11

THE METABOLISM OF ARSENIC IN MAN AND ANIMALS

S. A. Peoples
Professor of Pharmacology (Emeritus)
University of California Davis
Davis, CA 95616

Advances in our knowledge of the metabolic fate of inorganic arsenic administered to animals and man have followed the development of the necessary analytical methods. It has been established that man, dog and the cow can methylate both trivalent (As^{3+}) and pentavalent (As^{5+}) arsenic to form monomethylarsonic acid (MAA) and dimethylarsinic acid (DMAA). There is evidence that the steps are $As^{3+} \rightarrow As^{5+} \rightarrow$ MAA$\rightarrow$ DMAA. The reduction of $As^{5+} \rightarrow As^{3+}$ occurs in the dog only when given intravenously in a dose which damages the kidneys and probably takes place in the damaged renal tubules.

1. Introduction

The present safety standards for arsenic set by various governmental agencies for water, food, hazardous wastes, and industrial exposure have been based on arsenic analyses which measure total arsenic. The results of such analyses are often reported as arsenic trioxide and this has led to setting standards based on the toxicity of trivalent inorganic arsenic.

During the last five years analytical methods have been developed which speciate the arsenic into its inorganic forms, trivalent and pentavalent, and the methylated compounds, monomethyl arsonic acid and dimethylarsinic acid. Since there is a wide range in the toxicity of these compounds it is now time to set these standards on the basis of the compound present.

2. The form of arsenic in the tissues of marine fish and shellfish

The first demonstration that tissues of marine fish and crustacea had high levels of organic arsenic was that of Chapman in 1926 [1][1].

[1]Figures in brackets indicate the literature references at the end of this paper.

In 1935 Coulson [2] fed shrimp which had high levels of arsenic (10 - 55 ppm) to rats and found that a dose of "shrimp arsenic" which would have been lethal to rats if given as arsenic trioxide, was non-toxic.

Charbonneau [3] made balance studies in monkeys comparing a single feeding of 1 mg/Kg of arsenic as fish and arsenic trioxide. After the ingestion of fish, the majority of the arsenic (57 - 84%) was excreted in the urine in 4 days, with little found in the feces. After the ingestion of inorganic arsenic the excretion was more rapid and entirely in the urine. More recently, Crecelius [4] fed a human subject an amount of Dungeness crab containing 2.0 mg. of arsenic and found control levels of trivalent arsenic (As^{3+}) and pentavalent arsenic (As^{5+}) and small amounts of monomethylarsonic acid (MAA) and dimethylarsinic acid (DMAA). However, after digesting the urine with sodium hydroxide, large amounts of DMAA were found indicating the rapid excretion of hydrolyzable organic arsenic compound. The chemical structure of shrimp and crab arsenic has been investigated but is unknown at the present time. The search for similar compounds in higher animals should be an interesting field of research.

3. The metabolism of inorganic arsenic in animals and man

The daily intake of arsenic by man and animals varies with the diet but can amount to several mg. per day [5]. The metabolic fate of this arsenic or that absorbed during industrial exposure or poisoning has received considerable study since the development of analytical methods which speciate the arsenic.

In 1975, Peoples [6], using a method previously developed for determining inorganic and methylated arsenic [7], found that when dogs and cows were fed either trivalent or pentavalent arsenic the arsenic was excreted rapidly in the urine; about 50% as inorganic and 50% as methylated arsenic.

In 1973 Braman [8], using a method for speciating arsenic into arsenite (As^{3+}), arsonate (As^{5+}) monomethylarsonic acid (MAA) dimethylarsinic acid (DMAA), found that DMAA was the predominant form of arsenic in human urine in individuals on a normal diet. Using the same analytical procedure, Crecelius 4 fed a human subject wine containing both As^{3+} and As^{5+} and analyzed his urine at intervals for 100 hours. There was a small sharp peak of As^{3+} and As^{5+} at 5 hours, falling to normal in 20 hours. There was a slow rise in the concentration of MAA and DMAA lasting 80 hours with DMAA predominating. When the individual ingested well water high in As^{5+}, there was a low sharp peak at 5 hours, little change in MAA and prepondorance of DMAA which was excreted for 70 hours.

Charbonneau [9] in 1979 fed dogs carrier-free ^{74}As in a dose of .2 - .6 micrograms/dog and followed the blood and urine concentrations of inorganic arsenic (As^{3+} + As^{5+}) MAA and DMAA in plasma, red cells and urine over a 400 minute period. The inorganic arsenic in the erythrocytes and plasma rose slightly above the base line value at 50 minutes and fell quickly to near the base line. The arsenic was rapidly methylated to DMAA, appearing in the plasma, erythrocytes and urine in 50 minutes after ingestion. No significant amounts of MAA were detected in the blood or urine. The total radioactivity in the urine which was DMAA reached 90% in 90 - 190 minutes and 95 - 100% in 100 - 180 minutes. The lack of MAA may be an intermediate which is quickly methylated to DMAA.

3.1. The metabolism of arsenic by dogs fed sawdust from CCA-C treated wood

Sawdust from wood treated with CCA-C was mixed with canned dog food and fed to two beagle dogs weighing about 10 Kg. They were placed in metabolism cages arranged to collect urine and feces separately. The urine was analyzed by a method developed in our laboratory [10]. Daily examinations showed no signs of toxicity and daily clinical tests of blood and urine were normal. The results of the urine analyses are given in tables 1 and 2.

It will be seen that the only form of arsenic in a dog on a normal diet is DMAA. During the feeding period the main form of arsenic is DMAA. There is no As^{3+} and negligible amounts of MAA. These results are consistent with excretion studies previously reported.

Table 1. The urinary excretion of arsenic by dog #4323 fed 6.00 mg. of pentavalent arsenic/day as CCA-C sawdust.

Day of feeding	mg./day of arsenic in urine			
	As^{5+}	As^{3+}	DMAA	MAA
0	.00	.00	.05	.00
0	.00	.00	.06	.00
1	.28	.00	2.09	.04
2	.21	.00	1.74	.03
3	.26	.00	2.76	.07
4	.25	.00	1.95	.05
5	.25	.00	2.23	.05
6	.25	.00	2.46	.05
7	.23	.00	2.12	.06
8	.19	.00	1.38	.03

Table 2. The urinary excretion of arsenic by dog #4324 fed 13.2 mg. of pentavalent arsenic day as CCA-C sawdust.

Day of feeding	mg/day of arsenic in urine			
	As^{5+}	As^{3+}	DMAA	MAA
0	.00	.00	.05	.00
0	.00	.00	.05	.00
1	1.20	.00	7.18	.08
2	.93	.00	7.35	.13
3	.85	.00	9.91	.17
4	.82	.00	7.41	.18
5	.93	.00	8.44	.17
6	.99	.00	6.85	.09
7	.85	.00	8.68	.17
8	.70	.00	6.76	.13

3.2 Renal clearance studies using As^{5+} in beagle dogs

Ginsburg reported in 1965 [11] a series of renal clearances of As^{5+} in dogs which showed that As^{5+} is reduced to As^{3+}, presumably in the renal tubules since the amount formed is roughly proportional to the amount of As^{5+} reabsorbed by the tubules. Since little As^{3+} was found in the blood, the As^{3+} must then be secreted by the tubules and it then appears in the urine. When he infused As^{3+}, there was an increase of As^{5+} in both plasma and urine indicating an in vivo oxidation of As^{3+}.

This experiment has never been repeated by Ginsburg or anyone else and yet it is the strongest evidence that As^{5+} is reduced in the body to As^{3+}.

Recent studies done in my laboratory in collaboration with Hidekazu Tsukamoto, D.V.M., M.S. and Harold Parker, D.V.M., Ph.D. have shown that when a clearance study is done using the same concentration of As^{5+} in the i.v. infusion as that used by Ginsburg, there is kidney damage as shown by finding protein and glucose in the urine and histopathological changes in the renal tubules. These changes were not found when the concentration of As^{5+} in the infusion was lowered to 1/10 and 1/50 of that used by Ginsburg. This strongly suggests that As^{5+} is reduced to As^{3+} only by tubules which are injured by the high plasma concentration of As^{5+}. These are preliminary experiments and will be repeated. The reason that no one has shown As^{3+} in the urine of any animal in feeding studies is that the absorption from the gastro-intestinal tract prevents the high peak plasma levels needed to damage the renal tubules.

4. Summary

It appears to be well established that both animals and humans have a biochemical mechanism for oxidizing As^{3+} to As^{5+} and then methylating these organic compounds to give MAA and finally DMAA. Since these methylated compounds are about 1/100 as toxic as As^{3+}, the process can be termed a detoxification. The rapidity of the process reduces the possibility of chronic toxicity from As^{3+} and As^{5+} since neither can exist in the body for very long, especially when small doses from food, water and moderate industrial exposure are absorbed. The experiments of Ginsburg and my group should be repeated using more sensitive analytical methods permitting lower plasma levels which do not damage the kidney. There is evidence that there may be complex organic compounds in man and animals which may have an essential biological function.

References

1 Chapman, A., On the presence of compounds in arsenic in marine crustaceans and shellfish, Analyst 51, 548- (1926).

2 Coulson, E. V., Remington, R. V., Lynch, Metabolism in the rat of the naturally occurring arsenic of shrimp as compared with arsenic trioxide, J. Nutrit. 10, 255-270 (1935).

3 Charbonneau, S. M., Spencer, K., Bryte, F., Sandi, E., Arsenic excretion by monkeys dosed with arsenic containing fish or with inorganic arsenic, Bull. Environm. Contam. Toxicol. 20, 470-477 (1978).

4 Crecelius, E. A., Changes in the speciation of arsenic following ingestion by man, Environ. Health Prospect. 19, 147-150 (1977).

5 Schroeder, H. A., Balassa, J. J., Abnormal trace elements in man: Arsenic, J. Chron. Dis. 19, 85 (1966).

6 Lakso, J. U., Peoples, S. A., Methylation of inorganic arsenic by mammals, J. Agric. Food Chem. 23, 674-676 (1975).

7 Peoples, S. A., Lakso, and Lais, T., The simultaneous determination of methylarsonic acid and inorganic arsenic in the urine, Proc. West. Pharmacol. Soc. 14, 178-182 (1971).

8 Braman, R. S., Foreback, C. C., Methylated forms of arsenic in the environment, Science 182, 1287 (1973).

9 Charbonneau, S. M., Tam, G. K. H., Bryce, F., Zadwidzka, Z., Sandi, E., Metabolism of orally administered inorganic arsenic in the dog, Toxicol. Lett. 3, 107-113 (1979).

10 Lakso, J. U., Rose, L. J., Peoples, S. A., Shirachi, D. Y, A colorimetric method for the determination of arsenite, arsenate, monomethylarsonic acid and dimethylarsinic acid in biological and environmental samples, J. Agric. Food Chem. 27, [6], 1229-1233 (1979).

11 Ginsburg, J. M., Renal mechanism for excretion and transformation of arsenic in the dog, Am. J. Physiol, 208, [5], 832-840 (1965).

DISCUSSION

A. Furst: I noticed you talked about cows and dogs, but you omitted rats. You should not neglect them because I often think that Sprague-Dawley rats grow up to become small mules, so they are fairly large animals. What then about the rat? How can they metabolize the arsenic compounds?

S. A. Peoples: I pointedly left rats out of my life--as far as arsenic is concerned. As I said before starting this discussion, when I gave my first paper I was the only one that did not use rats. Rats have the péculiar habit of storing 90-some percent of their arsenic in their red cells. No other animal does this. I, therefore, object to using the rat as a subject because he sequesters the arsenic and it doesn't give the kidney and the liver and other metabolic sites a chance to work on it. There have been some studies on the rates of excretion in different animals--the chicken probably excretes arsenic faster than any living creature, with a half life of about 3 hours. In man the half-life is 24-36 hours. The rat has a very slow rate of excretion because it has to wait for its red cells to recycle. Therefore, I don't think rats are suitable for metabolic studies. It also leads people to the strange belief that arsenic is stored in the body. People frequently tell you about the storage of arsenic in hair and nails. I said they are not part of the living body and it's an excretory mechanism, not a storage mechanism.

V. L. Zaratzian: I'm concerned about the aromatic arsenical compounds that are used for poultry and swine, and I'd like to know if there is any evidence, that the carbon-arsenic bond is broken during metabolism. I have not found any information that the carbon-arsenic bond is broken, and of course, that's a concern of ours.

S. A. Peoples: As far as I know it is not broken--arsanilic acid and the other phenyl arsenates, are not metabolized appreciably in animals.

V. L. Zaratzian: They are metabolized to the phenylarsonic acid or its acetylated derivatives.

S. A. Peoples: There are changes on the ring.

V. L. Zaratzian: But is the ring broken in any way?

S. A. Peoples: No.

W. L. Marcus: I want to confirm what Dr. Peoples said. As a matter of fact, the rat is quite an improper model to use for exactly those reasons. A study done by Dutkowitz in 1977* showed that red cell binding and arsenic excretion in rats is quite different than any other animal that he looked at including monkeys, dogs and man. Therefore, we firmly agree that the rat is the improper model to use.

S. Lamm: Recognizing the high prevalence of renal disease in man and your explanation for the findings in Ginsburg's work, have you considered examining dogs that already have damaged kidneys to see whether in the presence of damaged kidneys the question may be valid?

S. A. Peoples: I haven't done that, but your suggestion is pertinent and Dr. Parker and I will try such a study. Renal damage is fairly common in dogs.

E. A. Woolson: What was the lowest dose where dogs did not have renal damage and arsenite was not found?

S. A. Peoples: The infusion rate was 17.5 μg/min. Trivalent arsenic was found in the urine when the infusion rate was 109 μg/min and 202 μg/min.

W. H. Lederer: I was wondering if you could comment on the technique that Ginsburg used in speciation of the arsenic.

S. A. Peoples: Ginsburg used a method which separated the arsenic by extraction with sodium xanthate. I consider this a relatively crude method. It was the only method which was available at that time for separation of pentavalent and trivalent arsenic. I think that this is another reason why Ginsburg's work should be repeated, not necessarily by me but by someone using more modern methods with a more exact technique. I tried this method and I found I could not achieve the accuracy that he claimed he got.

E. A. Woolson: I'm not a biomedical person so I'm not exactly sure how the absorption of arsenic occurs in the body. Does it absorb from the intestinal tract into the blood and then is purged from the blood at the kidney or liver?

S. A. Peoples: Yes.

E. A. Woolson: So that if reduction or metabolism occurs at the liver, it would have already essentially passed through the body.

S. A. Peoples: That's true. If the arsenic is given orally, it goes through the liver before it gets to the rest of the body. If you give it intravenously as we did, it gets everywhere at once.

---

*1977 published, "Experimental Studies on Arsenic Absorption Routes in Rats"; Environmental Health Perspectives, Vol. XIX, pp. 173-177.

E. A. Woolson: Would it go through the liver and be metabolized to cacodylic acid before going to the bloodstream?

S. A. Peoples: It could be. Sometimes the liver is not 100% effective, especially at higher doses, so that it may take two or three passes through the liver before the biochemical process in the liver is complete. This is why often early after giving a dose orally you may get some pentavalent arsenic in the urine, but not very much. The liver will take care of most of it. By the way, it has never been shown definitely that the liver does this. Those of you who like original research can work on this problem.

E. A. Woolson: I believe Marie Vahter in Sweden has or is about to publish some work showing that the site of metabolism is the liver. What I was trying to establish was if one ingests arsenite, arsenate, or one of the organic forms, does the metabolism occur before it gets to the bloodstream and the skin and various organs, or is it detoxified first?

S. A. Peoples: Actually, anything that you eat, regardless of whether it's arsenic or not, must enter first through the portal system of the liver. It's a separate circulatory internal matter, you might say, between the intestinal tract and the liver. Then the liver puts it back into the bloodstream through the hepatic vein and then into the main bloodstream. Determining whether it would all get metabolized depends on the dose. If the dose is small, the liver will handle it all, but as you increase the dose, it may get through the liver and out before it is fully metabolized.

G. D. Rosebery: With regard to this rather peculiar habit of rats to store arsenic in their cells, would you say that the use of rodents in chronic feeding oncogenicity, teratology, etc., studies would be inappropriate for most arsenic species?

S. A. Peoples: I think it would be inappropriate for arsenical compounds.

G. D. Rosebery: In that regard, with your knowledge of the metabolism of arsenic in general, what sort of animal would you recommend for long-term chronic studies?

S. A. Peoples: We have found that you could use guinea pigs and hamsters. Hamsters are very good experimental animals, and of course, they are completely inbred. I think those would be about the best small animals.

12

TOXICOLOGY OF PRENATAL EXPOSURE TO ARSENIC

Ronald D. Hood
Developmental Biology Section
Department of Biology
The University of Alabama
P. O. Box 1927
University, Alabama 35486

The effect on the offspring of maternal arsenic exposure during pregnancy is influenced by the form of arsenic involved, species and individual differences in susceptibility, dose level and exposure route, peak level attained in the conceptus, maternal metabolism and excretion, and timing of exposure during gestation. Studies have shown that prenatal death, malformation, and inhibition of growth can result from exposure to arsenite, arsenate, cacodylate, and methylarsonate during the susceptible stages of gestation. Such effects generally are seen, however, only at dose levels causing or approaching maternal toxicity and above those typically encountered by man. Malformation associated with arsenic treatment typically involves maternal exposure by the intraperitoneal or intravenous rather than the oral administration routes. Also, the developing mouse appears to be less susceptible to arsenic effects than is the hamster or rat. In addition, valence state influences the relative toxicity of inorganic arsenic. Arsenite is more acutely toxic to both mother and conceptus, but arsenate is more likely to cause malformation of fetuses. Presumably, these differences in effect are due to different mechanisms of action, although both forms are believed to disturb energy metabolism. Inorganic arsenic is also known to be methylated by mammals. The methylated metabolites are much less toxic than the inorganic parent forms, although high doses of mono- or dimethylated arsenic can be teratogenic.

1. Introduction

Arsenicals were among the first chemical agents tested for teratogenic potential, although the early studies, dating back to the 1930ies, were done with chick embryos [1,2,3,4,5].[1]

The initial mammalian studies with arsenic involved arsenate and are summarized in table 1. Although James *et al.* [6] had earlier described stunting of one of four lambs whose mothers had been fed potassium arsenate during gestation, the first detailed report of arsenic teratogenicity

---

[1] Figures in brackets indicate the literature references at the end of this paper.

in a mammal was that of Ferm and Carpenter in 1968 [7]. They injected hamsters iv with high doses of sodium arsenate (20 mg/kg) on gestation day eight. Such treatment resulted in a high percentage of malformed fetuses, as well as increased prenatal mortality. Doubling the dose resulted in completely resorbed litters, while a lower dose (5 mg/kg) had no effect. In later studies with hamsters, Holmberg and Ferm [8] described a protective effect of injected selenium against both arsenate- and cadmium-induced malformations, and Ferm *et al*. [9] outlined the spectrum of malformations resulting from treatment on different times during day eight of pregnancy. Defects observed in the latter study included skeletal and urogenital tract anomalies, as well as exencephalies. Similarly, Wilhite [10] reported exencephaly and related defects (cranioschisis occulta and encephalocoele) following intravenous injection of sodium arsenate in day-eight pregnant hamsters.

TABLE 1

Prenatal Effects of Inorganic Arsenate in Laboratory Rodents

Species	Fetal Stunting	Prenatal Mortality	Malformations	Reference
Mouse	+	+	+	Hood and Bishop [12], Hood and Pike [13] , Hood *et al*. [14].
Hamster	?[a]	+	+	Ferm and Carpenter [7], Holmberg and Ferm [8], Ferm *et al*. [9], Wilhite [10], Ferm and Kilham 11 .
Rat	+	+	+	Beaudoin [16], Burk and Beaudoin [23].

[a]Fetuses not weighed.

In an additional study, Ferm and Kilham [11] presented evidence that if pregnant hamsters are given 10 mg/kg sodium arsenate intraperitoneally and placed in a hyperthermia-inducing environment at 40° C for 50-60 minutes, both prenatal mortality and the incidence and severity of fetal malformations are increased in comparison with the effect of either treatment alone.

The initial study in our laboratory in 1972 [12] involved ip injection of sodium arsenate in mice on one of gestation days 6-12. The day a copulation plug was seen was considered to be day one (all descriptions of timing in gestation in this review have been adjusted to conform to this terminology). As was the case with the hamster, arsenate treatment during development with a relatively high dose (45 mg/kg) was associated with both gross and skeletal malformations and prenatal deaths. The defects observed involved a wide variety of structures. A lower dose (25 mg/kg) was ineffective.

We also tested the ability of BAL, an arsenic chelating agent, to protect against the effects of arsenate on the mouse fetus [13]. Treatment of pregnant mice on gestation day nine with BAL given subcutaneously resulted in a protective effect against arsenate given ip on the same day. This occurred with regard to fetal growth and gross malformations, regardless of whether the BAL was given four hours before or after, or concurrently with the arsenate. Skeletal malformations were prevented only by concurrent treatment, however.

In a more recent study, we [14] compared the outcome of administration of oral versus intraperitoneal sodium arsenate to pregnant mice on one of gestation days 7-15. Results of 40 mg/kg ip treatment were similar to those discussed previously, while treatment by gavage at 120 mg/kg resulted in a lesser rate of fetal stunting and malformation. Lower oral doses had little if any effect, and results of multiple doses of up to 80 mg/kg on gestation days 7-9, 10-12, or 13-15 were negative.

In an additional study by Hood *et al.*, [15] pregnant mice were fed throughout gestation with a diet containing 10% sawdust from wood treated with the arsenical-containing wood preservative, CCA. The wood had been impregnated with CCA at a level of 0.66 lb/ft^3. No adverse effects were seen in the offspring, although the amount of arsenic pentoxide ingested was estimated to be 136 mg/kg/day. This lack of adverse effect was presumably due to the unavailability of the arsenic bound to wood fibers.

The rat has also been a subject for arsenate teratogenicity testing. In 1974, Beaudoin [16] administered sodium arsenate ip to rats on one of days 8-13 of pregnancy. Malformations were observed at doses of 20 to 40 mg/kg, and consisted mainly of defects of the head, urogenital tract, and skeleton. Fetal weights were decreased in some cases at the higher doses, as was prenatal survival.

Results of mammalian teratogenicity studies with arsenite are summarized in table 2. The initial such test was published by us in 1972 [17]. In this study, sodium arsenite was administered ip to mice on one

TABLE 2

Prenatal Effects of Inorganic Arsenite in Laboratory Rodents

Species	Fetal Stunting	Prenatal Mortality	Malformations	Reference
Mouse	+	+	+	Hood [17], Matsumoto *et al.* [18,19], Baxley *et al.* [21].
Rat[a]	-	-	-	Kojima [20].
Hamster	+	+	+	Harrison and Hood [22], Wilhite [10].

[a]Chronic low dose exposure (100 ppm or less in the diet).

of gestation days 7-12 at doses of 10 or 12 mg/kg. Such treatments resulted in increased prenatal mortality. In some cases, decreased fetal weight and increased incidences of gross and skeletal malformations were also observed.

Subsequently, Matsumoto *et al*. [18,19] described decreased fetal weights and survival in mice given sodium arsenite orally at doses of up to 40 mg/kg for three consecutive days. A few minor malformations were noted, but their incidences did not appear to be dose-related.[2] According to Kojima [20], however, feeding a diet containing up to 100 ppm arsenite to rats throughout pregnancy (with or without subsequent feeding during the lactation period or for 12 weeks thereafter) had no effect on the offspring.

Our [21] recent results with single oral doses of sodium arsenite indicate that levels of 40-45 mg/kg are needed to cause adverse effects on developing mice. Such treatments on one of gestation days 8-15 resulted in some increased rates of prenatal death, but had only occasional effects on growth. Only a suggestive but nonsignificant increase in gross malformations was observed.

In addition, we [22] reported that in the hamster, ip injection with 5 mg/kg arsenite or oral treatment with 25 mg/kg on one of gestation days 9 or 10 resulted in totally resorbed litters. Later, more complete results [Hood and Harrison, unpublished] show fetal stunting and an apparent decrease in fetal survival following maternal oral treatment with a dose of 25 mg/kg. Treatment by ip injection with 5 mg/kg on one of days 8, 11, or 12, or with 2.5 mg/kg on day 10, resulted in apparently increased prenatal mortality, and the high dose was also associated with fetal stunting. Low rates of gross and skeletal malformations were seen in treated groups, but the incidences observed were nonsignificant. According to Wilhite [10], however, iv injection with 5 or 10 mg/kg sodium arsenite on gestation day eight caused malformations in the litters of treated hamsters.

The relative effects of tri- and pentavalent arsenic on the mouse conceptus are compared in table 3. The data indicate that both valence states are frequently fetotoxic, with treatment resulting in prenatal growth inhibition or mortality. Both gross and skeletal malformations are seen more frequently, however, following arsenate treatment. Even at comparably toxic doses, however, the two valence states produce somewhat dissimilar effects. For example, both forms of arsenic can induce rib defects in mice, but the pattern of such defects is strikingly different [12,17]. Thus it can be seen that studies of the biological effects of arsenite do not predict the effects of exposure to arsenate. The two forms differ in their relative toxicities, in many of their presumed mechanisms of action, and in their prenatal effects.

---

[2]The title of the reference from *Teratology* [18] suggests that the study involved arsenate, but it apparently involved arsenite [19, and A. L. Schmidt, personal communication].

TABLE 3

Comparative Effects on Mouse Development of Arsenate and Arsenite[a]

Prenatal Effect	Arsenate	Arsenite
Fetal Stunting	Severe	Moderate
Prenatal Mortality	Severe	Severe
Gross Malformations	Severe	Moderate
Skeletal Malformations	Severe	Slight
Visceral Malformations	Rare	Rare

[a]From references given in tables 1 and 2.

Species differences between the mouse, rat, and hamster in susceptibility of the offspring to exposure to pentavalent arsenic during pregnancy are outlined in table 4. The available experimental evidence indicates that the mouse conceptus may be the least sensitive of the three to arsenate, in terms of minimum effective dose. The apparent differences are not large, however, and may be artifacts of experimental protocol, strains employed, and similar factors.

TABLE 4

Species Effects on Susceptibility to Prenatal Exposure to Inorganic Arsenate[a]

Prenatal Effect	Species		
	Mouse	Rat	Hamster
	Severity of Effect (Effective dose in mg/kg)		
Fetal Stunting	Severe (40 ip)	Moderate (30 ip)	Unknown
Prenatal Mortality	Severe (40 ip)	Severe (30 ip)	Severe (17.5 iv)
Gross Malformations	Severe (40 ip)	Moderate (30 ip)	Severe (17.5 iv)
Skeletal Malformations	Severe (40 ip)	Severe (30 ip)	Severe (20 iv)
Visceral Malformations	Rare	Severe (30 ip)	Moderate (20 iv)

[a]From references given in table 1.

The types of malformations observed tend to be similar across species, with the exception of visceral malformations. These have not been observed in the mouse [14], but urogenital tract defects, particularly renal agenesis, are seen in both the rat and the hamster [23,9].

Only the mouse and hamster have been subjected to high doses of inorganic arsenite during pregnancy. As is shown in table 5, both species can be affected by such treatment. The only apparent species differences are the greater resistance of the mouse to induction of visceral malformations and its apparent greater sensitivity to arsenite-induced skeletal defects.

TABLE 5

Species Effects on Susceptibility to Prenatal Exposure to Inorganic Arsenite[a]

Prenatal Effect	Species	
	Mouse	Hamster
	Severity of Effect (Effective dose in mg/kg)	
Fetal Stunting	Moderate (12 ip)	Moderate to Severe (25 po)
Prenatal Mortality	Severe (10 ip)	Severe (10 iv)
Gross Malformations	Moderate (12 ip)	Moderate (5 iv)
Skeletal Malformations	Moderate (12 ip)	Slight (5 iv)
Visceral Malformations	Rare	Rare[b]

[a]From references given in table 2.

[b]Insufficient data to establish effective dose.

Since the summarized effects are seen in all three species at dose levels close to the maternal lethal dose, the responses seen should be at least roughly comparable. It must be kept in mind, however, that such data can be greatly influenced by a number of variables in addition to dose. The relative sensitivity of the developing human to arsenic is as yet unknown, as there is no evidence in the literature regarding human exposure at potentially sensitive stages of development. Therefore, we are left with only animal models as predictors of effects.

One example of a major factor influencing the outcome of toxic insult during gestation is the developmental stage during which exposure occurs. Samples of the differences in response that can be obtained from the same treatment given on different gestation days are shown in table 6. For

TABLE 6

Effect in the Mouse of Developmental Stage on the Outcome of Prenatal Exposure to Arsenate[a,b]

Prenatal Effect	Gestation Day of Arsenic Exposure							
	6	7	8	9	10	11	12	Solvent
Prenatal Mortality (%)	51	37	56	60	51	69	78	4
Fetal Weight (g ± SE)	0.88 ± 0.02	0.67 ± 0.03	0.87 ± 0.02	0.61 ± 0.03	0.79 ± 0.03	0.94 ± 0.03	1.05 ± 0.02	1.05 ± 0.01
Exencephaly (%)	0	3	31	54	0	0	0	0
Open Eye (%)	5	19	14	20	2	0	0	0
Umbilical Hernia (%)	0	18	0	9	0	0	0	0
Malformed Limb (%)	0	0	0	2	5	3	0	<1
Malformed Tail	2	0	0	9	25	5	0	0
Malformed Skeleton (%)	0	0	28	100	73	0	0	0

[a]Maternal Dose = 45 mg/kg ip.

[b]Adapted from Hood and Bishop [12].

example, although treatment of pregnant mice with a high dose of sodium arsenate on one of gestation days 6-12 decreased prenatal survival, the mortality rates ranged from a low of 37%, following day 7 treatment, to a high of 78% after day 12 treatment. Similar results were seen with regard to fetal weights, which ranged from 0.61 g to 1.05 g, a value no different from that of the solvent (H_2O) treated controls. In addition, it can be seen that the spectrum of malformations produced was highly treatment-day dependent.

Such a developmental stage dependence of effects of teratogens on the conceptus is seen in man as well as in experimental animals. In most cases, human exposure to a potential teratogen in the first trimester of pregnancy, during the period of early organogenesis, is more likely to result in malformation than would similar exposure at other times. This is particularly true of gestation days 15-40 [24]. In the case of agents affecting the brain or the genital tract, however, the critical periods are thought to extend later in development. This is because the brain and genitalia are among the last organs to mature.

Another significant variable in treatment protocol is maternal treatment route. This is shown in table 7, which compares the outcome of prenatal exposure of CD-1 mice (Charles River Breeding Labs, Wilmington, Massachusetts) to arsenate and arsenite given as single doses orally or intraperitoneally to the pregnant mother. With regard to arsenate, treatment effects were similar, with the exception of malformation rates. Maternal oral exposure resulted in significantly lower fetal malformation rates than did parenteral treatment. With arsenite, however, oral treatment was less hazardous to the offspring in every category examined.

Compared on the basis of elemental arsenic, the arsenate doses involved for oral and ip dosing are found to be approximately 48.4 and 16 mg/kg, respectively. For arsenite, the elemental arsenic doses for oral and for ip treatment were 23 and 5.8 - 6.9 mg/kg, respectively.

The point must be made, however, that the observed lesser effects on the embryo and fetus following maternal oral exposure were not associated with observable differences in maternal toxicity. For example, the incidences of maternal deaths following both ip and po arsenate treatments were virtually identical, averaging approximately 16% across treatment days. In the case of arsenite treatment, the oral dose was even somewhat more toxic than the ip dose, with maternal mortality ranging from 19 to 36% for the former and 7 to 24% for the latter.

It has been speculated that such differences are due to a too rapid uptake of arsenic in the blood and transfer to the fetus following maternal injection. This could largely bypass or overwhelm the maternal detoxification mechanism, arsenic methylation. Methylation would normally tend to reduce the degree of fetal insult following maternal arsenic exposure, especially when exposure is by the oral route. This supposition is supported by our data showing that a substantial percentage of arsenic taken up by the mouse fetus is in the methylated form, particularly as dimethylarsonate [Hood, Vedel, Zaworotko and Tatum, unpublished].

That the degree of methylation is decreased when arsenate is injected, rather than given orally, is supported by the results of Odanaka _et al._ [25]. They recovered 66% of urinary arsenic in methylated forms

TABLE 7

Influence of Maternal Arsenic Exposure Route of Mice on Insult to the Developing Offspring[a]

Agent	Treatment Route and Dose (mg/kg)	Prenatal Mortality	Fetal Stunting	Malformation	Maternal Mortality
Arsenate	ip:40	Moderate to Severe	Moderate	Moderate	16.2%
	po:120	Moderate to Severe	Moderate	Slight	16.1%
Arsenite	ip:10-12	Severe	Moderate	Moderate	6.67 - 24.4%
	po:40-45	Moderate	Slight	Slight	19 - 36%

[a]Adapted from Hood *et al*. [14], Hood [17], and Baxley *et al*. [21].

versus only 34% as inorganic arsenic following oral dosing of mice with arsenic acid (As^{+5}). These results are in contrast to the values of 45% methylated and 55% inorganic arsenic following intravenous injection. Similar results they obtained with hamsters confirmed these observations.

Additional support for the premise that a much greater amount of total arsenic reaches the fetus following intraperitoneal dosing than is the case with oral treatment has also been recently obtained in our laboratory. The data is shown in table 8 and was presented at the past meeting of the Teratology Society [26]. In this study, 18 day pregnant mice were treated with sodium arsenate at a dose of either 20 mg/kg by intraperitoneal injection or 40 mg/kg by gavage. We then sacrificed the mice at a variety of sampling times up to 24 hours. We determined total arsenic in fetuses and placentas by hydride generation atomic absorption spectrophotometry.

TABLE 8

Fetal Arsenic Content (as % of Dose) Following ip (20 mg/kg) or po (40 mg/kg) Sodium Arsenate Given to Mice on Gestation Day 18[a]

	Tissues Assayed and Treatment Mode					
	Fetuses		Litters		Placentas	
Time (hr)	ip	po	ip	po	ip	po
0.5	0.20	0.02	2.45	0.28	0.83	0.10
1	0.87	0.06	8.43	0.67	1.97	0.30
2	1.65	0.06	16.50	0.72	1.84	0.27
4	1.12	0.06	12.73	0.59	1.13	0.16
6	0.35	0.08	3.80	1.01	0.25	0.23
12	0.26	0.05	2.68	0.53	0.14	0.17
18	0.09	0.04	0.83	0.43	0.06	0.09
24	0.09	0.06	0.36	0.65	0.00	0.10

[a]Each value is the mean of samples from 3 or more litters.

At peak levels of uptake, a much higher percentage of the dose administered was found in the fetuses and in their placentas following maternal arsenate injection than was the case after oral treatment. We have obtained equally clear cut results in similar experiments involving arsenite. This indicates that peak levels of arsenic in the fetus are much higher after maternal injection than following oral treatment. Such data clearly show the dangers inherent in attempting to extrapolate arsenic effects following oral exposure from data derived from animals given arsenic by injection. These results also support the relative teratogenicity data obtained comparing the two exposure routes, as it is generally believed that teratogenic effects are highly correlated with peak exposure levels for the conceptus.

Methylated arsenicals have also been tested for teratogenicity, and the results are summarized in table 9. For example, Wilhite [10] reported no significant effect from iv injection of hamsters with relatively low doses of methylarsonic acid or cacodylic acid (dimethylarsinic acid) on day eight of gestation. Rogers _et al_. [27], however, obtained cleft palates in mice following maternal gavage with 400 or 600 (but not 200) mg/kg/day doses of cacodylic acid on gestation days 7-16, along with dose-dependent decreases in fetal weight and fetal survival. They described similar fetotoxicity in rats at doses of 40-60 mg/kg/day, with no cleft palates. Misaligned palatine rugae were seen, however, at doses of 30-60 mg/kg/day in these rats.

We have observed fetotoxicity and malformations in litters from mice and hamsters treated ip with high doses of sodium cacodylate. Disodium methanearsonate had similar effects in mice, but our results with hamsters are still preliminary [28, and Hood and Harrison, unpublished]. The doses employed were much higher than those of inorganic arsenic, confirming the relative reduction in toxicity due to arsenic methylation.

According to the reasoning of Johnson [29], our major concern should be those teratogenic agents to which the developing offspring is uniquely susceptible, rather than those that are teratogenic only at doses toxic to the mother. The doses of both inorganic and methylated arsenic required to produce fetal malformations are very near or within the range of maternally lethal doses. In general, the animal data suggest few prenatal effects from acute arsenic intoxication at doses that are not toxic to the mother. Thus, the similarity of the teratogenic and maternal toxic doses in animal models indicates that our greatest concern with arsenic should be its general toxicity rather than its potential as a teratogen, assuming the same relationship may hold true for man. The possible effects of chronic exposure to arsenic, however, have not been adequately tested.

In table 10, major factors affecting arsenic teratogenicity are listed. From these, it can be seen that for a teratogenic response to inorganic arsenic, the valence state is a significant consideration. With regard to both inorganic and methylated arsenicals, additional factors to consider include evidence that (1) dose level generally must be close to or above the maternal minimum lethal dose, (2) a steep dose-response curve is typical, (3) exposure route is important, as oral exposure tends to be significantly less effective than parenteral treatment, (4) methylation decreases arsenic toxicity, (5) both species of animal and timing of exposure during gestation influence the effect on the conceptus, and (6) insufficient data are available on the prenatal effects of chronic exposure to arsenic.

TABLE 9

Prenatal Effects in Rodents Following Maternal Treatment with Methylated Arsenicals

Agent	Test Species	Dose (mg/kg)	Fetal Stunting	Prenatal Mortality	Malformations	References
Sodium Cacodylate	Mouse	800 - 1,200 (ip)	-	+	+	Harrison *et al*. [28]
	Hamster	800 - 1,000 (ip)	+	+	+	Hood and Harrison, unpublished
Cacodylic Acid	Hamster	20 - 100 (iv)	?[a]	-	±	Wilhite [10]
	Mouse	200 - 600 (po)[b]	+	+	+	Rogers *et al*. [27]
	Rat	7.5 - 60 (po)[b]	+	+	+	Rogers *et al*. [27]
Disodium Methanearsonate	Mouse	1,200 - 1,500 (ip)	+	+	+	Harrison *et al*. [28]
	Hamster	700 - 800 (ip)	+	+	?	Hood and Harrison, preliminary data
Methylarsonic Acid	Hamster	20 - 100 (iv)	?[a]	-	-	Wilhite [10]

[a]Weights not reported.

[b]Chronic dosing.

TABLE 10

Summary of Factors Affecting Arsenic Teratogenicity

Valence State:	Arsenic (III) more toxic than Arsenic (V)
Dose Level:	Effective dose typically near the maternal toxic dose
Dose Response:	Typically steep response curve
Route of Exposure:	iv or ip >> po
Inorganic or Methylated:	Inorganic far more toxic than methylated arsenic
Timing During Gestation:	Greatest susceptibility during early organogenesis
Species Exposed:	Rat ≃ Hamster > Mouse
Single or Chronic Dosing:	Little data available on chronic exposure to teratogenic dose levels

References

[1] Ancel, P., and Lallemand, S., Sur l'arrêt de développement du bourgeon caudal obtenu expérimentalement chez l'embryon de poulet, Arch. Physique Biol. 15, 27-29 (1941).

[2] Ancel, P., Recherche expérimentale sur la spina bifida, Arch. Anat. Microsc. Morphol. Exp. 36, 45-68 (1946).

[3] Ancel, P., Recherche sur la réalisation expérimentale de la celosomie, Arch. Anat. 30, 6-15 (1947).

[4] Landauer, W., Le probléme de l'électivité dans les epériences de tératogenése biochimique, Arch. Anat. Microsc. Morphol. Exp. 38, 184-189 (1949).

[5] Ridgway, L. P., and Karnofsky, D. A., The effects of metals on the chick embryo: Toxicity and production of abnormalities in development, Ann. N. Y. Acad. Sci. 55, 203-215 (1952).

[6] James, L. F., Lazar, V. A., and Binns, W., Effects of sublethal doses of certain minerals on pregnant ewes and fetal development, Am. J. Vet. Res. 27, 132-135 (1966).

[7] Ferm, V. H. and Carpenter, S. J., Malformations induced by sodium arsenate, J. Reprod. Fert. 17, 199-201 (1968).

[8] Holmberg, R. E., Jr. and Ferm, V. H., Interrelationships of selenium, cadmium, and arsenic in mammalian teratogenesis. Arch. Environ. Health 18, 873-877 (1969).

[9] Ferm, V. H., Saxon, A., and Smith, B. M., The teratogenic profile of sodium arsenate in the golden hamster, Arch. Environ. Health 22, 557-560 (1971).

[10] Wilhite, C. C., Arsenic-induced axial skeletal (dysraphic) disorders, Exp. Mol. Pathol. 34, 145-158 (1981).

[11] Ferm, V. H. and Kilham, L., Synergistic teratogenic effects of arsenic and hyperthermia in hamsters, Environ. Res. 14, 483-486 (1977).

[12] Hood, R. D. and Bishop, S. L., Teratogenic effects of sodium arsenate in mice, Arch Environ. Health 24, 62-65 (1972).

[13] Hood, R. D. and Pike, C. T., BAL alleviation of arsenate-induced teratogenesis in mice, Teratology 6, 235-238 (1972).

[14] Hood, R. D., Thacker, G. T., and Patterson, B. L., Prenatal effects of oral versus intraperitoneal sodium arsenate in mice, J. Environ. Pathol. Toxicol. 1, 857-864 (1978).

[15] Hood, R. D., Baxley, M. N., and Harrison, W. P., Evaluation of chromated copper arsenate (CCA) for teratogenicity, Teratology 19, 31A (1979).

[16] Beaudoin, A. R., Teratogenicity of sodium arsenate in rats, Teratology 10, 153-158 (1974).

[17] Hood, R. D., Effects of sodium arsenite on fetal development, Bull. Environ. Contam. Toxicol. 7, 216-222 (1972).

[18] Matsumoto, N., Okino, T., Katsunuma, H., and Iijima, S., Effects of Na-arsenate on the growth and development of the foetal mice, Teratology 8, 98 (1973).

[19] Matsumoto, N., Okino, T., Katsunuma, H., and Iijima, S., Effects of Na-arsenite on the growth and development of the foetal mice, Congenit. Anom. Curr. Lit. 13, 175-176 (1973).

[20] Kojima, H., Studies on developmental pharmacology of arsenite. 2. Effect of arsenite on pregnancy, nutrition and hard tissue, Fol. Pharmacol. Japon. 70, 149-163 (1974).

[21] Baxley, M. N., Hood, R. D., Vedel, G. C., Harrison, W. P., and Szczech, G. M., Prenatal toxicity of orally administered sodium arsenite in mice, Bull. Environ. Contam. Toxicol. 26, 749-756 (1981).

[22] Harrison, W. P. and Hood, R. D., Prenatal effects following exposure of hamsters to sodium arsenite by oral or intraperitoneal routes, Teratology 23, 40A (1981).

[23] Burk, D. and Beaudoin, A. R., Arsenate-induced renal agenesis in rats, Teratology 16, 247-260 (1977).

[24] Tuchmann-Duplessis, H., in Drug Effects on the Fetus, pp. 38-46 (ADIS Press, Sydney, NSW, 1975).

[25] Odanaka, Y., Matano, O., and Goto, S., Biomethylation of inorganic arsenic by the rat and some laboratory animals, Bull. Environ. Contam. Toxicol. 24, 452-459 (1980).

[26] Hood, R. D., Vedel, G. C., Zaworotko, M. J., Tatum, F. M., and Fisher, J. G., Arsenate distribution in po or ip treated pregnant mice and their fetuses, Teratology 23, 42A (1981).

[27] Rogers, E. H., Chernoff, N., and Kavlock, R. J., The teratogenic potential of cacodylic acid in the rat and mouse, Drug Chem. Toxicol. 4, 49-61 (1981).

[28] Harrison, W. P., Frazier, J. C., Mazzanti, E. M. and Hood, R. D., Teratogenicity of disodium methanearsonate and sodium dimethylarsinate (sodium cacodylate) in mice, Teratology 21, 43A (1980).

[29] Johnson, E. M., Screening for teratogenic hazards: Nature of the problems, Ann. Rev. Pharmacol. Toxicol. 21, 417-429 (1981).

DISCUSSION

E. A. Woolson: Do you know what EPA's guideline is relative to the maternal toxicity in determining whether a teratogenicity study is valid? Is it 10% maternal death?

R. D. Hood: That's my impression, according to the FIFRA guidelines (Federal Register, Vol. 43, No. 163, August 22, 1978). The IRLG guidelines employ the same criteria.

E. A. Woolson: At the 10% maternal death treatment levels, do you see any teratogenicity?

R. D. Hood: In many cases, you have to go to 10% or greater. In others, we've seen teratogenicity at less than 10% maternal mortality. It's typically in the range of maternal toxic effects, and in most cases it seems to have been in the range of maternal lethal effects--at least at the lower end of the range.

W. L. Marcus: The EPA Office of Drinking Water sets the Maximum Contaminants Levels (MCLs) for chemicals in drinking water. When we look at a chemical that has a teratogenic response, or any response for that matter, maternal toxicity would not be allowed above 10% body weight loss. We would not consider significant death rates in the upper dose levels as appropriate data in an experiment. For the purpose of setting MCLs, a 10% weight loss is our upper endpoint for teratology and not mortality.

R. D. Hood: Those criteria appear to apply only to the Office of Drinking Water, but they seem to be a reasonable approach.

M. L. Roy: Since we're now aware that teratogenic effects can be manifested through the male and most of the workers in arsenic tend to be male workers, are you aware of any epidemiological studies involving workers and their wives?

R. D. Hood: There have not been studies, to my knowledge, involving the husband's exposure other than in perhaps living near an environmental point source. There have been some epidemiological studies, but the only ones that I am aware of were confounded with exposure to so many other pollutants, sulfur compounds, other metals, and so forth, that I'm not sure how reliable they could be with regard to singling out arsenic effects.

M. L. Roy: That's right. They would have to be compared to another group of miners or somebody who wasn't exposed to arsenic ore.

R. D. Hood: Right.

W. L. Marcus: Are you aware of any toxicological studies in respect to male exposure?

R. D. Hood: Not with regard to arsenic teratogenesis. There have been some studies done with other agents, but none that I know of with arsenic.

W. H. Lederer: Would you classify pentavalent arsenic as teratogenic, using the criterion of 10% maternal mortality?

R. D. Hood: It would certainly only be borderline, using that kind of a criterion, at least with acute exposure. Now, with regard to low-dose chronic exposure, there have been some studies like that. The doses were so low that I don't think they could have established a no-effect level. I don't believe there are any good studies in the literature using doses high enough to establish an obvious no-effect level with chronic exposures, so I couldn't say in that case. The literature is mainly on acute exposure.

G. D. Rosebery: With regard to the routes of exposure, it's rather unlikely that human beings would become exposed via any I.V. or I.P. route so the oral route would be the most common. However, are you aware of any inhalation teratology that's being done? This is another significant route of exposure that one could consider perhaps equivalent to I.V. or I.P. application.

R. D. Hood: I'm not aware of any inhalation exposure studies involving arsenic and teratogenicity and I'll grant that the comment about the exposure routes is quite valid. In fact, that's why I mentioned that a lot of inferences sometimes have been made on the basis of the injection experiments, but when we looked at oral exposure, we had to give much higher doses and saw much lower rates of malformation in most cases than were seen with the other exposure routes.

13

A NEW LOOK AT ARSENIC CARCINOGENESIS

Arthur Furst
Institute of Chemical Biology
University of San Francisco
San Francisco, CA 94117

For several decades arsenic has been considered the classical human carcinogen. Many epidemiological studies seem to bear this out. If, however, one evaluates the total picture, one must come to a different conclusion. Regulatory agencies have set three criteria which must be met before a substance is declared to be a carcinogen; these are epidemiological evidence, results of animal experiments, and information from genotox studies. Only epidemiological studies implicate arsenic as a human carcinogen, and these studies are not without major confounding factors. The animal experiments have been uniformly negative even when arsenic has been coadministered with known rodent carcinogens such as polyaromatic hydrocarbons or nitrosamines. The genotox aspect will not be treated in detail here, but the evidence for the mutagenicity of arsenic is tenuous. It is postulated in this review that arsenic is not a primary carcinogen. If it is carcinogenic at all, the mechanism of action must be completely different from that of any known carcinogen. Conclusions can then be drawn that arsenic may replace essential trace elements (like selenium) which may be protective agents, or that arsenic may require an unusual number of co-factors before the carcinogenic action is manifested, or that arsenic, rather than carcinogenic, is a powerful growth stimulant. Once the initiation and promotion by other agents have begun, arsenic may accelerate the growth so rapidly that it appears to be the etiological agent.

1. Introduction

Seldom are metals or metaloids mentioned in reviews or compilations on carcinogenic compounds. If inorganic agents are included, arsenic is the most frequently listed element. The exception to this, of course, are reviews by Furst [1][1], Flessel et al [2], Luckey and Venugopal [3], Radding and Furst [4], and Sunderman [5,6] who have published extensively on the topic of metal carcinogenesis. An entire issue of Environmental Health Perspectives is devoted to this topic [7]. That arsenic is a primary carcinogen appears to be part of the folk lore of the science of carcinogenesis, and, as a result, few have challenged this "fact". The reason is that a number of authors dating back to Hutchinson in 1888 [8] have associated arsenic with human cancer. The International Agency for Research on Cancer (IARC) [9] concluded that sufficient evidence exists that implicates inorganic arsenic compounds as carcinogens for the skin and lungs of humans.

---

[1] Figures in brackets indicate the literature references at the end of this paper.

Leitch and Kennaway [10] in 1922 reported that one mouse developed a metastasizing squamous cell carcinoma of the skin about six months after 100 mice were painted with a solution of potassium arsenite three times a week. Only 33 mice survived three months. This experiment also gave evidence to the idea that arsenic is carcinogenic even though Neubauer [11] discusses various unsuccessful attempts to confirm these results.

A critical review of the world literature on the carcinogenic action of arsenic reveals that the evidence which includes arsenic as a carcinogen is not at all definitive; it is possible to conclude that arsenic is not a primary[2] carcinogen and in fact, it may not be a carcinogen at all.

This review will elaborate on this concept, and will present hypotheses which will attempt to explain how arsenic can act in the neoplastic process. Furthermore a critical experiment will be proposed to strengthen or refute the hypothesis.

2. Criteria for Designating an Agent as a Carcinogen

There are no set criteria to follow for designating an agent as a carcinogen. Furst and Haro [12] listed some steps which must be considered before a metal or metaloid is to be designated an animal carcinogen. These include: tumors must be induced in more than one species; neoplasms must appear at a site distant from that where the agent was applied; the tumor should have the ability to metastasize; the neoplasm must be transplantable in an unconditioned host. Since no animal experiment is positive when arsenic is tested as a carcinogen, these criteria, obviously can not be one.

A more inclusive set of standards has been published to assist various regulatory agencies in making decisions to declare what compounds or substances can be listed as carcinogens. Arsenic will be considered in light of these recommendations. The report of the Interagency Regulatory Liaison Group (IRLG), CPSC, EPA, FDA, and OSHA; on the Scientific Basis for Identifying Potential Carcinogens and Estimating Their Risks states that:

"Evidence of carcinogenicity can be obtained from three sources:

1) Epidemiological evidence from exposed human populations;
2) Experimental animal evidence from long-term bioassays;
3) Suggestive evidence, derived from studies of chemical structure, reactivity, DNA damage and repair, mutagenicity, neoplastic transformation of cells in culture, induction of pre-

---

2 Apparently metals and metaloids do not follow the two-stage carcinogenesis mechanism; thus a primary carcinogen should both initiate and promote the neoplastic process.

> neoplastic changes, or from other short-term tests that correlate with carcinogenicity."

Using these criteria, it is much more difficult to classify elemental arsenic or any of its compounds as a primary carcinogen.

3. Evidence from Epidemiology

The only evidence that implicates arsenic as a human carcinogen comes from epidemiological studies; epidemiology can implicate an agent as a potential carcinogen, but cannot prove cause and effect. However, a detailed analysis of the published epidemiological studies reveal that a number of complicating confounding factors make the conclusion that arsenic is the true etiological agent rather tenuous. Perhaps the most quoted study is by Lee and Fraumeni, Jr. [13] who found at least a threefold excess of lung cancer among white male smelter workers who handled arsenic trioxide compared with the white male population of the same states. In their paper they note that "SO_2 exposure and respiratory cancer mortality were also positively related" and they "could not distinguish the influences of arsenic from SO_2 or unknown agents...". More cancers were also found among men exposed to ferromanganese dust than was expected. In another study, Blot and Fraumeni, Jr. [14] compared the lung cancer rates among people living in the vicinity of copper, lead, and zinc smelters and concluded the excess mortality for both men and women could not be explained by their occupational exposure alone.

Ott, Holder and Gordon [15] conducted an epidemiology study of workers involved in formulating and packaging insecticides containing either lead arsenate, calcium arsenate, copper acetoarsenite and magnesium arsenate. The concentrations varied. Respiratory tract neoplasms were found among 16.2% of those exposed compared to 5.7% of the controls. No other site of malignant neoplasms was significant. Four of the decedents were exposed to asbestos prior to working with arsenic.

No statisically significant differences in cancer deaths over controls was found among an exposed group who worked in a coal-burning power plant. The fuel had a concentration of one kg of arsenic per ton of coal. [16]

Studies on the vineyard pesticide sprayers was made by Roth [17]; a discussion of this group will be found in section 5.1.

The Taiwan area of high arsenic content in water has been extensively studied. Levels of arsenic have been reported by Tseng et al [18] to vary between 0.53 and 1.19 ppm. He also reported that soluble salts were also present. No complete mineral analysis appears to be available. Later he related a dose response relationship between skin cancer and blackfoot disease with arsenic [19]. What complicates the Taiwan studies is that Lu, Tsai and Ling [20] found that the water fluoresced and that some derivatives of ergotamine were present. The peripheral blood vessel contractions can easily be associated with the alkaloid, and hence account for the gangrene-like symptoms. Also Tsang [18] failed to include a control population which did not drink arsenic-enriched water.

The epidemiology will be covered more extensively by other speakers at this symposium.

4. Genotox Assays-Short Term Tests.

Arsenic as (III) or (V) was not found to be mutagenic in the Ames test [21]. Nishioka [22] did publish that rec effects in _B. subtilis_ were positive for both As (III) and As(V); however, there is a question about the technique used by Nishioka [22] and also Nishioka's [22] report on the sodium arsenite induced reversion by tryptophan in _E. coli_ has been challenged by Rossman et al [23] who could not reproduce the effect.

A more sensitive test for genotoxic activity can be the effect of a metal on the fidelity of DNA synthesis [24]. No apparent change in accuracy of DNA synthesis was noted when either As_2O_5, or Na_2HAsO_4 were added to the medium [25].

A more detailed report on the lack of relationship between mutagenesis, or genotox, and arsenic will be reported at this symposium.

5. Animal Studies

Studies in which attempts were made to induce any cancer in experimental animals have proven uniformly negative. The very few experiments which propone to show positive effects do not stand up to any statistical analysis. Details have been summerized by the IARC monograph #23 [9] and by Neubauer [11]. Kraybill [26] analyzed the dose levels of the arsenic compounds used in the various carcinogenic studies. The skin painting studies were conducted over a dose range of 0.4% to 1.8%; the oral doses were from 0.0004 ppm to 646 ppm. There is no advantage of repeating the parameter of each experiment which was noted in the IARC monograph [9]; a summary of the nature of the animals tested, the routes of administration, and the types of compounds are given in Tables 1, 2, and 3.

TABLE I

ANIMALS TEST FOR ARSENIC CARCINOGENESIS

Mouse Strains:

C57Bl/6 [27]	STS [33]
Swiss [28, 29, 39]	DBA [34]
NMR.I [30]	BALB/c [34]
B6C3F-1 [31]	
B6AKF-1 [31]	Rockland [33]
C3H [32]	S [37]
	Unknown [10, 38]

TABLE 1 (CON'T)

Rat Strains:

Osborne-Medel [36,41]
Long Evans [42]
Wistar-King [47,48]
Unknown [35,40,43]
Bethesda Black [27]
Wistar [45]
BDIX [51]

Dogs -- Beagle [36]

Rabbit [41]

Swine [44]

Chickens [44]

TABLE 2

ARSENIC COMPOUNDS TESTED FOR CARCINOGENICITY

Arsenic trioxide	[10,27,28,30,32,34,43,47,48]
Sodium arsenite	[29,32,36,38,39]
Fowlers solution	[30]
Dimethylarsenic acid	[31]
Arsenilic acid	[33,44]
Potassium arsenite	[33,37]
Lead arsenate	[35,45]
Calcium arsenate	[40,51]
Sodium arsenate	[28,36,39]
Arsenic metal	[41]

TABLE 3

ROUTE OF ADMINISTRATION OF ARSENIC COMPOUNDS

Oral: Drinking water [27,28,29,30,31,32,34,42]
Food [31,33,35,36,44,45]

Skin: Painting [10,28,33,37]

Respiratory track: Inhalation [38]
Intratracheal [43,47,48,51]

Parenteral: Intravenous [39]
Subcutaneous [39,40]
Intramedullary [41]

In addition to the attempt to induce cancer with an arsenic compound, attempts were made to see if arsenic could augment the carcinogenicity of known carcinogens. Mice were painted with potassium arsenite and croton oil or dimethylbenzanthracene (DMBA) with no indication of a promotion effect [33]. A similar experiment was carried out with feeding mice arsenilic acid prior to painting with DMBA, again with no result [33]. Kroes et al [45] treated rats with lead arsenate or sodium arsenate with and without N-nitrosodiethylamine and no differences were found in the incidence of tumors or time of appearance.

5.1 Two Recent Publications

Pershagen [47] mentions two publications which may bear on the possible carcinogenic action of arsenic in rats. These need a more detailed analysis. The first is by Ishimishi et al [47], who elaborated on their previous paper [48]. In the 1976 paper, the authors reported no malignant changes after administering arsenic trioxide by stomach tube into Wistar-King rats at a daily dose level of 2 mg/kg or 10 mg/kg for 40 days. At the higher dose levels hyperkeratoacanthosis of the epidermis was noted; these did not become malignant. In 1977 these authors reported that following 15 intratracheal instillations of a series of dusts in the same strain of rats with and without benzo(a)pyrene a few lung tumors were noted.

Table 4 gives the data.

Table 4

Lung Tumor Yield Following Instillation of Some Dust Agents

Agent	No. instilled	Tumors per survivors
Copper ore	14	0/10
Flue dust	23	1/7
Arsenic trioxide	14	0/8
B(a)P	21	1/7
Copper ore + B(a)P	20	2/10
Flue dust + B(a)P	25	3/10
Arsenic Trioxide + B(a)P	21	3/7
Saline	23	0/7

It is difficult to conclude that arsenic trioxide acts more than just a carrier for B(a)P. Saffiotto et al [49] used "hematite" (an iron oxide) as carrier for B(a)P and reported more respiratory tract tumors in hamsters given the combination of the iron oxide and B(a)P than the hydrocarbon alone. Other oxides also have been used as carriers for B(a)P. Among these are aluminum bronze powder [50].

The second publication often cited as a possible proof that arsenic is a carcinogen for rats is that of Ivankovic, Eisenbrand and Preussmann [51].

Following the intratracheal instillation of 0.1 ml of a mixture of copper sulfate, calcium hydroxide and calcium arsenate in 25 BDIX rats, 10 died. Of the fifteen rats which survived, nine developed adenocarcinomas of the lungs; two were found only after serial sections were made. Pulmonary adenomas were found in the vehicle controls. No squamous cell carcinomas were reported. Missing from the publication were data on rats instilled with copper sulfate alone, calcium hydroxide, or calcium oxide alone, or the mixture of copper sulfate and calcium hydroxide. Information on the treatment of these rats with a commercial Bordeaux mixture would also be useful. Until these controls are carried out, it is difficult to evaluate the Ivankovic paper.

The logic of the Ivankovic [51] study is based upon the finding of Roth [17] in 1956 of 24 autopsies of Mosel vineyard workers who were exposed to lethal doses of arsenic in the sprays. Of fourteen that had cancer, seven were brochiogenic carcinomas. In 1957 Roth [52] reviewed the previous data and now reported three liver tumors from the previously exposed workers who were found to have "arsenic-induced cirrhosis" of the liver. Denk et al [53] reviewed this latter report. The IARC group [9] did not believe the evidence associating arsenic and liver cancer was adequate.

Omitted in these publications is that the original Bordeaux mixture did not contain arsenic, and thus the exact nature of the commercial mixture which was used then is not known. In Portugal, men who used the Bordeaux mixture without added arsenic also seem to have a higher lung cancer incidence than the population as a whole [54]. Final judgment on the Ivankovic study must await the results of the logical controls, and to see what happens at doses below the LD40 level.

6. Discussion

That arsenic is toxic is not in doubt; its carcinogenicity must be questioned. There are a large number of papers on the presumed carcinogenic action of arsenic on humans without adequate questioning [55]. However, a detailed analysis of the publications where exposure has been to pesticides [56], inorganic compounds as such [57,58] or in the environment of smelters [59,60] reveals that too many compounding factors are present. Not all studies agree [61]. In favor of the possible action of arsenic on humans is that skin cancer may be possible [62,63],but this has not been adequately proven. Not enough information is given on the recovery rate of men who develop skin ulcers following contact with arsenic.

Speciation is important; it is known that As(III) is more toxic than AS(V). It may also be possible that if an arsenic compound is eventually found to be a carcinogen, As(III) will be the culprit. To date, however, all oxidation states of arsenic have been tested; perhaps the trend of research should be toward finding an animal model if possible.

7. Hypotheses as to the Nature of Arsenic as a Carcinogen

The following hypotheses and notes are suggested to help interpret the information available to date on the question of the carcinogenicity of arsenic.

1. Arsenic is not a primary carcinogen.

   Note: Since the epidemiology is complicated by a variety of confounding factors, and the mutagenic studies are not conclusive, and above all, none of the animal experiments are positive from the IRLG guide lines, arsenic would have to be classified as a non-carcinogen.

2. The wrong animal model has been used.

   Note: After over 10 strains of mice, 5 strains of rats, one strain of dog, rabbit, chicken and swine have been evaluated for their carcinogenic reaction and none have reported, it is difficult to accept hypothesis #2 as valid.

3. Humans are species specific for arsenic carcinogenesis.

   Note: No precedent is set for this, but it is still a tenable hypothesis.

4. Arsenic is a cocarcinogen if not a primary carcinogen.

   Note: The fact that arsenic did not alter the outcome of either hydrocarbon or nitrosamine carcinogenic results mitigates against this hypothesis. Leonard and Lauwerys [64] believe arsenic is a carcinogen.

5. Arsenic can produce cancer only if it is applied with at least two other compounds.

   Note: So often when epidemiology studies are analyzed, the humans have been exposed to arsenic, a sulfur compound, and another "heavy" metal; is a triad necessary for action?

6. Arsenic is not a primary carcinogen, but is a powerful growth factor.

   Note: It is possible that arsenic is not involved in the initiation or promotion of cancer, but after the cancer cells are established even at a microscopic level, arsenic accelerates the growth of the neoplastic process. The growth promotion properties of arsenic for fowl and swine are well documented.

8. A Critical Experiment

To test some of these hypotheses, it will be necessary to coadminister arsenic as (III) or (V) with a sulfur compound and another heavy metal. A procedure to mimic the smelter workers' exposure would be helpful; here the experimental animal should be exposed by inhalation to SO_2, an arsenic oxide and another agent such as copper metal powder.

A more important crucial experiment is to study a species other than rats or mice (hamster?, guinea pig?) for a detailed metabolic pathway of arsenic trioxide.

These metabolic products should be found in the urine and feces using HPLC techniques. Similar studies should be made on humans; in both species major and minor excretion products should be matched.

That species which is closet to man should be tested for arsenic trioxide carcinogenicity by oral (drinking water) and also by inhalation (or intratracheal) routes.

9. Conclusion

From the available evidence it is not possible to conclude that arsenic is a primary carcinogen. This conclusion covers all forms of arsenic tested to date: inorganic As(III), As(V) and organoarsenic compounds. A hypothesis is presented that arsenic is a powerful growth factor, and after the neoplastic processes are initiated and established, arsenic, especially As(III), accelerates the rate of growth of the cancer, making it appear that arsenic is the carcinogenic factor. A critical experiment is proposed to strengthen or refute the hypothesis.

10. Acknowledgement

The support of the Carrie Baum Browning Trust Fund from the Co-trustees of the Bank of America NTSA is gratefully acknowledged.

References

[1] Furst, A., Inorganic agents as carcinogens, Advances in Modern Toxicology, H.F. Kraybill and M. A. Mehlman, eds., Vol. 3, pp. 209-29 (Hemisphere Publishing Co., Washington, D. C., 1977).

[2] Flessel, C. P., Furst, A., and Radding, S. B., A comparison of carcinogenic metals, Metal Ions in Biological Systems, H. Siegel, ed., Vol. 10, Chapter 2 (Marcel Dekker, New York, N.Y., 1980).

[3] Luckey, T. D. and Venugopal, B., Metal Toxicity in Mammals, Vol. I, pp. 129-160 (Plenum Press, New York, N. Y., 1977).

[4] Radding, S. B. and Furst, A., A review of metals and carcinogenicity, Molecular Basis of Environmental Toxicity, R. S. Bhatnagar, ed., pp. 359-72 (Ann Arbor Science, Ann Arbor, MI, 1980).

[5] Sunderman, F. W., Metal carcinogenesis, Advances in Modern Toxicology, R. A. Goyer and M.A. Mehlman, eds., pp. 257-95 (John Wiley & Sons, New York, N. Y., Vol. 2, 1977).

[6] Sunderman, F. W., Jr., Carcinogenic effects of metals, Fed. Proc. 37, [1], 40-46 (1978).

[7] Environmental Health Perspectives 40,(August 1981).

[8] Hutchinson, J., On some examples of arsenic-keratoses of the skin and of arsenic cancer, Trans. Path. Soc. London, 39, 352-263 (1888).

[9] IARC (International Agency for Research on Cancer), IARC Monographs on the Evaluation of the Carcinogenic Risk of Chemicals to Humans, Vol. 23, Some Metals and Metallic Compounds, (IARC, Lyon, France, 1980).

[10] Leitch, A and Kennaway, E. L., Experimental production of cancer by arsenic, Br. Med. J., ii, 1107-1108 (1922).

[11] Neubauer, O., Arsenical cancer: a review, Br. J. Cancer 1, 192-251 (1947).

[12] Furst, A. and Haro, R. T., A survey of metal carcinogenesis, Progr. Exp. Tumor Res. 12, 102-133 (1969)

[13] Lee, A. M. and Fraumeni, J. F. Jr., Arsenic and respiratory cancer in man: an occupational study, J. Natl. Cancer Inst. 42, 1045-1052 (1969).

[14] Blot, W.J. and Fraumeni, J. F. Jr., Arsenical air pollution and lung cancer, Lancet 142-144 (1975).

[15] Ott, M. G., Holder, B.B. and Gordon, H. L., Respiratory cancer and occupational exposure to arsenicals, Arch. Environ. Health 29, 250-255 (1974).

[16] Bencko, V., et al, Rate of malignant tumor mortality among coal burning power plant workers occupationally exposed to arsenic, J. Hyg., Epidemiol., Microbiol., Immunol. 24, 278-284 (1980).

[17] Roth, F., Ueber die chronische Arsenvergiftung der Moserwinzer unter besonderer Beruecksichtigung des Arsenkrebses, Z. Krebsforsch. 61, 287-319 (1956).

[18] Tseng, W.P., Chu, H.M., How, S. W., Fong, J. M., Lin, C. S., and Yeh, S., Prevalence of skin cancer in an endemic area of chronic arsenicism in Taiwan, J. Natl, Cancer Inst. 40, 453-463 (1968).

[19] Tseng, W. P., Effects and dose-response relationships of skin cancer and blackfoot disease with arsenic. Environ. Health Perspect. 19, 109-119 (1977).

[20] Lu, F. J., Tsai, M. H. and Ling, K. H., J. Formosan Med. Assoc. 76, 209-217 (1977).

[21] Löfroth, G. and Ames, B.N., Mutagenicity of inorganic compounds in Salmonella typhimurium arsenic, chromium, and selenium, Mutat. Res. 53, 65-66 (1978).

[22] Nishiok, H., Mutagenic activities of metal compounds in bacteria, Mutat. Res. 31, 185-189 (1975).

[23] Rossman, T. G., Stone, D., Molina, M. and Troll, W., Absence of arsenite mutagenicity in Escherichia coli and Chinese hamster cells, Environ. Mutagen. 2, 371-379 (1980).

[24] Sirover, M. A., Effects of metals in in vivo bioassays, Environ. Health Perspect. 40, 163-172 (1981).

[25] Tkeshelashvili, L. K., Shearman, C. W., Zakour, R. A., Koplitz, R. M., and Loeb, L. A., Effects of arsenic, selenium, and chromium on the fidelity of DNA synthesis, Cancer Res. 40, 2455-2460 (1980).

[26] Kraybill, H. F., Carcinogencity of arsenic experimental studies, Health Effects of Occupational Lead and Arsenic Exposure, B. W. Carnow, ed., pp. 272-283 (U.S.D.H.E.W., Center for Disease Control, Contract No. 210-75-0026, 1976).

[27] Hueper, W. C. and Payne, W. W., Experimental studies in metal carcinogenesis. Chromium, nickel, iron, arsenic, Arch. Environ. Health 5, 445-462 (1962).

[28] Baroni, C., van Esch, G. J., and Soffiotti, U., Carcinogenesis tests of two inorganic arsenicals. Arch. Environ. Health 7, 668-674 (1963).

[29] Kanisawa, M. and Schroeder, H. A., Life term studies on the effects of arsenic, germanium, tin, and vanadium on spontaneous tumors in mice, Cancer Res. 27, 1192-1195 (1967).

[30] Knoth, W., Arsenic treatment, Arch. Klin. Exp. Dermatol. 227, 228-234 (1966).

[31] Innes, J. R. M., Ulland, B. M., Valerico, M. G. etal, Bioassay of pesticides and industrial chemicals for tumorigenicity in mice: a preliminary note, J. Natl. Cancer Inst. 42, 1101-1113 (1969).

[32] Schrauzer, G. N. and Ismael, D., Effects of selenium and of arseniic on the genesis of spontaneous mammary tumors in inbred C3H mice, Ann. Clin. Lab. Sci. 4, 441-447 (1974).

[33] Boutwell, R. K., A carcinogenicity evaluation of potassium arsenite and arsenilic acid, J. Agri. Food Chem. 11, 381-384 (1963).

[34] Milner, J. E., The effect of ingested arsenic on methylcholanthrene-induced skin tumors in mice, Arch. Environ. Health 18, 7-11 (1969).

[35] Fairhall, L. T. and Miller, J. W., A study of the relative toxicity of the molecular components of lead arsenate, Publ. Health Rep. 56, 1610-1625 (1941).

[36] Byron, W. R., Bierbower, G. W., Bronwer, J. B., and Hansen, W.H., Pathological changes in rats and dogs from two-year feeding of sodium arsenite or sodium arsenate, Toxicol. Appl. Pharmacol. 10, 132-147 (1967).

[37] Salaman, M. H. and Roe, F. J. C., Further tests for tumour-initiating activity: N,N-di-(2-chloroethyl)-p-aminophenylbutyric acid (CB1348) as an initiator of akin tumour formation in the mouse, Br. J. Cancer 10, 363-377 (1956).

[38] Berteau, P. E., Flom, J. O., Dimmich, R. L., and Boyd, A. R., Long-term study of potential carcinogenicity of inorganic arsenic aerosols to mice, Toxicol. Appl. Pharmacol. 45, 323 (1978).

[39] Osswald, H. and Hoerttler, K., Arsenic-induced leucoses in mice after diaplacental and postnatal application, Verh. Dtsch. Ges. Pathol. 55, 289-293 (1971).

[40] Arkhipov, G.N., Carcinogenicity of chemical poisons containing arsenic, Vopr. Pitan. 27, 77 (1968).

[41] Hueper, W. C., Experimental studies in metal carcinogenesis. VI. Tissue reactions in rats and rabbits after parenteral introduction of suspensions of arsenic, beryllium, or asbestos in lanolin, J. Natl. Cancer Inst. 15, 113-124 (1954).

[42] Kanisawa, M. and Schroeder, H. A., Life term studies on the effect of trace elements on spontaneous tumors in mice and rats, Cancer Res. 29, 892-895 (1969).

[43] Growth, D. H., Scheel, L. D. and Mackay, C. R., Comparative pulmonary effect of beryllium and arsenic compounds in rats, Lab. Invest. 26, 477-478 (1972).

[44] Frost, D. U., Arsenicals in biology -- retrospect and prospect, Fed. Proc. 26, 194-208 (1967).

[45] Kroes, R., van Logten, J. M., Berkvens, J. M., de Vries, T., and van Esch, G. J., Study on the carcinogenicity of lead arsenate and sodium arsenate and on the possible synergistic effect of diethylnitrosamine, Food Cosmet. Toxicol. 12, 671-679 (1974).

[46] Pershagen, G., The carcinogenicity of arsenic, Environ. Health Perspect. 40, 93-100 (1981).

[47] Ishinishi, N., Kodama, Y., Nobutomo, K., and Hisanaga, A., Preliminary experimental study on carcinogenicity of arsenic trioxide in rat lung, Environ. Health Perspect. 19, 191-196 (1977).

[48] Ishinishi, N., Osato, K., Kodama, Y., and Kunitake, E., Skin effects and carcinogenicity of arsenic trioxide: a preliminary experimental study in rats, Effects and Dose-Response Relationships of Toxic Metals, G. F. Nordberg, ed., pp. 471-479 (Elsevier, Amsterdam, 1976).

[49] Saffiotti, U., Cefis, F., and Kolb, L. H., A method for the experimental induction of bronchogenic carcinoma, Cancer Res. 28, 104-124 (1968).

[50] Blair, W. H., Sharma, U., and Daco, R., Lung neoplasms induced with aluminum bronze powder and benzo(a)pyrene, Proc. Am. Assoc. Cancer Res. 15, 136 (1974).

[51] Ivankovic, S., Eisenbrand, G., and Preussmann, R., Lung carcinoma induction in BD rats after single intratracheal instillation of an arsenic-containing pesticide mixture formerly used in vineyards, Intl. J. Cancer 24, 786-788 (1979).

[52] Roth, F., Arsen-Leber-Tumoren (Haemangioendothelism), Z. Krebsforsch. 61, 468-503 (1957).

[53] Denk, R., Holzmann, H., Lange, H.-J., and Greve, D., Ueber Arsenspatschaeden bei obduzierten Moselwinzern, Med. Welt 20 (N. F.), 557-567 (1969).

[54] Villar, Vineyard sprayer's lung, Am. Rev. Respir. Dis. 110, 545-555 (1974).

[55] Axelson, O., Arsenic compounds and cancer, J. Toxicol. Environ. Health 6, 1229-1235 (1980).

[56] Mabuchi, K., Lilienfeld, A. M. and Snell, L.M., Lung cancer among pesticide workers exposed to inorganic arsenicals, Arch. Environ. Health 34, 312-320 (1979).

[57] Perry, K. et al, Studies in the incidence of cancer in a factory handling inorganic compounds of arsenic. II. Clinical and environmental investigations, Br. J. Ind. Med. 5, 6-15 (1948).

[58] Hill, A. B. and Faning, E. L., Studies in the incidence of cancer in a factory handling inorganic compounds of arsenic. I. Mortality experience in the factory, Br. J. Ind. Med. 5, 1-6 (1948).

[59] Kuratsune, M., Occupational lung cancer among copper smelters, Intl. J. Cancer 13, 552-558 (1974).

[60] Axelson, O., et al, Arsenic exposure and mortality: a case-referent study from a Swedish copper smelter, Br. J. Ind. Med. 35, 8-15 (1978).

[61] Pinto, S. S. et al, Mortality experience in relation to a measured arsenic trioxide exposure, Environ. Health Perspect. 19, 127-130 (1977).

[62] Braun, W., Carcinoma of the skin and the internal organs caused by arsenic: delayed occupational lesions due to arsenic, Ger. Med. Monthly 3, 321-324 (1958).

[63] Sanderson, K. V., Arsenic and skin cancer, Trans. St. John Hosp. Derm. Soc. 49, 115-122 (1963).

[64] Leonard, A. and Lauwerys, R. R., Carcinogenicity, teratogenicity and mutagenicity of arsenic, Mutat. Res. 75(1), 49-62 (1980).

DISCUSSION

I. Harding Barlow: How would you feel about arsenic not as a co-carcinogen, but as a co-promoter?

A. Furst: In other words, you suggest that arsenic may not be a primary carcinogen or a promoter per se, but a co-promoter. This means there would be some other promoter present. In a sense that would be consistent with the idea of there being another growth factor, although I hadn't thought in terms of a co-promoter per se. I think that is possible.

M. A. Schneiderman: Nick Day at IARC and Charles Brown at the National Cancer Institute, in looking at their mathematical model of multi-stage carcinogenesis and looking at the human data, seemed to come to conclusions rather similar to yours. They find that the human data would imply that arsenic was not an initiator, not a complete carcinogen, but was operating somewhere in the late stages of the development of carcinogenesis. It would seem to me that's amenable to experimental verification.

A. Furst: I certainly think it is. Just a word or two about experiments from a personal point of view. I did 119 experiments for the National Cancer Institute in metal carcinogenesis and, although these were done a few years ago, they were not anywhere near as expensive as the figures industry has been quoting. I think it incumbent upon some of the Institutes to fund experiments so we can settle the question of whether arsenic is a carcinogen, without spending a half a million dollars per experiment.

14

IS ARSENIC A MUTAGEN?

Vincent F. Simmon
Genex Corporation
6110 Executive Boulevard
Rockville, MD 20852

Genex has undertaken a review of the literature as well as laboratory experiments to investigate the mutagenic potential of inorganic arsenicals. The majority of the data indicates that these compounds are not capable of inducing point mutations in mammalian or microbial cells. The evidence in the literature that arsenicals cause DNA damage in *B. subtilis* is contradictory and not supported by studies conducted at Genex. Although evidence available from the literature suggests that arsenic is a potent clastogen; (that is, it causes chromosome breakage) analysis of reports of humans exposed to arsenic for therapeutic purposes provides conflicting data on the effect of such treatment on human chromosomes.

1. Introduction

A number of reports in the published and unpublished literature indicate that arsenic is a mutagen [1,2]. However, there are equally compelling reports that directly or indirectly refute the evidence that arsenic is mutagenic [3,4]. The purpose of this review is to analyze the conflicting data.

2. What Is a Mutagen?

A mutation is defined as any heritable change in genetic material. It is generally assumed that new mutations are deleterious and some investigators estimate that as much as 10 per cent of human disease may be attributed to mutation. However, there are no data which demonstrate that any agent (chemical or physical) has induced any mutations in human germ cells, despite many studies designed to detect such mutations.

A mutagen is an agent (chemical or physical) that can be demonstrated to cause a mutation. While the concern is that a mutagen will cause a change in human germ cells, most of the assays to identify mutagens are not conducted in germ cells. Assays that detect germ cell mutations are generally expensive and take many weeks or months to conduct. In contrast, assays that detect agents that can damage DNA (deoxyribonucleic acid, the hereditary material) are relatively rapid (one day or one or two weeks) and inexpensive using microorganisms or mammalian cells in culture. Chemicals that cause a mutagenic response in these short-term assays are considered to be suspected or presumptive human mutagens, unless adequate germ cell assays demonstrate that the chemical has no measurable effect.

3. Assay Systems

A number of bacterial assays have been developed to provide evidence (but not proof) that a chemical interacts with DNA. Some of these assays compare the relative toxicity of a chemical to two strains of bacteria; one strain is proficient in repairing damage to its DNA and the second strain is impaired in one or more of the pathways used to repair damaged DNA. The presumption is that a chemical that damages DNA will be more toxic to the repair-deficient strain than to the repair-proficient strain.

Two reports in the literature indicate that arsenicals are more toxic to *Bacillus subtilis* deficient in one of the DNA repair enzymes (*rec*) than to a repair-proficient *B. subtilis* [1,5]. Using the same strains and procedures, we were unable to reproduce the results of either of these studies [7]. Assays in *Escherichia coli* to measure the effect of arsenic on DNA synthesis (errors in DNA replication can lead to mutation) indicated no effect [6]. In other assays in *E. coli*, arsenic appeared to have an antimutagenic effect by inhibiting an error-prone DNA repair system [8].

Assays that measure the induction of mutations in *E. coli* and in other systems have also been conducted. In 1975, Nishioka reported that sodium arsenite induced tryptophan reversion (mutation from a tryptophan requirement to tryptophan non-requirement) in *E. coli* [1]. The procedure used is known to be subject to artifacts [9]. Subsequent attempts to reproduce the observed mutagenic effects by Rossman *et al.* [3] Kanematsu *et al.* [5] and Peirce and Simmon [10] have been unsuccessful. Perhaps the most widely used assay system for detecting chemical mutagens is the Salmonella/microsome assay. Loefroth and Ames report that neither As^{+3} nor As^{+5} was mutagenic in this assay [4].

Casto reported an increase in 8-azaguanine resistance in Chinese hamster cells exposed to arsenic [2]. In one experiment, the increased mutation frequency was observed only when survival was 5% or less. Although it is generally desirable to demonstrate some toxicity in such experiments, this level of survival is generally regarded as too low to be meaningful. In another experiment, an increase in mutation frequency was reported at only one dose (30% survival). Rossman *et al.* measured the effect of sodium arsenite on two genetic end points (mutation to ouabain resistance and 6-thioguanine-resistance) in V79 Chinese hamster cells and were unable to demonstrate any mutagenic activity under a variety of assay conditions [3]. Amacher and Paillet observed no mutagenic effect of arsenic in L5178Y mouse lymphoma assays [12]. Therefore, there is no clear evidence that arsenic is a mutagen in *in vitro* mammalian cell assays.

Jung *et al.* [12] and Jung and Trachsel [13] have reported one toxic effect of arsenic. The investigators observed a reduction in UV-induced tritiated thymidine incorporation into the DNA of epithelial cells treated with sodium arsenate. Based on this observation, they hypothesized that arsenic inhibits the excision repair process responsible for the UV-induced DNA repair. While their data may support this hypothesis, they do not preclude numerous other explanations for the observed reductions. These investigators have failed to consider that the amount of tritiated thymidine incorporation into the DNA of the irradiated cells is dependent not only on the activity of the repair enzymes, but also on the amount of tritium-labeled precursor available to the enzymes. Several general toxic phenomena can reduce the availability of the tritiated thymidine. For example, reduction of thymidine kinase activity would reduce the cells' ability to scavenge the nucleotides carrying the label, or the inhibition of oxidative phosphorylation could reduce the ATP available for transport and phosphorylation of the thymidine, thus reducing the proportion of the labeled precursor in the intracellular pool. In the absence of more definitive evidence for the inhibition of DNA repair (e.g., reduced excision of UV-induced pyrimidine dimers) there is inadequate evidence to suggest that the observations of Jung *et al*. reflect anything other than an acute toxic effect.

Unfortunately, no investigations have been reported in mammalian cells that might support or contradict the hypothesis of a DNA-repair-specific inhibitory effect of arsenic. Studies have, however, been reported in which the bioeffects of arsenic have been observed using *E. coli*. Tkeshelashvili *et al*. [6] reported *E. coli* DNA polymerase activity and fidelity in a cell-free system were unaffected by sodium arsenate. This enzyme is the bacterial equivalent of the mammalian enzyme that Jung *et al*. speculated was arsenic-sensitive. Rossman *et al*. [3] reported an enhancement of UV toxicity to *E. coli* deficient in excision repair capabilities (*uvrA*⁻) by treatment with sodium arsenite. They suggest that such an enhancement should not have been observed in the *urvA*⁻ strain (lacking excision repair) if the excision repair enzymes are the target of the arsenic. Thus the studies with *E. coli* do not lend support to the hypothesis of Jung *et al*.

Rossman *et al*. [3] observed that the UV sensitivity of *E. coli* deficient in inducible, error-prone DNA repair (generally believed to be associated with the process of mutagenesis) was unaffected by sodium arsenite and suggested that this repair system may be a target for arsenic inhibition. This hypothesis was further supported by their observation that the mutagenic effect of UV light could be reduced by treatment of the *E. coli* with sodium arsenite following UV irradiation.

In contrast with the conflicting results reported in microbial and mammalian *in vitro* assay systems to detect chemicals that cause point mutations or DNA damage, reports that arsenicals may cause damage to chromosomes are in general agreement. Agents which are known to cause chromosome damage are called clastogens, and, in *in vitro* assays, arsenicals are clastogenic. Petres and his colleagues have reported positive effects on human lymphocytes in a number of studies [14-16]. These results have been confimred by Paton and Allison [17], and by Oppenheim and Fishbein [18].

In view of the conflicting data in other mutagenicity assays, the mechanism of action of arsenic as a clastogen might be reasonably questioned. For example, in *in vitro* assays, arsenic has been shown to reduce the mitotic index [14,19], inhibit RNA, DNA and protein synthesis [20,21], and to inhibit oxidative phosphorylation [16]. Thus the chromosome damage caused by arsenic may be a manifestation of the toxic effect of arsenic. Such general toxic effects have an acute influence on cell survival, but may have little, if any, chronic significance.

Humans have been exposed occupationally and therapeutically to arsenic. Petres *et al.* has reported that such exposure results in chromosome aberrations [16,22], while Burgdorf observed an increase in sister chromatic exchange rates, but no increase in chromosome aberrations [23].

There are only two reports of *in vivo* experiments with arsenic [24,25]. In both studies, dominant lethal assays were performed with mice. Both studies report that arsenic by itself did not increase dominant lethality in either acute or chronic exposures.

Conclusion

The published reports of arsenic causing mutations are flawed because the results are not reproducible in other laboratories, are subject to known artifacts, or the conclusions are drawn from experiments in which there was extreme toxicity. Furthermore, arsenic has not been found to be mutagenic in a large number of *in vivo* assays, although arsenic treatment of cells in culture clearly leads to chromosome aberrations. However, even this clear-cut clastogenic effect should be interpreted cautiously because many agents not considered to be human mutagens (e.g., EDTA) are positive in *in vitro* cytogenetic assays. This is not to say that arsenic is not a toxic chemical; however, even though it is a toxic chemical, the evidence for it being a hazard because of its mutagenic activity is very poor.

References

1. Nishioka, H. (1975) Mutagenic activities of metal compounds in bacteria. Mutat. Res. 31, 185-189.

2. Casto, B. (1977) Letter, No title, dated May 31, 1977 from Bruce Casto, Sc.D., Biolabs, Inc. to Dr. Bill Waugh, EPA.

3. Rossman, T. G., Stone, D., Molina, M. and Troll, W. (1980) Absence of arsenite mutagenicity in Escherichia coli and Chinese Hamster Cells. Environ. Mutagen. 2, 371-379.

4. Loefroth, G. and Ames, B. N. (1978) Mutagenicity of inorganic compounds in Salmonella typhimurium: arsenic, chromium and selenium. Mutat. Res. 53, 65-66.

5. Kanematsu, N., Hara, M. and Kada, T. (1980) Rec assay and mutagenicity studies on metal compounds. Mutat. Res. 77, 109-116.

6. Tkeshelashvili, K., Shearman, C. W., Zakour, R. A., Koplitz, R. M. and Loeb, L. A. (1980) Effects of arsenic, selenium, and chromium on the fidelity of DNA synthesis. Cancer Res. 40, 2455-2560.

7. Peirce, M. V. and Simmon, V. F. (1981) An evaluation of the comparative genotoxic activities of five test articles in the Bacillus subtilis differential DNA repair assay. Study No. 3002-8. Contract report to Koppers Company.

8. Rossman, T. G., Meyn, M. and Troll, W. (1977) Effects of Arsenite on DNA Repair in Escherichia coli. Environ. Health Perspect. 19, 229-233.

9. Witken, E. M. (1976) Ultraviolet mutagenesis and inducible DNA repair in Escherichia coli. Bact. Rev. 40, 869-907.

10. Peirce, M. V. and Simmon, V. F. (1981) An evaluation of the comparative genotoxic activities of five test articles in the Escherichia coli WP2, WP2 uvrA, and CM571 tryptophan reversion assay. Study No. 3002-9. Contract report to Koppers Company.

11. Amacher, D. E. and Paillet, S. C. (1980) Induction of trifluoro-thymidine-resistant mutants by metal ions in L5178Y TK^{+}/- cells. Mutat. Res. 78, 279-288.

12. Jung, E. G., Trachsel, B. and Immich, H. (1969) Arsenic as an inhibitor of the enzymes concerned in cellular recovery (dark repair). Germ. Med. Mth. 14, 614-616.

13. Jung, E. G. and Trachsel, B. (1970) Molecular-biological studies on arsenic carcinogenesis. Arch. Klin. Exp. Derm. 237, 819-826.

14. Petres, J. and Hundeiker, M. (1968) Chromosome pulverization induced by arsenic in cell cultures in vitro. Archiv. fur Klinische u. Exp. Derm. 231, 366-370.

15. Petres, J. and Berger, A. (1972) The effect of inorganic arsenic on DNA synthesis of human lymphocytes in vitro Arch. Derm. Forsch. 242, 343-352.

16. Petres, J., Baron, D. and Hagedorn, M. (1977) Effects of arsenic cells metabolism and cell proliferation: Cytogenetic and biochemical studies. Environ. Health Perspect. 19, 223-227.

17. Paton, G. R. and Allison, A. C. (1972) Chromosome damage in human cell cultures induced by metal salts. Mutat. Res. 16, 332-336.

18. Oppenheim, J. J. and Fishbein, W. N. (1965) Induction of chromosome breaks in cultured normal human leukocytes by potassium arsenite, hydroxylurea and related compounds. Cancer Res. 25, 980-985.

19. Tsuda, S. (1974) Effects of 2,4-dinitrophenol, sodium arsenate, and oligomycin on mitosis of mouse L cells growing in monolayer culture. Tokoshima J. of Exp. Med. 21, 49-59.

20. Baron, D., Kunick, I., Frischmuth, I. and Petres, J. (1975) Further in vitro studies on the biochemistry of the inhibition of nucleic acid and protein synthesis induced by arsenic. Arch. Derm. Res. 253, 15-22.

21. Skipper, H. E., Mitchell, J. H., Jr., Bennett, L. L., Jr., Newton, M. A., Simpson, L. and Edison, M. (1951) Observations on inhibition of nucleic acid synthesis resulting from administration of nitrogen mustard, urethan, colchicine, 2,6-diaminopurine, 8-azaguanine, potassium arsenite, and cortisone. Cancer Res. 11, 145-149.

22. Petres, J., Schmid-Ullrich, K. and Wolf, U. (1970) Chromosomeraberrationen an menschlicken lymphocyten bei chronischen arsenschaden. Dtsch. Med. Wschr. 95, 79-80.

23. Burgdorf, W., Kurvink, K. and Cervenka, J. (1977) Elevated sister chromosome exchange rate in lymphocytes of subjects treated with arsenic. Hum. Genet. 36i, 69-72.

24. Hodge, H. C. (1977) Estimation of the mutagenicity of arsenic compounds utilizing dominant lethal mutations in mice. S.E.R.A. Project 7625.

25. Sram, R. J. and Bencko, V. (1974) Evaluation of the genetic risk of exposure to arsenic. Cesk. Hyg. 19, 308-315.

DISCUSSION

E. P. Radford: It is my understanding that in the pentavalent form as arsenate, one mechanism by which arsenic can interfere with metabolism is by substituting for phosphate. On this basis, wouldn't you anticipate that arsenates could lead to abnormalities in the DNA chain, by substituting for phosphates?

V. F. Simmon: If that were to occur and arsenic became part of the DNA chain, then one could assume this might have an effect. But I don't believe that it has ever been demonstrated that arsenic ends up as part of a DNA or RNA or any polymer. It would have to, at one stage, be the first phosphate on a nucleotide, and would then have to have two other phosphates added to it. What I'm trying to point out is that there are a number of steps which it would have to go through. Presumably we have evolved in such a way that if arsenic does substitute for phosphate at any step, it cannot go successfully through the next step.

15

CONSEQUENCES OF ARSENIC DEPRIVATION IN LABORATORY ANIMALS

E.O. Uthus, W.E. Cornatzer and F.H. Nielsen

United States Department of Agriculture
Agricultural Research Service
Grand Forks Human Nutrition Research Center

and

Department of Biochemistry
University of North Dakota

Grand Forks, North Dakota 58202

Arsenic deprivation signs have been described for the chick, goat, minipig and rat. However, studies with chicks indicated that the extent, severity, and direction of many signs were affected by the arginine, zinc and manganese status of the animal. Moreover, arsenic deprivation markedly influenced the effects of high levels of dietary arginine, and/or zinc deprivation. Parameters affected by dietary variations in arsenic, arginine and zinc included growth, kidney arginase, plasma alkaline phosphatase, plasma urea, plasma uric acid and hematocrit. For example, arsenic deprivation apparently elevated kidney arginase, but did not affect plasma urea, in chicks fed a normal level of arginine. When dietary arginine was increased to 34 mg/g by a supplement of 20 mg/g, kidney arginase and plasma urea were substantially elevated. Zinc deficiency alleviated the elevation in these two parameters in arsenic-deprived chicks. On the other hand, zinc deficiency exacerbated the elevation in kidney arginase and plasma urea in arsenic-supplemented chicks. The findings to date suggest that arsenic has a role that strongly influences arginine metabolism. Furthermore, because arginine metabolism is affected by arginine, manganese and zinc nutriture, these nutrients may modify the role of arsenic, thus influencing the signs of arsenic deprivation. Perhaps arsenic, arginine, manganese and zinc interact to affect the conversion of arginine into urea and ornithine which is catalyzed by the enzyme arginase.

1. Introduction

Early findings which suggested arsenic may be an essential element included a favorable effect of arsenicals on hemoglobin production [1], tadpole metamorphosis [2], and tissue culture growth [3]. It was found that growth was stimulated in chickens, turkeys, swine, rats and calves when they were fed derivatives of phenyl arsenic acid (0.005% to 0.01% of the diet) [4].

Early attempts to prove the essentiality of arsenic were unsuccessful. Hove, Elvehjem and Hart [5] found diets furnishing as little as 2 μg of arsenic per day per rat had no significant effect on growth, hemoglobin concentration and red blood cell number or fragility. They did observe, however, that dietary arsenic caused a delay in the rate of fall of hemoglobin levels in rats fed whole milk without mineral supplementation subsequent to being fed a mineral-supplemented milk diet compared to controls not given arsenic. Schroeder and Balassa [6] found rats and mice grew and developed normally on diets containing as little as 0.053 μg of arsenic/g. They observed, however, that arsenic supplementation resulted in a healthier appearance to the skin and hair of rats and mice.

Subsequent findings from the chick, goat, minipig, and rat stimulated the concept that arsenic is an essential dietary nutrient. These findings and their possible nutritional significance will be described here.

2. Evidence of Essentiality for Goats and Minipigs

Anke et al. [7,8] fed an arsenic-deficient diet to growing goats and minipigs that were allowed to reproduce. Arsenic deprivation did not adversely affect the growing animals. The first signs of arsenic deprivation appeared in F_1 offspring and also, for goats, lactating adults. Arsenic-deprived goats and minipigs exhibited depressed birth weights and skeletal ash, impaired fertility, elevated perinatal mortality, and elevated copper and manganese in some tissues. Other signs of deprivation in goats were depressed serum triglycerides, and death during lactation [9]. Histological examination revealed myocardial damage in the lactating goats that died.

3. Evidence of Essentiality for Rats

While investigating the essentiality of nickel, Nielsen et al. [10] observed that in third-generation pregnant rats fed a diet based on skim milk powder, whether nickel-deficient or -supplemented, perinatal death of pups was 90%. They rebred a group of ten nickel-supplemented third-generation dams and gave five a supplement of 4.0 μg of arsenic as sodium arsenate/g and 0.5 μg of arsenic as sodium arsenite/g. Loss of pups was 80% in the group not receiving the arsenic supplement and 40% in the group receiving the arsenic supplement. In the arsenic-supplemented group most of the deaths occurred in one litter, apparently as a result of maternal neglect. At weaning (24 days) mean body weight was 26 g for the six surviving arsenic-deprived pups and 42 g for the 26 surviving arsenic-supplemented pups.

The essentiality of arsenic for rats was tested in two additional experiments [11]. The offspring of Sprague-Dawley dams fed an arsenic-deficient (30 ng/g) diet from day 3 of gestation were examined for signs of deficiency. Controls received 4.5 μg of arsenic/g. During the suckling period, pups did not differ in mortality or appearance. Possibly, perinatal mortality was not elevated and growth was not depressed because the dams had not been on the arsenic-deficient regimen for sufficient time. After weaning, the growth of the arsenic-deprived offspring was slower than that of the supplemented offspring. At 40 to 44 days, control males weighed significantly more than arsenic-deprived males. At this time the deprived rats appeared less thrifty than the supplemented rats; their coats were rougher and yellowish. Other signs of arsenic deprivation were elevated erythrocyte osmotic fragility, elevated spleen iron and splenomegaly. Initially, the second of these two experiments was to continue through three generations with the objective of confirming the original finding that arsenic-deprivation impaired perinatal survival. However, the stock of low-arsenic skim milk powder for the diet became depleted, and a new supply could not be located. Thus, the initial findings from studies using a skim milk powder based diet were never duplicated.

In 1979, we successfully formulated a new diet which routinely contained less than 15 ng of arsenic/g. This diet was composed of high-protein casein (16%), acid-washed ground corn (68%), corn oil (7.5%), and supplemental vitamins, minerals and amino acids to meet the nutritional requirements of the laboratory rat (table 1). The major problem in developing this diet was finding mineral sources low in arsenic. The minerals used and their sources are shown in table 2.

Thus far, we have completed one experiment in which time-pregnant dams were placed on the newly formulated diet within 2 days after being bred. Controls received a supplement of 2 μg of arsenic as sodium arsenate/g of diet. The pups were weighed every 4 days starting with day 8. At age 24 days, the pups were weaned, and at age 40-44 days, separated according to sex. F_2 and F_3 generations were obtained by mating rats from different litters at age 75 days. Animals not used for reproduction were used to determine the effect of dietary arsenic on the following parameters: growth; hematocrit; hemoglobin; plasma alkaline phosphatase, glucose, urea, cholesterol, uric acid and triglycerides; ratios of heart wt/body wt, spleen wt/body wt, liver wt/body wt; and liver content of copper, iron, manganese and zinc.

The most consistent effect of arsenic deprivation observed throughout all three generations was growth depression. Examples of the growth depression are shown in table 3. Other less consistent and less marked changes caused by arsenic deprivation included: (a) depressed hematocrits and hemoglobins in F_2 and F_3 generation males; (b) elevated plasma alkaline phosphatase in F_2 and F_3 offspring; and (c) elevated manganese content in liver.

Recently, we initiated another experiment to confirm the findings of our first experiment. However, the findings of the first experiment will not be confirmed because only 2 of 12 arsenic-deprived F_1 females

Table 1. Composition of basal arsenic-deficient diet used since 1979.[1]

Ingredient	g/kg
Casein, high protein[2,3]	160.00
Corn, ground, acid-washed[4]	699.46
Corn oil[5]	75.00
DL-α-tocopherol[6]	0.20
Vitamin mix[7]	4.59
Choline chloride[6]	0.75
Arginine[8]	5.00
Mineral mix[9]	55.00
	1000.00

[1]Diet routinely contains less than 15 ng of arsenic/g.

[2]Teklad Division, Harlan Industries, Madison, WI.

[3]Mention of a trademark or proprietary product does not constitute a guarantee or warranty of the product by the U.S. Department of Agriculture, and does not imply its approval to the exclusion of other products that may also be suitable.

[4]U.S. Biochemicals Corp., Cleveland, OH. See reference [12] for the procedure for acid washing corn.

[5]ICN Nutritional Biochemicals, Cleveland, OH.

[6]Grand Island Biological Co., Grand Island, NY.

[7]The vitamin mix contained (in mg): retinyl acetate (500,000 IU/g), 16; retinyl palmitate (250,000 IU/g), 32; menaquinone, 1; thiamine•HCl, 10; pyridoxine•HCl, 7.5; niacin, 30; D-pantothenic acid, calcium salt, 48; vitamin B_{12} (0.1% triturate), 20; folic acid, 2; biotin, 1; riboflavin, 27; inositol, 50; para-aminobenzoic acid, 5; vitamin D_3 (400,000 IU/g), 3.75; dextrose, 4,336.8.

[8]Sigma Chemical Co., St. Louis, MO.

[9]See table 2.

Table 2. Composition of mineral mix for arsenic-deficient diet used since 1979.

Ingredient	g/kg
$CaCO_3$[1,2]	15.0000
KH_2PO_4[3]	15.0000
NaCl[4]	3.5000
$Mg(C_2H_3O_2)_2 \cdot 4H_2O$[5]	3.5000
$Mn(C_2H_3O_2)_2 \cdot 4H_2O$[6]	0.2250
$CuSO_4 \cdot 5H_2O$[7]	0.0300
KI[4]	0.0004
$Zn(C_2H_3O_2)_2 \cdot 2H_2O$[5]	0.0504
Na_2SeO_3[8]	0.0003
$(NH_4)_6Mo_7O_{24} \cdot 4H_2O$[7]	0.0040
$Cr(C_2H_3O_2)_3 \cdot H_2O$[6]	0.0020
NH_4VO_3[7]	0.0002
$NiCl_2$[4]	0.0020
Iron sponge[7] dissolved in HCl[9]	0.0400
$Na_2SiO_3 \cdot 9H_2O$[5]	0.5100
NaF[4]	0.0022
H_3BO_3[5]	0.0057
Corn, ground, acid-washed[10]	17.1278
	55.0000

[1]MCB Manufacturing Chemists, Inc., Cincinnati, OH.

[2]See footnote 3, table 1.

[3]"ULTREX" grade, J.T. Baker Chemical Co., Phillipsburg, NJ.

[4]"ULTRAPURE" grade, Alfa Div., Ventron Corp., Danvers, MA.

[5]"REAGENT" grade, J.T. Baker Chemical Co., Phillipsburg, NJ.

[6]"CERTIFIED" grade, Fisher Scientific Co., Fair Lawn, NJ.

[7]"PURATRONIC" grade, Alfa Div., Ventron Corp., Danvers, MA.

[8]Alfa Div., Ventron Corp., Danvers, MA.

[9]"INSTRA-ANALYZED" grade, J.T. Baker Chemical Co., Phillipsburg, NJ.

[10]See footnote 4, table 1.

Table 3. Effect of arsenic-deprivation on rat growth.

Dietary Arsenic	No. of Pups		Body Weight at 42 Days	
	Female	Male	Female	Male
μg/g			g	
		First Generation		
0	8	15	102 ± 12[1]	112 ± 12
2	20	14	126 ± 9	143 ± 16
		Second Generation		
0	25	22	100 ± 11	109 ± 15
2	14	12	112 ± 14	132 ± 16
		Third Generation[2]		
0	12	9	59 ± 15	73 ± 16
2	20	21	68 ± 7	81 ± 12

[1]Mean ± SD.

[2]Body weight at 30 days. Some pups were used for experimental purposes before age 42 days.

became pregnant after being mated with arsenic-deprived males for 10 days. Only six pups survived from one of the two litters which were delivered. This contrasts with the arsenic-supplemented group in which 9 of 12 females became pregnant. A similar finding may have occurred in our first experiment because 6 of 8 arsenic-deprived, but 8 of 8 arsenic-supplemented, F_2 females became pregnant after being mated for 12 days. Furthermore, the number of pups/litter was smaller in the arsenic-deprived group. The findings suggested that arsenic deprivation impairs fertility.

The findings to date strongly suggest that arsenic is an essential element for the rat. Several signs of deprivation have been described, but these need to be confirmed and extended. At present, depressed growth appears to be the most consistent sign of arsenic deprivation in the rat. The studies also show that further refinement of experimental conditions are needed before arsenic deprivation in rats can be adequately investigated.

4. Evidence of Essentiality for Chicks

Nielsen and Shuler [13] reported that arsenic might be an essential nutrient for growing chicks. In four experiments, day-old cockerel chicks were fed a diet based on skim milk powder and acid-washed ground corn that was supplemented with arginine (20 g/kg). The basal diet contained about 20 ng of arsenic/g in experiments 1 and 2, 35 ng/g in experiment 3, and 45 ng/g (no low-arsenic skim milk powder available) in experiment 4. Controls were fed a supplemental 1.0 μg of arsenic/g. In the first three experiments, the arsenic-deprived chicks weighed significantly less than controls at 28 days. The growth difference might have been more marked if the experiment had been prolonged, because arsenic deprivation did not affect growth until days 17 to 20 of the experiment. Chicks fed the basal diet containing 45 ng of arsenic/g grew as well as controls. Chicks fed the basal diet containing 20 ng of arsenic/g had larger, darker livers than controls. When the basal diet contained 35 ng of arsenic/g, the elevation in size of livers was less but still significant. When the basal diet contained 45 ng of arsenic/g, liver weight was not elevated. The arsenic-deprived chicks exhibited elevated levels of zinc in the liver except in experiment 4, in which the basal diet contained 45 ng of arsenic/g. Other possible signs of arsenic deprivation observed only in experiment 3 were depressed plasma alkaline phosphatase, depressed white cell count, elevated erythrocyte osmotic fragility. As with rats, the unavailability of low-arsenic skim milk powder prevented any further studies with chicks fed a skim milk powder-based diet.

After formulating the new arsenic-low diet for rats (tables 1 and 2), minor modifications were made so that it met the requirements of the chick. Using the modified diet, two experiments were done with day-old cockerel chicks which were assigned to groups of 24 of similar weight. In experiment 1, arsenic as $Na_2HAsO_4 \cdot 7H_2O$ at levels of 0 and 2 μg/g was supplemented to the basal diet containing 5 ng of arsenic/g. Experiment 2 was a fully crossed, two way, two-by-two design with supplemental arsenic and zinc as the variables. The supplements were arsenic as $Na_2HAsO_4 \cdot 7H_2O$, 0 and 2 μg/g; zinc as $Zn(C_2H_3O_2)_2 \cdot 2H_2O$, 5 (zinc was left out of basal diet so this diet was zinc-deficient) and 40 μg/g. After 28 days in experiment 1 and 32 days in experiment 2, the following parameters were determined: growth, leg scores (1 for normal through 5 for severely abnormal), liver wt/body wt ratio, spleen wt/body wt ratio, hematocrit, hemoglobin, plasma alkaline phosphatase, plasma uric acid, and kidney arginase.

In experiment 1, chicks fed the arsenic-deprived diet exhibited depressed growth, leg abnormalities, and elevated hematocrit and hemoglobin (table 4). In experiment 1, kidney arginase was not determined, and arsenic deprivation did not markedly affect the other parameters examined.

In experiment 2, an interaction between arsenic and zinc affected growth, hematocrit, plasma alkaline phosphatase activity and kidney arginase activity (table 5). When dietary zinc was 40 μg/g, the arsenic-

Table 4. Effect of arsenic deprivation on chick growth, leg development, hematocrit, and hemoglobin at age 4 weeks.

Dietary Arsenic	Weight	Leg Score[1]	Hematocrit	Hemoglobin
µg/g	g		%	g/100 ml blood
0	747 ± 25[2]	3.5	32.4 ± 0.5	9.32 ± 0.15
2	894 ± 25	2.4	30.5 ± 0.4	8.82 ± 0.13
P values	0.004		0.01	0.02

[1]1 = normal; 5 = severely abnormal.

[2]Mean ± SEM.

deprived chicks exhibited leg abnormalities that differed from those of the zinc-deficient chicks. The legs were twisted and appeared weak but were not extremely short and thick with swollen hocks. Thus because the scoring system rated degree but not type of leg abnormalities, the effect of the arsenic-zinc interaction on legs was not evaluated statistically. If arsenic supplementation had any affect on leg abnormalities of zinc-deficient chicks, it was that it made them slightly worse.

The effect of dietary arsenic on growth and hematocrits was dependent upon the zinc status of the chick. When dietary zinc was 40 µg/g, arsenic-deprived chicks exhibited depressed growth and elevated hematocrits. During zinc deficiency, however, growth was more markedly depressed and hematocrits more markedly elevated in arsenic-supplemented than in arsenic-deprived chicks.

In contrast to effects on growth and hematocrits, effects of dietary zinc on plasma alkaline phosphatase activity and kidney arginase activity apparently depended upon the arsenic status of the chick. Plasma alkaline phosphatase activity was elevated in zinc-deficient chicks that were fed supplemental arsenic. On the other hand, zinc deficiency slightly depressed plasma alkaline phosphatase activity in arsenic-deprived chicks. Zinc deficiency elevated kidney arginase activity of arsenic-supplemented chicks. When the chicks were arsenic-deprived, zinc-deficiency did not affect kidney arginase activity.

Table 5. Effects of arsenic, zinc and their interaction on chick growth, leg development, hematocrit, plasma alkaline phosphatase activity and kidney arginase activity.

Treatment					Plasma	
		Body[1]	Leg[2]		Alkaline Phosphatase	Kidney Arginase
As	Zn	Weight	Score	Hematocrit	Activity	Activity
μg/g	μg/g	g		%	units[3]	units[4]
0	5	783	3.1	31.3	1.381	4.11
0	40	904	2.6	30.9	1.632	4.08
2	5	746	3.6	32.8	2.012	5.84
2	40	979	1.5	29.8	1.464	3.18
Analysis of Variance - P Values						
Arsenic effect		NS		NS	NS	NS
Zinc effect		0.0001		0.01	NS	0.007
Arsenic x zinc		0.06		0.05	0.003	0.009

[1]Weight of chicks at age 32 days.

[2]1 = normal; 5 = severely abnormal.

[3]Units were μmoles p-nitrophenyl phosphate split/min/ml of plasma.

[4]Units were μg urea formed/min/mg protein (x 100).

The finding that arsenic deprivation apparently elevated, or depressed, kidney arginase activity with the direction depending on the zinc status of the chick, and the earlier finding that arginine supplementation of chicks made arsenic deprivation signs appear earlier and more severe [13], prompted us to do a three-way, two-by-two-by-two arranged experiment with supplemental arsenic, arginine, and zinc as the variables. The supplements were arsenic as $Na_2HAsO_4 \cdot 7H_2O$, 0 and 2 μg/g; arginine, 0 and 20 mg/g; zinc as $Zn(C_2H_3O_2)_2 \cdot 2H_2O$, 0 and 50 μg/g. Day-old cockerel chicks were used and assigned to groups of 15 of similar weight. At 12 days, chicks in the high-arginine, zinc-deficient groups began dying, so the zinc-deficient diet was then supplemented with 5 μg of zinc/g. The additional zinc stopped the mortality in these groups, but not before 11 of 15 in the arsenic-deficient, and 7 of 15 in the arsenic-supplemented, group had died. In addition to the parameters examined in the previous two experiments, the following were determined: plasma glucose, heart wt/body wt ratio, plasma urea, plasma cholesterol, and liver content of copper, iron, manganese and zinc. Only selected data will be presented here.

The most interesting findings were those which showed that dietary arsenic markedly influenced arginine metabolism (table 6). In this experiment, arsenic deprivation apparently elevated kidney arginase activity and plasma uric acid in chicks fed a normal level of arginine. When dietary arginine was increased to 34 mg/g by a supplement of 20 mg/g, kidney arginase and plasma urea were substantially elevated. However, the elevation of these two parameters was markedly influenced by dietary arsenic and zinc. Zinc deficiency alleviated the elevation in plasma urea and kidney arginase activity in arsenic-deprived chicks. On the other hand, zinc deficiency exacerbated the elevation in these two parameters in arsenic-supplemented chicks. The effects of a high level of dietary arginine was less marked on plasma uric acid. Only the zinc-deficient, arsenic-supplemented chicks exhibited a substantial increase in plasma uric acid upon feeding a high level of dietary arginine. In contrast, high dietary arginine tended to depress plasma uric acid in the zinc-adequate, arsenic-deprived chicks.

The finding with plasma uric acid, plasma urea, and kidney arginase show that the signs of arsenic deprivation may be greatly influenced by other nutrients such as zinc and arginine. Furthermore, although the data are not shown, this influence was also suggested by the findings with growth, hematocrit, hemoglobin, plasma alkaline phosphatase activity, and liver iron.

5. Biological Function and Biochemistry of Arsenic

From the preceding, it should be evident that no biological role for, or specific action of, arsenic has been found. Thus, the mode of action of arsenic is open to conjecture. The findings do suggest, however, that arsenic does have a role that strongly influences arginine metabolism. Furthermore, because arginine metabolism is affected by arginine, manganese and zinc nutriture, these nutrients may modify the role of arsenic, thus affecting the extent, severity, or direction of the signs of arsenic deprivation.

Perhaps arsenic, arginine, manganese and zinc interact to affect the reaction:

$$\text{Arginine} \xrightarrow[\text{arginase}]{} \text{urea + ornithine}$$

The enzyme arginase is composed of four subunits, to each of which is bound one atom of Mn^{2+} and is specific for L-arginine.

Arginase levels in chick kidney can be depressed by dietary excess of the amino acids glycine, threonine, and α-aminoisobutyric acid

Table 6. Effects of arsenic, arginine, zinc and their interaction on plasma uric acid, plasma urea, and kidney arginase activity.

Treatment[1]			Plasma Uric Acid	Plasma Urea	Kidney Arginase
As	Arg	Zn			
μg/g	mg/g	μg/g	mg/100 ml	mg/100 ml	units[3]
0	0	0-5[2]	6.99	0.98	5.80
2	0	0-5	5.83	0.99	3.48
0	0	50	6.66	1.07	4.30
2	0	50	4.71	0.86	3.79
0	20	0-5	7.64	5.27	21.97
2	20	0-5	8.24	15.60	37.14
0	20	50	6.00	9.00	28.00
2	20	50	5.24	9.05	27.12
			Analysis of Variance - P Values		
Arsenic effect			0.001	0.001	0.01
Arginine effect			0.004	0.0001	0.0001
Zinc effect			0.0001	NS	NS
As x Zn			0.03	0.0008	0.002
As x Arg			0.003	0.0007	0.0002
Zn x Arg			0.002	NS	NS
As x Zn x Arg			NS	0.001	0.0001

[1]Amounts of arsenic (sodium arsenate), arginine, and zinc (zinc acetate) supplemented to basal diet containing approximately 15 ng of arsenic, 5 μg of zinc and 14 mg of arginine/g.

[2]Initially, no supplemental zinc was given to the zinc-deficient groups. After age 12 days, a supplement of 5 μg of zinc (zinc acetate) was given.

[3]Units were μg urea formed/min/mg protein (x 100).

[14,15], and elevated by dietary excess of the amino acids ornithine, histidine, lysine, isoleucine, tyrosine, and, of interest to this report, arginine [15-17]. Chu and Nesheim [18] suggested that dietary amino acids such as arginine might alter arginase activity in chick kidney by altering the rate of arginase degradation.

Because manganese is an integral part of arginase, dietary manganese is likely to affect the activity of the enzyme. Thus, it is not surprising that there are reports showing that manganese deficiency depressed liver arginase activity in rats [19], rabbits [20], and pigs [21]. Moreover, manganese-deprived animals exhibited depressed liver manganese levels, and there is evidence that, in the liver, manganese is mainly in the arginase extract [22].

The interrelationship between arginase and zinc is not well defined. However, Hoekstra [23] noted that in the serum of zinc-deficient chicks fed isolated soybean protein there was about twice as much free arginine as in similar chicks fed dried egg white. Although the isolated soybean protein contained more arginine than the dried egg white, the isolated soybean protein-diet did not contain an extraordinary amount of arginine. This finding suggested that zinc deficiency affected arginine metabolism. Most likely, this effect was due to an effect on some arginine metabolic pathway other than that catalyzed by arginase because we found elevated arginase levels in kidneys of zinc-deficient chicks. Perhaps, zinc deficiency depressed the amount of arginine shunted through another pathway which resulted in elevated amounts of arginine in the serum, thus simulating the effect obtained by feeding elevated amounts of arginine. As mentioned, elevated dietary arginine elevated kidney arginase activity.

The mechanism of action through which arsenic affects arginase activity is open to speculation. Most likely, the effect is indirect because, in our experiments, arsenic deprivation elevated or depressed kidney arginase activity with the direction determined by the zinc and arginine status of the chick. Perhaps through an effect on manganese metabolism, arsenic affects arginase activity. As described, arsenic deprivation elevated the manganese level in several tissues from minipigs and goats, and in liver from rats. Unfortunately, we did not measure the effect of arsenic deprivation on the level of manganese in chick kidney where the majority of the arginase activity is located in that species. On the other hand, arsenic may directly, or indirectly, affect arginase levels in chick kidney, or rat liver, and thus affect their manganese content. Regardless of the mechanism of action, the findings do show that arsenic affects arginine metabolism in some way. Thus, dietary arsenic, like arginine, manganese and zinc affects arginase activity.

Another reaction through which arginine can be shunted is:

$$\text{Arginine} + \text{glycine} \rightarrow \text{guanidoacetic acid} + \text{ornithine}$$

which is catalyzed by glycine amidinotransferase in the chick kidney. Although we have not determined the effect of arsenic deprivation on glycine amidinotransferase, we have some preliminary indirect evidence (unpublished) that dietary arsenic may affect that enzyme. Arsenic deprivation in the rat depressed the activities of the following phosphatidylcholine biosynthetic enzymes: phosphatidylethanolamine methyltransferase, phosphatidyldimethylethanolamine methyltransferase and choline phosphotransferase. Uthus [24] found that feeding rats guanidoacetic acid, a methyl depleter through its conversion to creatine, depressed phosphatidylcholine biosynthetic enzymes in liver. Perhaps, glycine amidinotransferase activity is elevated during arsenic deprivation, then more guanidoacetic acid is formed and this depresses the phosphatidylcholine biosynthetic enzymes. Of course, if guanidoacetic acid is elevated, so would be ornithine. As mentioned previously, ornithine elevates arginase activity in chick kidney, and we found that arsenic deprivation elevated arginase activity in kidneys of chicks fed normal levels of arginine. The findings to date show that the effect of arsenic deprivation on glycine amidinotransferase activity should be ascertained.

The possible role of arsenic might be more than just the regulation of some enzyme involved with the metabolism of arginine. Excess amino acid nitrogen is usually eliminated from the chick as uric acid. We found that arsenic deprivation elevated plasma uric acid in chicks fed normal levels of arginine. Furthermore, we have preliminary findings which indicated arsenic deprivation depressed the protein level in the liver of rats. This suggests that arsenic may have some role in the utilization of amino acids in the synthesis of protein, or is necessary for the control of protein degradation.

Of course, the biological role of arsenic may be something other than one of those described. However, the possibilities suggested appear to be viable and thus will be investigated in our future arsenic studies.

6. Concluding Comments

Evidence has emerged which shows that arsenic is an essential element for several animal species. Several signs of deprivation have been described. These signs need to be confirmed and extended, but this may be difficult. Findings to date indicate that several dietary nutrients can affect the extent, severity and direction of the signs of arsenic deprivation. Thus, unless dietary conditions are rigidly defined, apparent differences may occur between findings from different laboratories, or even from different experiments within a laboratory.

Also, although there is evidence that arsenic affects arginine metabolism, and, perhaps, protein metabolism in general, little is understood regarding the mechanism of action of physiological levels of arsenic at the cellular and molecular level. Thus, much work still must be done to assess the importance of arsenic in nutrition and metabolism.

References

[1] Skinner, J.T. and McHargue, J.S., Supplementary effects of arsenic and manganese on copper in the synthesis of hemoglobin, Am. J. Physiol. 145, 500-506 (1946).

[2] Godonneche, J. and Dastugue, G., Action of the arsenical mineral waters of Bourboule on the germination of seeds and the growth and metamorphosis of batrachian larvae, Bull. Soc. Chim. Biol. 16, 248-256 (1934).

[3] Sanjo, K., Pharmacological investigation on the pigment epithelial cells of the iris in cultures in vitro. VII. The influence of arsenious acid, sodium arsenite, sodium arsenate and arsenic acid on the growth of cultures in vitro of iris epithelium as well as the histological changes due to these poisons, Folia Pharmacol. Japon. 19, 151-164 (1934).

[4] Frost, D.V., Overby, L.R. and Spruth, H.C., Studies with arsanilic acid and related compounds, J. Agric. Food Chem. 3, 235-243 (1955).

[5] Hove, E., Elvehjem, C.A. and Hart, E.B., Arsenic in the nutrition of the rat, Am. J. Physiol. 124, 205-212 (1938).

[6] Schroeder, H.A. and Balassa, J.J., Abnormal trace metals in man: Arsenic, J. Chronic Dis. 19, 85-106 (1966).

[7] Anke, M., Grün, M. and Partschefeld, M., The essentiality of arsenic for animals, Book, Trace Substances in Environmental Health, Vol. 10, D.D. Hemphill, ed., pp. 403-409 (University of Missouri, Columbia, MO, 1976).

[8] Anke, M., Grün, M., Partschefeld, M., Groppel, B. and Hennig, A., Essentiality and function of arsenic, Book, Trace Element Metabolism in Man and Animals-3, M. Kirchgessner, ed., pp. 248-252 (Tech. Univ. München, Freising-Weihenstephan, BRD, 1978).

[9] Anke, M., Groppel, B., Grün, M., Hennig, A. and Meissner, D., The influence of arsenic deficiency on growth, reproductiveness, life expectancy and health of goats, Book, 3. Spurenelement-Symposium Arsen, M. Anke, H.-J. Schneider and Chr. Bruckner, eds., pp. 25-32 (Friedrich-Schiller-Universität, Jena, DDR, 1980).

[10] Nielsen, F.H., Myron, D.R. and Uthus, E.O., Newer trace elements - vanadium (V) and arsenic (As) deficiency signs and possible metabolic roles, Book, Trace Element Metabolism in Man and Animals-3, M. Kirchgessner, ed., pp. 244-247 (Tech. Univ. München, Freising-Weihenstephan, BRD, 1978).

[11] Nielsen, F.H., Evidence of the essentiality of arsenic, nickel, and vanadium and their possible nutritional significance, Book, Advances in Nutritional Research, Vol. 3, H.H. Draper, ed., pp. 157-172 (Plenum Publishing Corp., New York, NY, 1980).

[12] Nielsen, F.H., Myron D.R., Givand, S.H. and Ollerich, D.A., Nickel deficiency and nickel-rhodium interaction in chicks, J. Nutr. 105, 1607-1619 (1975).

[13] Nielsen, F.H. and Shuler, T.R., Arsenic deprivation studies in chicks, Fed. Proc., Fed. Am. Soc. Exp. Biol. 37, 893 abs. (1978).

[14] Shao, T.-C. and Hill, D.C., Effect of α-aminoisobutyric acid on arginine metabolism in chicks, J. Nutr. 98, 225-234 (1969).

[15] Austic, R.E. and Nesheim, M.C., Role of kidney arginase in variations of the arginine requirement of chicks, J. Nutr. 100, 855-867 (1970).

[16] Jones, J.D., Peterburg, S.J. and Burnett, P.C., The mechanism of the lysine-arginine antagonism in the chick: Effect of lysine on digestion, kidney arginase, and liver transamidase, J. Nutr. 93, 103-116 (1967).

[17] Nesheim, M.C., Kidney arginase activity and lysine tolerance in strains of chickens selected for a high or low requirement of arginine, J. Nutr. 95, 79-87 (1968).

[18] Chu, S.-H.W. and Nesheim, M.C., Analysis of the rates of dietary induced changes in chick kidney arginase activity, Proc. Soc. Exp. Biol. Med. 164, 347-350 (1980).

[19] Shils, M.E. and McCollum, E.V., Further studies on the symptoms of manganese deficiency in the rat and mouse, J. Nutr. 26, 1-19 (1943).

[20] Ellis, G.H., Smith, S.E. and Gates, E.M., Further effects of manganese deficiency in the rabbit, J. Nutr. 34, 21-31 (1947).

[21] Burch, R.E., Williams, R.V., Hahn, H.K.J., Jetton, M.M. and Sullivan, J.F., Tissue trace element and enzyme content in pigs fed a low manganese diet. I. A relationship between manganese and selenium, J. Lab. Clin. Med. 86, 132-139 (1975).

[22] Committee on Biologic Effects of Atmospheric Pollutants, Division of Medical Sciences, National Research Council, Medical and Biological Effects of Environmental Pollutants: Manganese, Chapter 4, pp. 77-82 (National Academy of Sciences, Washington, D.C., 1973).

[23] Hoekstra, W.G., The complexity of dietary factors affecting zinc nutrition and metabolism in chicks and swine, Book, Trace Element Metabolism in Animals-1, C.F. Mills, ed., pp. 347-353 (E. & S. Livingstone, Edinburgh & London, 1970).

[24] Uthus, E.O., Effects of Diet and Alcohol on Phosphatidylethanolamine Methyltransferase and Choline Phosphotransferase Activity in Liver Microsomes, M.S. Thesis (University of North Dakota, Grand Forks, ND, 1976).

DISCUSSION

R. D. Wauchope: Are you having any difficulty convincing nutritionists that arsenic is an essential element?

F. H. Nielsen: At times, I feel that only Dr. Anke from the GDR (German Democratic Republic), myself, and our research associates are the only ones who believe that arsenic is essential for animal life. However, other people who work in the field of trace element nutrition and metabolism apparently are beginning to accept the concept of arsenic essentiality. Evidence for this can be seen in reviews such as the one by Dr. Walter Mertz which appeared in the September 18, 1981, issue of *Science*. Today, the majority of the nutrition community does not regard arsenic as an essential nutrient for any animal. Hopefully, when some of the recent results obtained in my laboratory, especially those showing interrelationships among arsenic, zinc, arginine, and possibly manganese, are published, many of the disbelievers and uninformed will be convinced that arsenic is an element of nutritional concern.

R. D. Wauchope: There are reports in the literature that the methylarsenical herbicides affect amino acid levels in some of the plants that are treated, so maybe there's an effect on protein synthesis. I notice you obtained control diets by supplementing your arsenic-deficient diets with from one to two parts per million arsenic. This seems a little high to me since most plant matter is lower than that. I was wondering how your animals behave on your supplemented diets relative to perhaps a more natural diet?

F. H. Nielsen: Many natural feeds contain between 0.5 and 1.0 ppm of arsenic. Thus, I believe the arsenic-supplemented diets in our experiments do not contain an unusually high amount of arsenic. At most, the arsenic-supplemented diets probably contain twice the amount of arsenic found in many natural diets.

Actually, our arsenic-supplemented diets may contain less arsenic than some commercial feeds, especially if they contain fish meal or some other seafood by-products. I am sure you are aware that many fish and seafoods contain relatively high levels of arsenic. I realize that the arsenic in natural feeds may be present mainly as organic arsenic, and we supplement the animal diets with inorganic arsenate. Nonetheless, I believe we are using a reasonable arsenic supplement, and our findings would not be significantly altered if the arsenic-supplemented diets contained less, but an adequate level, of arsenic.

R. D. Wauchope: Yes, but there is some evidence that arsenic can stimulate growth. If you have two parts per million available arsenic, I would say you are more like ten times the normal. You might be getting some stimulation in your controls.

F. H. Nielsen: Perhaps the best way to respond to your question is by the following: We use chicks which grow to approximately 1,000 grams in four weeks when fed a commercial diet. In our experiments, the growth rate of the arsenic-supplemented, but not the arsenic-deficient, chicks was that which was expected. In our last experiment, the average weight at 4 weeks of an arsenic-deficient chick fed 25 ppm dietary zinc was about 750 g. Thus, the growth difference apparently was not due to the arsenic-deficient chick growing at an expected rate and arsenic supplementation stimulating a slight increase in growth. Furthermore, the difference in growth between the arsenic-deficient and -supplemented chicks was much greater than that usually observed in experiments in which arsenicals stimulated growth. There is no doubt in my mind that the growth depression was a true sign of arsenic deficiency.

E. A. Woolson: Would you care to speculate what the dietary intake minimum requirement might be if you could extrapolate your chick data or any of the other data to humans?

F. H. Nielsen: The only experiments which indicated a possible requirement level of arsenic for animals were those in which dried skim milk based diets were used. In those experiments, apparent deficiency signs were seen when the diet contained 35 ng of arsenic/g. No deficiency signs were seen when dietary arsenic was 45 ng/g. So, if we assume the dietary requirement for arsenic is near 40 ng/g - or 40 μg/kg, this would extrapolate to a human dietary requirement of 30-40 μg/day because, I believe, a person normally eats 700-1000 g of food per day.

E. A. Woolson: Okay, and that would be as inorganic arsenic. In our food, a significant portion of arsenic is going to be organically bound and nobody has examined what happens to that in the stomach.

F. H. Nielsen: I would like to say that I made the previous extrapolation only because I was requested to do so. I dislike making such extrapolations because it usually is inappropriate to extrapolate a human requirement for a nutrient from animal data.

E. A. Woolson: The point I'm making is that we probably get most of our inorganic arsenic from drinking water, a significant portion of it anyway.

16

BACTERIAL RESISTANCE TO ARSENIC COMPOUNDS

Simon Silver[1] and Hideomi Nakahara[2]
Department of Biology, Washington University, St. Louis, Mo. 63130 U.S.A.[1]; and Department of Hygiene, Jikei University School of Medicine, Tokyo, Japan[2]

Many bacterial species have variants that show resistance to arsenite and/or arsenate. The determination of arsenic resistance can be governed by genes on the bacterial chromosome or on extrachromosomal plasmids, which also contain genes governing resistances to antibiotics. *Escherichia coli* strains isolated from a Tokyo hospital showed a 61% frequency of $AsO_4{}^{3-}$resistance and approximately similar frequencies of resistance toward Hg^{2+}, Cd^{2+} and a range of antibiotics. Analogous strains isolated from a polluted river showed an 84% frequency of $AsO_4{}^{3-}$resistance, but 10 times lower antibiotic resistance levels.

The mechanism of chromosomally-governed arsenate resistance is an alteration in the properties of the phosphate transport system which is responsible for cellular accumulation of arsenate. Plasmid mediated arsenate resistance is due to a separate highly specific arsenate efflux "pump". The mechanism of plasmid-determined arsenite resistance is distinct from that for arsenate. The mechanism of plasmid-determined arsenite resistance is not understood, but does not involve extracellular chelation by thiol compounds or detoxification. Presumably chromosomally-determined arsenite resistance in *Alcaligenes* strains is due to oxidation of arsenite to arsenate by an inducible enzyme system.

1. Introduction

Bacteria appear to have mechanisms of coping with all toxic heavy metal pollutants with which they may come into contact. We have worked extensively on mercury (and organomercurial) resistance, cadmium resistance and arsenic resistance in studies that range from surveys of the frequencies of heavy metal resistance in strains of clinical and environmental sources to the basic biochemistry and genetics of the mechanisms of the resistances. These studies have been published [1-4, plus others] and reviewed [5,6]. For this symposium, we will review what is currently known about bacterial resistances to arsenic compounds.

2. Frequency and distribution of arsenic resistances

Much of the work on the genetics and biochemistry of arsenic resistance in bacteria has been done with bacterial strains isolated from hospital sources [1-4,7,8]. Thus it is important to ask whether such resistances occur equally in bacteria from non-hospital sources where environmental pollution with arsenic compounds might be expected.

Table 1 contains comparative data for a single bacterial species from isolates from a Tokyo hospital and from the Tama River, a river toward the western suburbs of Tokyo with a significant industrial pollution problem [7]. The bacteria were initially screened for resistance to three heavy metals and five antibiotics. One might note that the frequency of heavy metal resistant E. coli was about the same for Hg^{2+},

Table 1. Frequency of Heavy Metal Resistances and Drug Resistances in Isolates of Escherichia coli from a Hospital and a River

	Clinical strains (1974-1977)	River strains (1979-1980)
Metal Resistance		
Hg^{2+}	323 (57)[a]	170 (49)
Cd^{2+}	524 (93)	258 (74)
AsO_4^{3-}	342 (61)	292 (84)
Drug Resistance		
Streptomycin	377 (67)	31 (8.9)
Tetracycline	344 (61)	15 (4.3)
Chloramphenicol	315 (56)	16 (4.6)
Kanamycin	186 (33)	16 (4.6)
Gentamicin	9 (1.8)	0 (0)
Total no. of strains	564	347

[a]Figures in parentheses represent % of total isolates.

Cd^{2+} and AsO_4^{3-}(49-93%) for bacteria isolated from hospital and from river samples. The slightly higher frequency of arsenate resistance and the slightly lower frequency of resistance to cadmium among the river isolates does not seem of consequence, although these are statistically significant differences. What is striking, however, is that the river and hospital E. coli showed about the same frequencies of heavy metal resistances, although the river isolates showed ten-times lower frequencies of resistances to antibiotics. It is as if both the hospital and the river contained equal sources of selection for resistances to Hg^{2+}, Cd^{2+} and AsO_4^{3-}, but that antibiotic selection pressure was much less in the river system than in the hospital. We would expect analogous screening results with other types of common bacteria from soil or water sources.

From the 170 Hg^{2+} resistant E. coli of river origin, 117 (69%) were able to transfer mercury resistance by bacterial matings [7]. Such a result is indicative of genes carried on extrachromosomal "resistance plasmids". Since selection for recombinants in these bacterial matings was always for Hg^{2+} resistance, this was a common transferrable property. It can be seen in table 2 that 115/117 of these bacterial strains transferred arsenate resistance along with mercury resistance. The most common class of transferrable resistances in these experiments included only Hg^{2+} and $AsO_4^{3-}$and none of the antibiotic resistances. Thus we can conclude that arsenate (and other heavy metal) resistance can be found at a high frequency in environmental samples and that these resistances are often governed by genes on transferrable resistance plasmids. As far as

Table 2. Demonstration of Resistance Plasmids Carrying Mercury Resistance in E. coli Isolates and Their Patterns of Resistance to Arsenate and Drugs

Resistance pattern of R plasmids	Number of E. coli strains with R plasmid
Hg^{2+}, AsO_4^{3-}	92
Hg^{2+}, streptomycin	1
Hg^{2+}, AsO_4^{3-}, streptomycin	6
Hg^{2+}, AsO_4^{3-}, kanamycin	2
Hg^{2+}, AsO_4^{3-}, streptomycin, tetracycline	1
Hg^{2+}, AsO_4^{3-}, streptomycin, kanamycin	4
Hg^{2+}, AsO_4^{3-}, streptomycin, chloramphenicol	1
Hg^{2+}, streptomycin, chloramphenicol, kanamycin	1
Hg^{2+}, AsO_4^{3-}, streptomycin, tetracycline, chloramphenicol	2
Hg^{2+}, AsO_4^{3-}, streptomycin, chloramphenicol, kanamycin	3
Hg^{2+}, AsO_4^{3-}, streptomycin, tetracycline, kanamycin	1
Hg^{2+}, AsO_4^{3-}, streptomycin, tetracycline, chloramphenicol, kanamycin	3

is currently known, arsenite resistance is always found associated with arsenate resistance on plasmids [8,9]. The genes governing arsenate and arsenite resistances are separate, however [9,10].

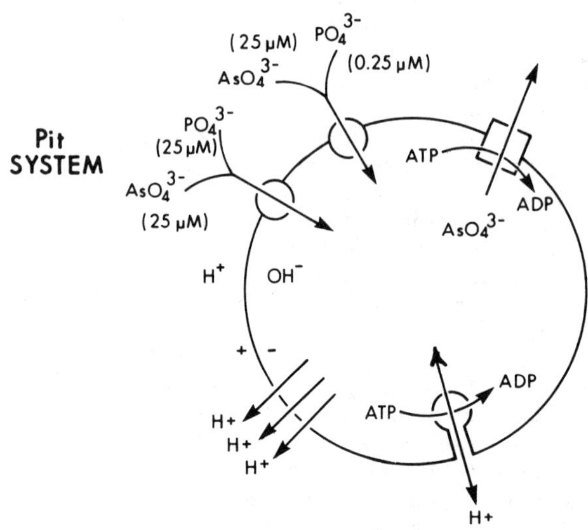

Figure 1. Model for phosphate (arsenate) transport systems and the arsenate efflux system. 0.25 μM or 25 μM, where indicated, are the K_m (for PO_4^{3-}) or the K_i (for AsO_4^{3-} as a competitive inhibitor of PO_4^{3-} transport) for the Pit and Pst phosphate transport systems of E. coli.

3. Arsenate uptake and chromosomal resistance

In all bacteria studied [11,12], AsO_4^{3-} is accumulated by highly specific, energy-dependent membrane "pumps" for which AsO_4^{3-} is the functional analog for the natural substrate PO_4^{3-}. Some bacteria (including E. coli) have two such phosphate pumps [11-13]. Transport systems for other anions, such as NO_3^- and SO_4^{2-}, are biochemically and genetically separate from the PO_4^{3-} transport systems [11]. Transport systems are thought to be composed of membrane proteins and their functioning is much analogous to that of enzymes. Thus, kinetic parameters, such as a V_{max}, a K_m, and K_i's are useful indicators of function. The K_m's of the two phosphate transport systems of E. coli are shown in Figure 1.

Of the two systems, one is relatively non-specific (called Pit, for inorganic Pi transport) and has approximately equal K_m for phosphate and K_i for arsenate of 25 μM [13]. The Pit system is synthesized and functions at all times (i.e. constitutively) at a rate with a V_{max} of 60 nmol/min per mg cells. The other system (called Pst, for phosphate specific transport) has a comparable K_i for arsenate as a competitive inhibitor of phosphate transport, but has a 100-times higher affinity for PO_4^{3-} with a K_m of 0.25 μM PO_4^{3-} [13]. Thus, surprisingly, the specificity of the Pst system is achieved by increasing the affinity for the natural substrate and not by reducing it for the toxic analogue, arsenate. As seems to be general for more highly-specific transport systems, the Pst system is synthesized only at times of need: its rate increases from about 15 to about 70 nmol/min per mg cells under conditions of phosphate starvation or arsenate toxicity [9,13]. When the Pit system is eliminated by mutational loss, the Pst system is synthesized and provides a more specific phosphate uptake system insensitive (relatively) to arsenate. Pit$^-$ Pst$^+$ bacterial cells are arsenate resistant.

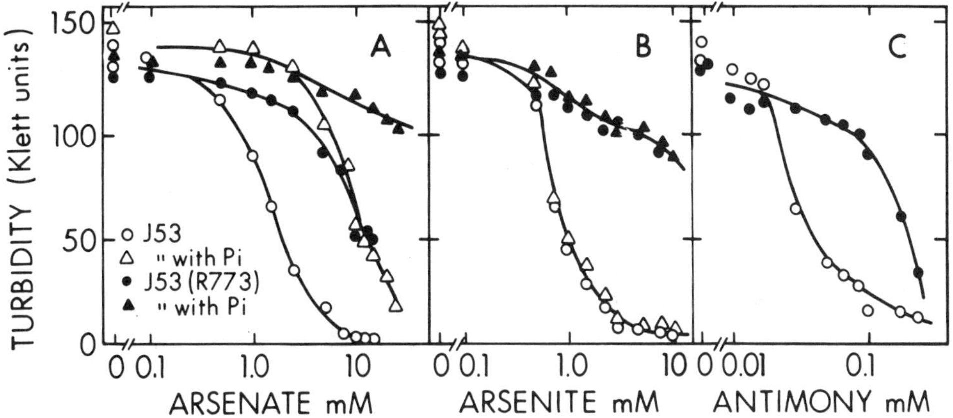

Figure 2. Resistances to arsenate, arsenite and antimony. Cultures of E. coli strain J53, with or without plasmid R773, were diluted 100-times into nutrient broth containing varying concentrations of Na_2HAsO_4, $NaAsO_2$ or $SbKOC_4H_6O_6$. After 6 h at 37°C, growth was measured as turbidity in a Klett-Summerson colorimeter [from ref. 9].

4. Plasmid-mediated resistance and induction

Plasmid gene-mediated resistance to arsenate is quite distinct and additive to that conferred by the chromosomal pit mutations [9]. Some bacterial plasmids (which are small extrachromosomal DNA molecules) have genes conferring resistances to toxic heavy metals as well as to antibiotics [1-6]. When tested, all arsenate-resistance plasmids also conferred resistances to arsenite and antimony(III) [8-10]. Figure 2 shows the resistance with one particular plasmid (R773) which we have studied most extensively. As far as we have determined, the basic properties of the arsenic-antimony resistance system are the same in both Gram negative and Gram positive bacteria [9,14].

For all three ions, the sensitive strain (without a plasmid) was inhibited at lower concentrations than required to inhibit the resistant strain (with plasmid R773). The addition of 6 mM PO_4^{3-} to the cells growing in less than 1 mM PO_4^{3-} broth protected both the sensitive and the resistant cells from arsenate (fig. 2A) but not from arsenite (fig. 2B) or Sb(III) (data not shown). That is because phosphate competes for cellular uptake with arsenate, but not with arsenite. One surprise from our studies [9] was that the kinetic parameters of phosphate uptake and arsenate inhibition of phosphate uptake were unaltered by the presence of the resistance plasmid. The resistance mechanism did not depend upon properties of the uptake transport systems. (It was also true that chromosomal arsenate resistance was additive to plasmid-determined resistance.)

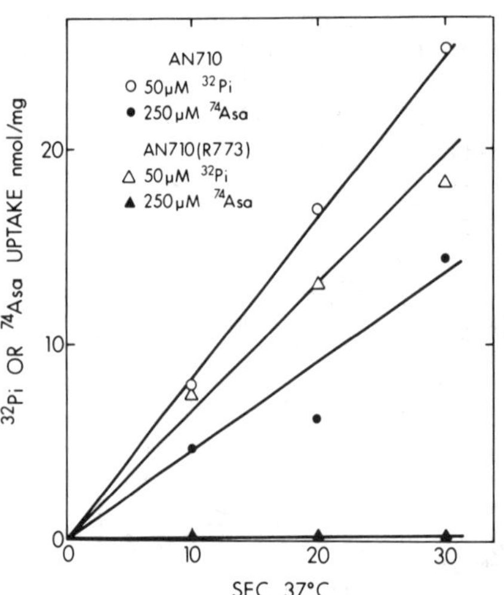

Figure 3. Arsenate and phosphate uptake by sensitive (AN710) and resistant [AN710(R733)] E. coli [from ref. 9].

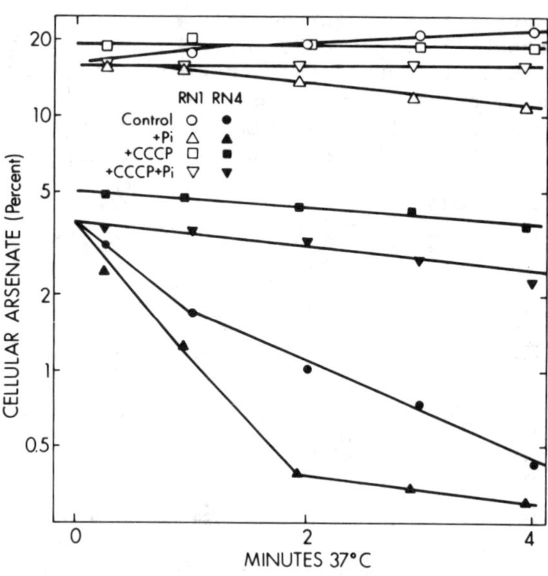

Figure 4. Accelerated energy-dependent efflux of arsenate by resistant S. aureus cells (RN4) but not by the sensitive cells (RN1). Cultures were grown, induced, "loaded" with $^{74}AsO_4$ and efflux was initiated by dilution as described in ref. 14. 40 μM CCCP and 5 mM PO_4^{3-} (Pi) were added as indicated [from ref. 14].

Plasmid-determined arsenate resistance was due to an accelerated efflux of arsenate from the cells by means of a highly specific arsenate-efflux "pump" (Fig. 1). We currently believe that this "pump" operates as a transport ATPase [14]. The evidence for this mechanism of arsenate resistance includes a striking reduction in the level of net arsenate accumulation by induced resistant cells (fig. 3) and a rapid energy-dependent efflux from induced resistant, but not sensitive cells (fig. 4). That figure 3 shows an experiment with Gram negative bacteria and figure 4 an experiment with Gram positive bacteria is to emphasize that the mechanism of arsenate resistance seems essentially the same in both cases and that equivalent experiments have been run. In addition to inhibition by CCCP (carbonyl cyanide m-chlorophenyl hydrazone, a small organic molecule that mobilizes H^+ movement across biological membranes), arsenate efflux was also inhibited by reduced temperature (4°C), by mercurial poisons and by agents that led to intracellular acidification [14]. Arsenate efflux was somewhat accelerated by extracellular phosphate (fig. 4), but did not depend upon this; nor did it require any other tested extracellular cation or anion [14]. The diagram in figure 1 summarizes what we believe to represent the situation for bacterial phosphate and arsenate uptake and exit. A membrane potential (negative inside) is generated by "pumping" of protons by the respiratory chain and a reversible membrane ATPase. This also leads to an acidity/alkalinity gradient, where the inside of the cell remains about pH 7.5 while the outside can be 1 or more units below that value. Phosphate (and arsenate) are accumulated by energy-dependent processes (two transport systems with "co-transport" of H^+). Intracellular phosphate is not exchanged, but intracellular arsenate is rapidly pumped out by a highly specific plasmid-governed system. The efflux pump must not transport phosphate, or else the cells would become phosphate starved, which would not have much advantage over being arsenate inhibited.

The arsenate-resistance, arsenate-efflux system is synthesized only in response to an environmental stimulus. This is referred to as "induction" and, in this case, low marginally toxic or subtoxic concentrations of heavy metal salts function as inducers. Since the same system of genes confers resistance to AsO_4^{3-}, AsO_2^- and SbO^+, it is not surprising that all three of these ions induce synthesis of the arsenate efflux system [6,9,14]. The rate of arsenate efflux increases by more than 10-fold after 1 h exposure to optimum inducing concentrations under conditions of protein synthesis [6,9,14]. BiO^+ is also an inducer of the arsenate resistance system, although these plasmid genes do not confer resistance to bismuth [6,14]. With some plasmids [10], there are still additional gene(s) and system(s) for bismuth resistance.

5. Arsenite sensitivity and resistance

Much less is known about the genetics and biochemistry of arsenite sensitivity and resistance than is known about arsenate sensitivity and resistance. It is clear that these systems are different, since phosphate protects against arsenate toxicity, but not against arsenite toxicity (fig. 2). Furthermore, studies with genetic mutants that have lost either arsenate or arsenite resistance, but retained the other [9,10] establish that different genes and therefore different cellular functions are involved.

There is both chromosomal [15] and plasmid mediated [9,10] arsenite resistance. The understanding of chromosomally controlled arsenite resistance is still at a very tentative, very preliminary stage. Smith and Myers [15] isolated a series of arsenite resistant mutants of E. coli and reported that 1 (out of 10) showed a defect in the membrane ATPase (diagrammed in fig. 1) that generates the membrane potential. The experiments leading to that conclusion look reasonable, but there is no current explanation as to why a mutant lacking this membrane enzyme should be arsenite resistant. Furthermore, most mutants lacking the ATPase activity remain arsenite sensitive and most arsenite resistant mutants retain the activity. Clearly, more understanding is needed. There are also arsenite hypersensitive chromosomal mutants in E. coli [8]. Arsenite resistance plasmids function effectively in these hypersensitive mutants, but we do not know the mechanisms responsible for either the hypersensitivity or the resistance.

There are two reports of isolation of soil or sewage isolates of Alcaligenes that can oxidize the more toxic oxyanion AsO_2^- to the less toxic form AsO_4^{3-}[16,17]. In both cases, the arsenite oxidizing Alcaligenes strains were arsenite resistant, but it is not known whether the genes governing this resistance are chromosomal or plasmid-borne. The arsenite-oxidizing Alcaligenes contained an intracellular enzyme, a cytochrome-linked oxido-reductase, that was synthesized only upon induction by added arsenite [16,17]. Non-oxidizing arsenite-resistant bacteria were also isolated from the same sources [17]. Apparently similar arsenite-oxidizing bacteria were isolated over the last 60 years from arsenical cattle and sheep dip solutions [original refs. discussed in refs. 5, 16 and 17]. There is no real basis today for telling whether arsenite-oxidizing or arsenite-nonoxidizing bacteria are more prevalent among arsenite-resistant natural isolates.

Plasmid-mediated arsenite resistance does not depend upon arsenite oxidation [9]; in fact, all extracellular detoxification mechanisms were eliminated by careful experiments in which medium containing arsenite, in which arsenite-resistant cells had been grown, retained the ability to inhibit growth of a sensitive, but not of a resistant strain [9]. That result means that an alternative hypothesis, that of excretion of an arsenite-binding dithiol compound was also eliminated [9]. We are left today with the unsatisfactory untested alternatives of (a) a transport barrier or efflux mechanism, (b) intracellular synthesis of arsenite binding compounds, or (c) a change in a critical intracellular target enzyme. These hypotheses will hopefully be tested during the coming year.

6. Summary

Arsenic resistant bacterial cells are found both in environmental and in clinical collections of bacteria. Arsenate and arsenite resistances are governed by different mechanisms; and the genes determining these mechanisms reside either on the cellular chromosome or on small resistance plasmids. The mechanisms of chromosomally-determined and plasmid-determined resistance are generally different. Arsenate resistance governed by chromosomal genes is due to a block on cellular uptake because of elimination of a relatively indiscriminate phosphate-arsenate membrane transport system. Plasmid-mediated resistance to arsenate is due to the synthesis of a highly specific arsenate efflux pump that eliminates intracellular arsenate as fast as it is accumulated. There appear to be two mechanisms of chromosomally governed arsenite resistance in different bacterial species: one involves oxidation of arsenite to arsenate by an inducible enzyme system. The second, non-oxidizing mechanism is not known. The mechanism of plasmid-mediated arsenite resistance does not involve extracellular detoxification, but it also is not known at this time.

The research in S.S.'s laboratory is being supported by research grants from the NSF and the NIH. H.N. wishes to thank Prof. H. Kozukue of Jikei University School of Medicine for support and encouragement.

References

[1] Summers, A.O. and Silver, S., Mercury resistance in a plasmid-bearing strain of *Escherichia coli*, J. Bacteriol. **112**, 1228-1236 (1972).

[2] Weiss, A.A., Silver, S. and Kinscherf, T.G., Cation transport alteration associated with plasmid-determined resistance to cadmium in *Staphylococcus aureus*, Antimicrob. Agents Chemother. **14**, 856-865 (1978).

[3] Kondo, I., Ishikawa, T. and Nakahara, H., Mercury and cadmium resistance mediated by penicillinase plasmid in *Staphylococcus aureus*, J. Bacteriol. **117**, 1-7 (1974).

[4] Nakahara, H., Ishikawa, T., Sarai, Y., Kondo, I., Kozukue, H. and Silver, S., Linkage of mercury, cadmium, and arsenate and drug resistance in clinical isolates of *Pseudomonas aeruginosa*, Appl. Envir. Microbiol. **33**, 975-976 (1977).

[5] Summers, A.O. and Silver, S., Microbial transformations of metals, Annu. Rev. Microbiol. **32**, 637-672 (1978).

[6] Silver, S., Mechanism of bacterial resistances to toxic heavy metals: arsenic, antimony, silver, cadmium and mercury, In *Environmental Speciation and Monitoring Needs for Trace Metal-Containing Substances from Energy-Related Processes*, F.E. Brinckman and R.H. Fish, eds., pp. 301-314 (National Bureau of Standards Special Publ. 618, Washington, D.C., 1981.

[7] Nakahara, H. and Kozukue, H., Volatilization of mercury determined by plasmids in *E. coli* isolated from an aquatic environment, In *Proc. 3rd Tokyo Symposium on Drug Resistance in Bacteria*, S. Mitsuhashi and H. Hashimoto, eds. (Japanese Scientific Societies Press, Tokyo, in press, 1981).

[8] Smith, H.W., Arsenic resistance in Enterobacteria:its transmission by conjugation and by phage, J. Gen. Microbiol. 109, 49-56 (1978).

[9] Silver, S., Budd, K., Leahy, K.M., Shaw, W.V., Hammond, D., Novick, R.P., Willsky, G.R., Malamy, M.H. and Rosenberg, H., Inducible plasmid-determined resistance to arsenate, arsenite, and antimony(III) in Escherichia coli and Staphylococcus aureus, J. Bacteriol. 146, 983-996 (1981).

[10] Novick, R.P., Murphy, E., Gryczan, T.J., Baron, E. and Edelman, I., Penicillinase plasmids of Staphylococcus aureus: restriction-deletion maps, Plasmid 2, 109-129 (1979).

[11] Silver, S., Transport of cations and anions, In Bacterial Transport, B.P. Rosen, ed., pp. 221-324 (Marcel Dekker, Inc., New York, 1978).

[12] Bennett, R.L., and Malamy, M.H., Arsenate resistant mutants of Escherichia coli and phosphate transport, Biochem. Biophys. Res. Commun. 40, 496-503 (1970).

[13] Rosenberg, H.R., Gerdes, R.G. and Chegwidden, K., Two systems for the uptake of phosphate in Escherichia coli, J. Bacteriol. 131, 505-511 (1977).

[14] Silver, S. and Keach, D., Energy-dependent arsenate efflux: the mechanism of plasmid-mediated resistance, manuscript in preparation (1981).

[15] Smiley, D.G. and Myers, J.W., An arsenite resistant mutant of Escherichia coli with a defective membrane-bound adenosine triphosphatase, Current Microbiol., in press (1981).

[16] Osborne, F.H. and Ehrlich, H.L., Oxidation of arsenite by a soil isolate of Alcaligenes, J. Appl. Bacteriol. 41, 295-305 (1976).

[17] Phillips, S.E. and Taylor, M.L., Oxidation of arsenite to arsenate by Alcaligenes faecalis. Appl. Environ. Microbiol. 32, 392-399 (1976).

DISCUSSION

E. A. Woolson: Have you looked at the arsenate that is pumped out in terms of whether or not it is self-metabolized to, for example, cacodylic acid, since this appears to be a detoxification mechanism?

S. Silver: It is not a detoxification mechanism. It is pumping out arsenic as a resistance mechanism. We have taken the arsenate that was intracellular and the extracellular arsenate after exposure of bacterial cells for 20 or 30 minutes to radioactive arsenate. With thin layer chromatographic analysis, it looks to be totally arsenate. There is no sign of any chemical alteration. Take, for example, mercury. Mercury is enzymatically changed from Hg^{2+} to Hg^0 and we have purified the enzyme responsible. There are speciation changes for mercury and for inorganomercurials. Phenylmercury is converted to benzene and Hg^0. But for arsenic, as far as we know, there is no chemical conversion. It's in every bacteria you can imagine. In the Chesapeake Bay, resistance occurs in pseudomonads and bacillus from the soil and elsewhere. We look at *E. coli* because that is where you have the best molecular genetics, which is such a powerful tool. We think the basic mechanism for arsenate resistance is the same wherever we have looked.

E. A. Woolson: A lot of bacteria seem to affect the ratio of arsenate to arsenite in solution.

S. Silver: I'm aware of a soil *Alcaligenes* species, which has an enzymatic system for oxidizing arsenite to arsenate. Now, that is a separate mechanism that does not seem to be plasmid mediated. Oxidation is not the mechanism amongst the bacteria we are studying from soil and water systems; but it is a mechanism that exists. I do not know if the methylation is ever a resistance mechanism, because my impression is that it occurs at a very slow rate. The systems that I have talked about today are widespread.

E. A. Woolson: The reason I was interested in the arsenite/arsenate conversion is that arsenite was used as cattle dip down at the Mexican border and they had a heck of a time keeping it as arsenite, because the bacteria would oxidize it as fast as they put it in.

S. Silver: I think the first arsenite resistant bacteria were found in cattle and sheep dips in South Africa and Australia. They were indeed arsenite oxidizing bacteria, so those do exist. We have not run into one, although we've looked at a fair range. Maybe it is because we start out looking for plasmid mediated resistances as a way of getting a genetic handle on this. Perhaps we are screening those out from our collections since we are not looking at *Alcaligenes*. I think the arsenite oxidizing ones are predominantly *Alcaligenes*.

SESSION IV - EPIDEMIOLOGY

Session Chairman: Dr. Edward P. Radford
University of Pittsburgh
Graduate School of Public Health
130 DeSoto Street, 517 Parran Hall
Pittsburgh, PA 15261

INTRODUCTION

It is my hope that this particular group of papers will receive extra attention because they deal with the question of human carcinogenesis or co-carcinogenesis. I would like to introduce this topic by asking the question: How would one know if a compound or material were a human carcinogen, as distinct from a cancer promoter in man? I have come up with a few thoughts that may help to set the stage for some of the discussion that is likely to occur. First, it is my perception that in contrast to some of the experimental animal species used in cancer research, the human animal is exposed to conditions in life that are hardly controlled experiments. There is reason to believe that there are a number of carcinogens and promoters naturally present in the human environment. Thus, development of human cancer may well be limited, in some circumstances, by the presence of a naturally occurring initiator, or in other circumstances by a naturally occurring promoter. For example, we are all exposed to naturally occurring radon and radon daughters, considered to be bronchial cancer initiators from the alpha radiation they produce. Indeed, the average radiation dose to the bronchial epithelium from alpha radiation from this source is the most significant background radiation exposure in any tissue of our body. Depending on various estimates, anywhere from 15% to 30% of all lung cancers in non-smokers may be due to exposure to background radiation of this type. Given this circumstance, development of a lung cancer may depend on the presence or absence of a potent promoter, because an initiator is normally present. The question is, how do you detect this?

Second, if a population is exposed to a carcinogen, let's say vinyl chloride, the probability of subsequently developing cancer may be enhanced by naturally occurring promoters such as viruses, irritant gases or trauma. It is not always easy to distinguish between an initiating or promoting agent, because both are naturally present in the environment. Of course, there is the possibility an agent can be both.

We might expect that a promoter could have an effect which would lead to an earlier onset of cancer in relation to age, compared to an unexposed population. In other words, by considering the age dependence of the observed cancers, we may get a clue; this is one of the factors that I think we can look at in the following studies. Nevertheless, I think we still have to develop models by which we will be able ultimately to distinguish when a chemical or physical agent is acting as a carcinogen or as a promoter. This question is pertinent because arsenic compounds have sometimes been considered as carcinogens and sometimes as promoters.

17

WHAT IS THE STATUS OF ARSENIC AS A HUMAN CARCINOGEN

I. Harding-Barlow
Consultant
3717 Laguna Ave
Palo Alto, CA 94306

To date no animal model exists for arsenic, hence its role as a supposed human carcinogen, rests entirely on epidemiological studies. Arsenic is known to interact with many elements and moieties which can increase or decrease its toxicity. This is discussed in terms of what is presently known experimentally and also in theoretical terms. Upon close examination of the epidemiological studies relating to respiratory cancer, where arsenic is apparently involved, it was found that copper and sulfur compounds may probably play a part with arsenic as co-promoters or co-carcinogens. Some enzyme systems which may be involved are discussed. With regard to studies on skin cancers where arsenic appears implicated, it would seem that the situation is complex, but again other factors such as immune response, dietary deficiencies and trace metal imbalances are of great importance, leading one to believe that arsenic is probably a co-promoter with other factors involved. It is also apparent that total amount of arsenic consumed is probably important and should be assessed when making predictions concerning toxicity of arsenic, rather than just consideration of intake due to inhalation or water alone.

When attempting to correlate the epidemiological data with the presumption that arsenic is a carcinogen, it was once again apparent that arsenic never seemed to be the only possible contributing cause. Thus, it is fruitful to consider what other factors might logically be involved.

In trace element content of human tissue studies conducted over twenty years ago [1], it had been apparent that interactions and hence synergisms and antagonisms, often occurred between three or four elements, not just two. Therefore, in an attempt to determine which elements might interact with arsenic, a listing was obtained of elements similar in size and/or binding power (as measured by ionization potential) to trivalent and pentavalent arsenic and the isoelectronic ions, quadravalent and sextavalent selenium, for which arsenic is known to substitute. It was found that the following trivalent ions were similar in size to arsenic, namely, aluminum, cobalt, chromium, iron and manganese, and elements similar in binding power were aluminum, chromium, iron, phosphorus and vanadium. Of the quadravalent ions none were similar in size to selenium, but lead and titanium were similar in binding power. Phosphorus was similar in size and binding power to pentavalent arsenic, whereas vanadium was similar in binding power. Sulfur was similar in size and binding power to sextavalent selenium, whereas chromium was similar in size. Thus in theory, under certain circumstances, arsenic should be able to interact with aluminum, cobalt, chromium, iron, manganese, vanadium and phosphorus, whereas selenium should be able to interact with chromium, titanium, lead and sulfur. At the present time under experimental conditions it is known that arsenic interacts with copper, iron, zinc, selenium, manganese, antimony, phosphorus, iodine, cadmium, lead and sulfur dioxide, whereas selenium is known to interact with copper, zinc, cobalt, arsenic, tellurium, tungsten, mercury, cadmium, lead, silver, thallium and sulfur. This does not mean that other interactions do not occur, but rather that they have not been tested. It is highly likely that for instance, arsenic and molybdenum interact, since it is known that arsenic can inhibit the molybdenum metalloenzyme xanthine dehydrogenase at the site of the molybdenum active center [2].

One of the problems which arises if one is to consider arsenic a complete carcinogen, with no threshold, is that arsenic is an essential element [3]. Thus, as with every essential element, one has concentration levels where deficiency occurs, concentration levels which are normal and concentration levels where toxic effects take place. When a deficiency occurs the body attempts to overcome this deficiency and there may be increased active absorption of the element or an attempt by the body to shift the balance of the other elements and perhaps find a substitute element to keep the key enzyme systems functioning. Under deficiency one may thus see one set of interactions.

Between deficient and toxic levels are the so-called normal levels, which will include the levels present under most usual conditions of exposure. At these normal levels a set of interactions will take place between the essential element under consideration and other elements which may overlap, but need not be identical to the interactions which occur under deficiency conditions.

At concentration levels at which toxic effects occur, the body attempts to accelerate detoxification mechanisms, but as concentration levels increase an overload point is reached, where the element of interest will be inhibiting systems it would not normally interfere with, thus creating a third set of interactions. These new interactions may include the creation of apparent deficiencies, where enzyme systems that are in the low normal or borderline deficient may now appear to act as if a defieciency exists. Thus if the action is on a system which is anti-tumorgenic, the net result will be to decrease the defense mechanism, even if the antitumorgenic agent is present in apparently normal amounts.

Another factor which is often overlooked is that dietary deficiencies can also affect the toxicity of elements. For example, it is thought that vitamin C and also methionine reduce the toxicity of arsenic, but that vitamin A deficiency increases the sensitivity to arsenic and high carbohydrate diets or high protein diets or even high fat diets protect against the toxic effects [4].

It should be noted that arsenic can act upon a large number of metabolic pathways; for example, the Krebs cycle, the electron transport chain and oxidative phosphorylation [5]. Arsenic may inhibit certain metabolic reactions at high concentrations, while promoting them at lower levels.

Turning to the epidemiological studies in which arsenic has been implicated as a carcinogen, it is seen that they fall into two groups, those in which respiratory cancer has been found and those in which skin cancer is diagnosed.

Respiratory cancer has been found in three sets of workers in the above studies, (a) copper smelter workers, (b) arsenical pesticide manufacturing workers and (c) arsenical pesticide applicators. The copper smelter workers are exposed to fairly large quantities of a multitude of substances including some which are thought to be carcinogens, however on further examination of the data, it is found that for the particular workers considered, in addition to arsenic, they were always exposed to high concentrations of copper and sulfur compounds - mainly sulfur dioxide On close examination of both the pesticide manufacturing workers and the pesticide applicators, in the above mentioned epidemiological studies, it was found that they would have been exposed to copper and sulfur compounds, in addition to arsenic and a nonhomogeneous group of other chemicals [6]. Individual work histories would have to be checked to see if there were any exceptions, but in these instances possible exposures to carcinogens such as asbestos and/or other potential carcinogens would have to be checked too.

Thus it would seem that in respiratory cancer supposedly caused by arsenic, there were co-exposures to at least two other substances at high concentrations, namely copper and sulfur compounds. Because of very imperfect knowledge of how metallo-enzyme systems function, particularly when dealing with an apparent multi-element toxicity, at present only guesses are possible as to which enzyme systems might be involved.

Selenium is thought to be an anticarcinogenic agent [7]. One of the enzymes which may be involved in the process is the glutathione peroxidase system which is selenium dependent. Arsenic can inhibit the action of selenium and this might result in an apparent deficiency if this enzyme system actually was present, but at low normal or borderline deficiency levels. The combined effects of high concentrations of sulfur compound as well as high concentrations of arsenic might be synergistic in enhancing this apparent deficiency.

Normally, the molybdenum containing enzyme, sulfite oxidase, functions in the detoxification mechanisms for sulfur dioxide and other sulfur compounds. It is well known that copper is an antagonist of molybdenum, in addition arsenic can inhibit sulfite oxidase. Thus it seems likely that high levels of arsenic combined with high levels of copper may at least partially, be capable of blocking the body's detoxification mechanism of some sulfur compounds.

The enzyme xanthine oxidase, another molybdenum containing enzyme, is probably involved in the three way balance interactions of copper, sulfur and molybdenum. In vitro it is known that both arsenic and copper can inhibit this enzyme. Thus, it is probable that this enzyme may also be affected when workers are exposed to high levels of copper, sulfur compounds and arsenic.

It seems probable that when workers are exposed to high levels of arsenic, copper and sulfur compounds, the affects on enzyme systems may well be different from those that would take place with exposure to only high levels of one or two of these three agents. Thus the combined affects of the three-arsenic, copper and sulfur compounds may either predispose the respiratory tissues to neoplastic changes or co-promote changes, once initiation has occurred.

With regard to the relationship of arsenic to skin cancer the epidemiological basis is not very consistent. On one hand there is a group of patients being treated with Fowler's solution for a number of diseases with very different etiology and on the other hand three groups of persons who used potable waters which were high in arsenic content.

Fowler's solution has been used for a couple of hundred years for treating a very large number of persons for a wide variety of long term chronic diseases. Of those persons treated with Fowler's solution a small number, some estimates have been as low as 0.1%, developed skin cancers. Some of these cases progressed via hyperpigmentation to keratosis to neoplastic growths, but some did not [8]. At no time is it clear what percentage were treated with other drugs and skin ointments, since the original condition being treated with Fowler's solution was often a skin condition. It is fairly certain however, that probably all these patients, because of their pre-existing chronic diseases, had disturbances in their copper metabolism and also other elements as well. The existence of non-normal levels of enzymes and immune system deficiencies is open for speculation. Also whether sulfur and/or zinc containing ointments were used by a large number of these patients for their treatments is again not known.

On the other hand one has three series of studies [9, 10, 11] in which the arsenic content of the water is stated to cause skin lesions of which a certain number then develop into skin cancers. In two of the study series, those in Chile and Argentina, it is impossible to calculate even approximately, what percentage of persons exposed to the arsenic containing waters developed and/or died of skin cancers.

In all three study series it is impossible at the present time to even roughly gage the total trace element picture (intake from food, water and air) for the populations exposed to high arsenic in their water supplies. It seems fairly certain that the water and food supplies in the endemic areas of Argentina and Chile could have been contaminated by wastes and tailings from copper mining, as well as natural contamination due to geological formations. It seems probably that the total copper and sulfur intakes were high in at least some instances. Little appears to be known concerning the geological origins of the deep aquifer that fed the Taiwanese wells. It should be noted that trace element concentrations in water supplies can vary considerably with time and that food supplies grown in areas high in trace element contamination can also contribute to total trace element intake.

It is not known to what organic toxic components the inhabitants of the endemic areas of Chile and Argentina were exposed in their water supplies. It is known that in the Taiwanese endemic area, ergotamines were found in the water supplies [12].

In the Taiwanese endemic area it is known that their diets were limited in protein, fats, and the sulfur containing amino acid, methionine, and also probably in vitamin C. All these factors would tend to increase the toxicity of arsenic and upset other mineral balances. In Antofagasta, Chile it is noted that it is a copper and saltpeter producing area, hence the levels of trace elements may be high in the food supplies. It is also probable that the children of this area might suffer from immune deficiencies, which could well have lead to abnormal trace element metabolism. In the Cordoba area of Argentina it is known that the vitamin C content of the inhabitants was often lower than normal. In addition the iodine and hence the thyroid metabolism of at least some of the inhabitants was abnormal.

Thus it would seem that in persons who developed skin cancers attributed to arsenic, other factors such as dietary deficiencies, immune deficiencies and/or major essential element metabolic imbalances co-existed with high doseages of arsenic. Thus it seems likely that arsenic is highly unlikely to be a complete carcinogen and would more likely be a co-promoter.

In conclusion it is of interest to note that the average person in the U.S. takes in 4.8 μg of arsenic in their drinking water and beverages per day, of which 4.1 μg is absorbed; 21.7 μg of arsenic in their food per day of which 14.7 μg is absorbed and breathes in 0.06 μg of arsenic from the air, of which 0.02 μg is absorbed. Therefore of a total intake of 26.5 μg of arsenic, about 18.8 μg is absorbed. If for argument sake this 18.8 μg of absorbed arsenic per day was considered to come totally via the inhalation route, it would be equivalent to an air concentration 3.3 μg As/m^3, and if the absorbed arsenic per day was considered to come totally via the water borne route the water concentration would be 14.7 μg/l.

The CAG group of EPA have indicated that life time probability (P) of getting lung cancer from arsenic should be given by the equation:-

$$P = 2.953 \times 10^{-3} X$$

where X is the concentration in $\mu g/m^3$ of arsenic in the air. Hence, if X were to be put equal to 3.3 μg As/m^3

$$P = 9.73 \times 10^{-3} \approx 10^{-2}$$

The CAG group of EPA have also indicated that the life time probability (P) of getting skin cancer should be given by the equation:-

$$P= \frac{2.41423\ X}{2.41423\ X + 6.02933}$$

where X is the average concentration of arsenic in the water in mg/l. Hence, if X were put equal to 0.0147 mg/l

$$P= 5.85 \times 10^{-3}$$

Thus it would seem appropriate that in the EPA calculations, due allowance be made for the total intake of arsenic not, just that by one route.

References

[1] Harding-Barlow, I., Studies on the trace element content of human tissues. Ph.D. Thesis, University of Cape Town, South Africa (1961).

[2] Johnson, J.L. and Rajagopalan K.V., The interaction of arsenite with the molybdenum center of chicken liver xanthine dehydrogenase, Bioinorg. Chem. 8, 439-44 (1978).

[3] Nielsen, F.H., Myron, D.R. and Uthus, E.O., Newer trace elements-vanadium and arsenic deficiency signs and possible metabolic roles. In: Trace Element Metabolism in Man and Animals, Vol. 3, M. Kirchgessner, ed. pp 244-7 (Arbeitskrese fur Tierernahrungsforschung, Weikenstephen, Germany, 1978).

[4] Calabrese, E.J., Nutrition and Environmental Health, Vol. 1 and Vol. 2 (John Wiley and Sons, New York, 1980).

[5] National Research Council of Canada, Effects of Arsenic in the Canadian Environment, N.R.C.C. Publication No. 15391 (1978).

[6] Harding-Barlow, I. to be published.

[7] Schrauzer, G.N., Trace elements, nutrition and cancer, Adv. Exp. Med. Biol. 91, 323-44 (1978).

[8] Fierz, U., Catamnestic investigations of the side effects of therapy of skin diseases with inorganic arsenic, Dermatologica 131, 41-58 (1965).

[9] Yeh, S., Skin cancer in chronic arsenicism, Hum. Pathol. 4, 469-85 (1973).

[10] Zaldivar, R. and Guillier, A., Environmental and clinical investigation of epidemic chronic arsenic poisoning in infants and children, Zentralbl. Bakteriol. Parasitenkd, Infektionskr. Hyg. Abt. 1, 165B, 226-34 (1977).

[11] Astolfi, E., Maccagno, A., Fernandez, J.C.G., Vaccaro, R. and Stimola, R., Relation between arsenic in drinking water and skin cancer, Biol. Trace Elem. Res. 3, 133-43 (1981).

[12] Lu, F., Tsai, M.H. and Ling K.H., Studies on fluorescent compound in drinking water of Blackfoot Disease endemic areas, J. Formosan Med. Ass. 76, 209-17 (1977).

18

AN EPIDEMIOLOGICAL STUDY OF ARSENIC IN DRINKING WATER IN MILLARD COUNTY, UTAH

J. W. Southwick, A. E. Western, M. M. Beck, T. Whitley, R. Isaacs, J. Petajan[1] and C. D. Hansen[1]

UTAH STATE DEPARTMENT OF HEALTH
150 West North Temple
P.O. Box 2500
Salt Lake City, Utah 84110

This epidemiological study evaluated the health effects of arsenic in drinking water at levels approximately four times the maximum allowed by the National Interim Primary Drinking Water Regulations. Study participants came from west Millard County, a desert area of Utah, which has a homogeneous, stable population with a predominantly "Mormon" lifestyle.

Physical examinations were given to 250 people and included evaluating dermatological and neurological health, sampling hair and urine for arsenic content and testing for anemia. Water consumption estimates were used to estimate arsenic ingestion. The 145 "exposed" participants came from the small communities of Hinckley and Deseret, where drinking water arsenic content averaged 0.18 and 0.21 mg/l respectively. The predominant arsenic species was arsenate (86% As^{+5}). A matched control group of 105 participants was selected from a larger, neighboring community named Delta, where drinking water arsenic averaged 0.02 mg/l.

A clear relationship was shown between the amount of arsenic consumed and the amount of arsenic present in hair and urine samples. Dermatological signs compatible with arsenic exposure were rare and, when found, were scattered singly among exposed and control participants rather than being clustered as multiple signs on individuals with higher arsenic exposure. Anemia was not found significantly more often among exposed participants. Nerve conduction slowing did not correlate significantly with arsenic exposure levels. Typical signs and symptoms of arsenic intoxication were not found in any of the study participants.

---

[1] University of Utah, College of Medicine, Salt Lake City, Utah

1. Introduction

About the same time in 1976 when the Utah State Department of Health was assessing the problem of naturally occuring arsenic in public drinking water in the town of Hinckley, we learned of the U.S. Environmental Protection Agency's (EPA) interest in gathering more epidemiological data on arsenic. Further checking showed that the level of arsenic in Hinckley's drinking water was midway between levels found in localities where epidemiological studies had been done.

A study in Taiwan (Tseng, et al., 1968) involved arsenic in water at about 0.6 mg/liter, while a Chilian study (Zaldivar, 1974) had arsenic at about 0.8 mg/liter. Both of these studies reported illness associated with consumption of these waters. By contrast, a study in Lane County, Oregon (Morton et al., 1976) showed no adverse health effects from ingestion of water containing 0.1 mg/liter of arsenic. Hinckley water contained approximately 0.2 mg/liter of arsenic.

With a grant from EPA we conducted an epidemiological study of arsenic in drinking water in west Millard County where Hinckley is located. The results were reported to EPA and published as an EPA report (Southwick, et al., 1980).

West Millard County is a desert area with low population density. Fortunately the area had a homogeneous, stable population with a predominantly "Mormon" lifestyle. The study drew its exposed participants from Hinckley and neighboring Deseret, while control participants came from the larger, near-by community of Delta (fig. 1).

We decided to evaluate dermatological effects of arsenic because cutaneous lesions appeared to be the more consistant signs of chronic arsenic exposure (National Academy of Science, 1977). Reports of neurological effects from chronic arsenic exposure (Vallee, et al., 1960 and Heyman et al., 1956) prompted us to do neurological testing. Anemia had been reported in some cases of arsenic poisoning (Mizuta et al., 1956 and Miyata et al., 1970) so we included hematocrit determinations in the study.

2. Methods

2.1. Selection of Study Participants

All eligible residents of Hinckley and Deseret were invited to be "exposed " participants. To be eligible a person needed to be five years of age or older, and be a resident of Hinckley or Deseret for at least the previous five years. Since Deseret residents all used private wells for drinking water, their individual wells had to show arsenic levels in excess of 0.150 mg/liter to be included in the study. Of the approximately 650 residents of Hinckley and Deseret, 223 were determined

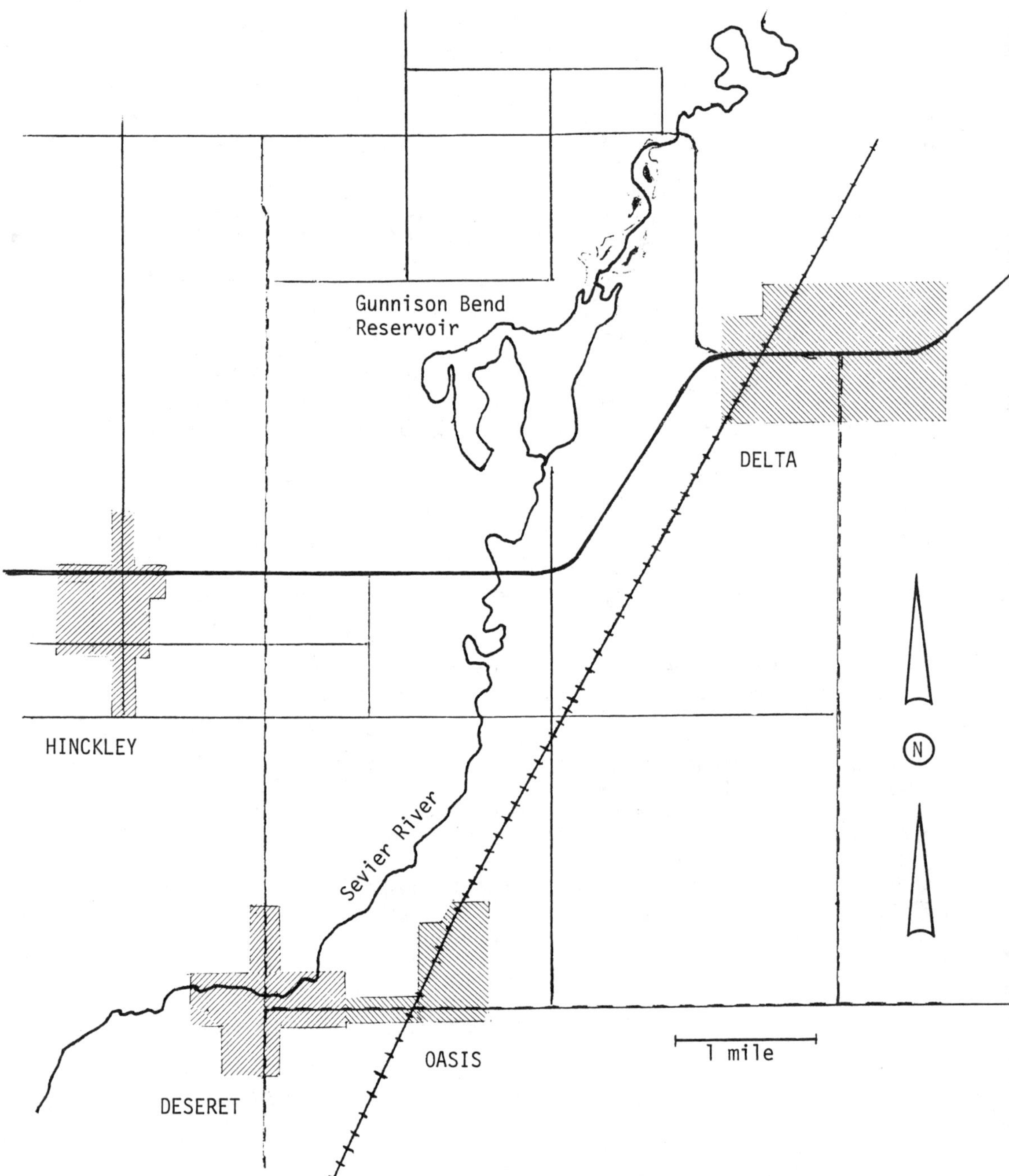

Figure 1. Map of West Millard County, Utah, showing the geographic relationship of the control community (Delta) to the arsenic exposed communities (Hinckley and Deseret).

to be eligible, 83 percent of the eligible agreed to participate and 65 percent actually participated.

Control participants were residents of Delta who had lived no part of their lives in communities where arsenic in drinking water exceeded the national standard of 0.05 mg/liter. Control participants were selected at random from age and sex categories in order to match the age and sex characteristics of the exposed group as closely as possible. Of the 226 control participants selected, 69 percent agreed to participate and 46 percent actually participated.

Table 1 shows the age and sex characteristics of the study population.

Table 1. Age and sex distribution of study participants from control (Delta) and exposed (Hinckley and Deseret) communities.

	Control				Exposed			
Ages	Male	%	Female	%	Male	%	Female	%
7-12	6	12.8	2	3.5	10	14.9	6	7.7
13-20	6	12.8	13	22.4	12	17.9	14	17.9
21-30	4	8.5	6	10.3	5	7.5	4	5.1
31-40	8	17.0	13	22.4	9	13.4	11	14.1
41-50	7	14.9	7	12.1	6	9.0	13	16.7
51-60	8	17.0	9	15.5	11	16.4	13	16.7
61-70	6	12.8	6	10.3	8	11.9	9	11.5
71+	2	4.3	2	3.5	6	9.0	8	10.3
TOTALS	47	44.8	58	55.2	67	46.2	78	53.8
GRAND TOTAL		105				145		

2.2. Assessment of Arsenic Exposure

All drinking water in the study communities came from wells. These wells were sampled periodically and analyzed for arsenic content. Urine samples were collected from participants during the study's August 1978 physicial examination. Hair samples were obtained by supplying each participant with a self-addressed, postage-paid envelope, with his name on it. Almost 75 percent of the envelopes were returned with hair samples.

Table 2 shows the number of hair and urine samples recieved and analyzed for arsenic content.

Study participants were closely questioned about their water consumption habits. Information obtained was used to calculate an annual arsenic dose. A crude estimated "total dose" was calculated from annual dose, times the number of years the participant had lived in the community.

Table 2. Number of physical examination participants from each study community showing number of hair and urine samples received.

Community	Number of Participants	Number of Samples Collected	
		Urine	Hair
Control (Delta)	105	99	68
Exposed communities	145	135	117
Hinckley	102	95	80
Deseret	43	40	37
TOTAL	250	234	185

2.3. Assessment of Health Status

Dermatological examinations were done by Dr. C. David Hansen (University of Utah College of Medicine, Division of Dermatology). He had no knowledge of any participant's community of residence until after his findings had been recorded. Body areas examined included face, back, abdomen, arms, legs, hands, and feet. No attempt was made to examine the chest, thigh or genital areas unless questionable lesions were indicated by the study participants. Notations were made of the individual's eye and hair color as an indication of the susceptibility to actinic damage and skin cancer.

Participants were examined for specific signs of arsenic toxicity, including palmar and plantar (palms and soles) keratoses, diffuse palmar and plantar hyperkeratoses, and skin tumors in non-sun exposed areas. All such tumors were recorded except for obviously benign lesions such as seborrheic keratoses and warts. Any histologically verified malignant tumors that were previously removed were noted. The location (i.e., sun-exposed vs palmar/plantar) and frequency of all keratoses and tumors were recorded. Any diffuse non-sun exposed hyperpigmentation was recorded, as were Mee's lines in nails. Arterial insufficiency was noted to assess any tendency toward "blackfoot disease" described for the Taiwan arsenic episode (Tseng, et al., 1968).

Neurological testing was done only on participants 47 years of age and younger. We focused on younger participants, in whom neuropathy was not generally expected, to see if an excess of neuropathy could be associated with high arsenic ingestion. In all, 150 participants received neurological examinations (Delta 67, Hinckley 53, and Deseret 30). Neurological testing was performed by a team from the University of Utah College of Medicine, Department of Neurology under the direction of Dr. Jack H. Petajan.

The neurological team was not apprised of the participants' residential data until after their testing was completed. A neurologist examined each participant for such things as threshhold for touch, hypalgesia, temperature sense and vibratory sense. Then each participant was taken to a different room where electrodiagnostic studies were performed to test motor and sensory nerves using a TE-4 electromyograph.

Hematocrit determinations were made on participants using a drop of blood obtained by finger prick.

3. Results

3.1. Assessment of Arsenic Exposure

3.1.1. Arsenic Consumption from Water

Monthly water samples taken from Hinckley and Delta (May 1976-May 1977) showed arsenic concentrations averaging 0.180 mg/l for Hinckley and 0.019 mg/l for Delta. Deseret well waters averaged higher in arsenic content than did Hinckley's or Delta's. Families in Deseret whose wells tested less than 0.150 mg/l were not included in this study. Arsenic in well water of participants from Deseret averaged 0.270 mg/l.

Estimates of arsenic consumed by each study participant showed a wide range (see table 3) within each community depending on the water consumption levels of each participant. Nonetheless, the mean arsenic consumption level was lowest in Delta (control) and highest in Deseret.

Table 3. Annual arsenic consumption from drinking water for study participants from three study communities.

Community	Number of Participants	Arsenic Consumption (mg)		
		Mean	Range	Median
Delta (control)	105	24.2	4 - 135	17
"Exposed"	145	152.4	12 - 853	119
Hinckley	102	135.5	12 - 853	115
Deseret	43	192.5	14 - 736	148

The species of arsenic in Hinckley water was determined to be predominantly arsenate (86% AS^{+5}) by Dr. Kurt Irgolic of Texas A & M University, College Station, Texas (Personal communication, 13 February 1980).

3.1.2. Arsenic in Urine

Table 4 shows participants from Deseret had the highest average arsenic-in-urine concentrations. These urinary arsenic data were transformed to logs and tested with the Duncan's Multiple Range Test. The results showed that Hinckley and Deseret did not differ from each other significantly, but both were significantly different from Delta. A t-test showed the two exposed communities to have more arsenic in urine than Delta at the P = < 0.0001 level of significance.

Table 4. Comparison of arsenic in urine of residents of three communities.

Community	Number of Samples	Arsenic Concentration (mg/l) Arithmetic Mean	Range	Median
Delta (control)	99	0.048	0.010 - 0.220	0.040
"Exposed"	135	0.185	0.025 - 0.660	0.158
Hinckley	95	0.175	0.025 - 0.580	0.150
Deseret	40	0.211	0.030 - 0.660	0.160

Table 5 shows a strong dose-response relationship between estimated annual dose of arsenic and arsenic in urine. The higher the estimated arsenic dose per year, the higher the average level of arsenic found in urine samples.

Table 5. Arsenic levels in urine compared to estimated annual arsenic dose

N	Annual Dose (mg)	Urine Arsenic Geometric Mean (mg/l)	Grouping*
10	0 - 9	.034	A
73	10 - 29	.044	A
63	30 - 99	.089	B
76	100 - 299	.152	C
12	300+	.302	D

*Means with the same letter not significantly different at P = .01 (Duncan's Multiple Range Test).

3.1.3. Arsenic in Hair

Table 6 shows Deseret residents to average somewhat less arsenic in hair than Hinckley residents but more than Delta residents. These arsenic in hair data were transformed to their logs and a Duncan's Multiple Range Test was performed. The results showed that Hinckley and Deseret did not differ from each other significantly but both were significantly different from Delta.

A t-test comparing Hinckley and Deseret (exposed) hair arsenic levels with Delta's (control) showed the difference to be statistically significant at P = < 0.0001.

Table 6. Comparison of arsenic in hair of residents of three communities.

Community	Number of Samples	Arsenic Concentration (mg/Kg) Arithmetic Mean	Range	Median
Delta (control)	68	0.32	0.10 - 3.10	0.20
"Exposed"	117	1.17	0.10 - 4.70	0.80
Hinckley	80	1.21	0.10 - 4.60	0.90
Deseret	37	1.09	0.10 - 4.70	0.65

3.2. Assessment of Health Status

3.2.1. Dermatological Findings

Results of the dermatological examination are shown in table 7. The finding of signs suggestive of arsenic toxicity was rare, with only 12 of 249 participants having any signs associated with arsenic ingestion. The fact that no participant had more than one sign (i.e. pigmentation and multiple cancers) suggested that the findings may have been incidental and not related to arsenic.

Table 7. Specific signs associated with arsenic ingestion as found in dermatological examinations.

Signs of Arsenic Ingestion	Control	Exposed
Palmar and plantar keratosis	1	2
Diffuse palmar or plantar hyperkeratosis	2	5
Tumors (nonsun-exposed)	0	0
Diffuse pigmentation (nonsun-exposed)	0	1
Arterial insufficiency	0	1
Mee's lines in nails	0	0
TOTAL	3	9
PERCENT	2.86%	6.25%

These twelve participants were not clustered among the more heavily exposed participants, as might be expected if high level arsenic exposure were responsible for the findings. For each indicator of exposure, the data from these twelve tended to fall on either side of the means for the respective communities. The ages of the twelve averaged 57.5 years (range 26-82). Half of the twelve were females, half were males. The signs of arsenic ingestion given in table 7, were regressed against annual arsenic dose and the log of annual dose, but no significant associations were found.

3.2.2. Neurological Findings

The mean values for nerve conduction velocity for any given nerve did not vary significantly with respect to age or community. However, some participants in each community and age group had below normal conduction velocities. For sensory nerves, velocities below 37 meters per second were considered abnormal, provided nerve temperature was above 30°C.

Sensory nerve conduction was observed to be primarily affected, with the sural nerve most often involved. Participants with low foot temperature explaining the slow conduction were also indicated. In such cases velocities were corrected according to the factor 1.8 meters per second per degree. For example, a sural velocity of 32 meters per second at 28°C could be corrected to 35.6 meters per second at 30°C. The corrected velocity remains below normal. This is an estimated value which will vary somewhat from participant to participant because the nerve temperature varies along its course depending upon the shape of the leg. Taking such participants into consideration, six exposed and six controls were found to have slowing of sural nerve conduction (table 8). In Hinckley, five participants had other nerves with abnormal conduction, as did three in Deseret and two in Delta.

Table 8. Study participants judged to have abnormal nerve conduction (temperature corrected).

Community	Participant Age Range	Number of Participants	Nerve With Abnormal Conduction
Delta (controls)	13-20	2	Sural
	21-30	3	Sural
	31-40	1	Sural
	31-40	1	Median Sensory
	41-47	1	Median and Ulnar Sensory
		8	(= 11.9% of 67 participants)
Hinckley	13-20	1	Ulnar Sensory
	13-20	1	Median and Ulnar Sensory
	21-30	1	Sural
	31-40	1	Sural
	31-40	1	Median Sensory
	31-40	1	Peroneal
	41-47	2	Sural
	41-47	1	Median Sensory&Peroneal
		9	(= 17.0% of 53 participants)
Deseret	7-12	1	Ulnar Sensory
	13-20	1	Sural
	31-40	1	Median Sensory
	31-40	1	Sural and Median Motor
		4	(= 13.3% of 30 participants)
Exposed (Hinckley and Deseret)		13	(= 15.7% of 83 participants)

The data can be viewed from the standpoint of number of nerves with abnormal conduction. More nerves were involved in exposed participants; median motor and peroneal nerves in addition to the median sensory, ulnar sensory, and sural which were also seen in control participants.

Participants with nerve conduction slowing were found in all three communities. The number of participants involved was small but the data indicate a slightly increased proportion of participants with slowing of nerve conduction among the exposed participants. Actual nerve conduction velocities were regressed against annual arsenic dose and log of annual dose, but no significant associations were found.

Neurological and physical findings are summarized in table 9. No trends were noted for any particular neurological finding or geographical location. In general, typical signs and symptoms of arsenic intoxication were not present in any of the participants examined, but slight impairment of sensation in the feet was found in two participants of each community.

Table 9. Neurological and physical findings.

Findings	Community Delta	Hinckley	Deseret
Cold feet	3	2	0
Cramps	8	6	2
Sweating	0	4	2
Decreased reflexes	3	0	0
Decreased sensation	2	2	2
Carpal tunnel syndrome	0	0	1

3.2.3. Hematocrit Findings

Table 10 shows results of hematocrit testing. The percentage of participants with anemia tended to be higher in the exposed communitities however, two of the four anemics in Deseret came from one family (mother and son, age 7).

Because of the apparent trend toward more anemia in communities with increasingly high arsenic levels in drinking water, a chi square test for a linear trend was made (Brown and Benedetti, 1979). No significant trend was found (P=.33).

Table 10. Anemia in study participants from three study communities.

Community	Number Tested	Anemia	
		Number	Percent
Control (Delta)	100	5	5.0
Exposed	137	10	7.3
Hinckley	95	6	6.3
Deseret	42	4	9.5

3.3. Statistical Analyses

For statistical treatment of the data, Hinckley and Deseret participants were combined since they constituted the "exposed" participants while Delta participants functioned as "controls."

A simple t-test (Goodnight, 1979) was performed to compare average health and exposure indicators from the control and exposed communities. Table 11 shows that only the exposure indicators were significantly different between exposed and control communities. None of the health indicators were significantly different between exposed and control communities.

Using health indicators for which incidence could be calculated, a chi square test (Sall, 1979) was performed to test the null hypothesis that the incidence in each community was the same. Table 12 shows the variables considered and that the health indicator incidences did not differ significantly between exposed and control communities, thus we were unable to reject the null hypothesis.

Table 11. A group analysis, comparison of means for health and exposure indicators in exposed and control communities.

Variable	t-Test Sig. Level	Control Number	Control Mean & Std. Dev.	Exposed Mean & Std. Dev.	Exposed Number
Urine arsenic	P <.0001*	99	48.1±30.7	185.3±124.3**	135
Annual dose	P <.0001*	105	24.2±22.1	152.4±130.8**	145
"Total dose"	P <.0001*	105	716±1112	4079±3807**	145
Hair arsenic	P <.0001*	68	.32±.49	1.17±1.09**	117
Urine/hair ratio	N.S. (.45)*	66	269±219	448±581**	111
Age	N.S. (.46)	105	38.7±19.3	40.6±21.5	145
Years in community	N.S. (.40)	105	26.2±15.4	28.0±17.6	145
Ulnar motor d.l.***	N.S. (.04)	24	2.29±0.31	2.53±0.59**	38
Ulnar motor v. ****	N.S. (.88)	23	63.8±8.2	64.1±8.5	38
Median motor d.l.	N.S. (.43)	67	2.98±0.46	3.05±0.55	82
Median motor v.	N.S. (.13)	67	62.6±6.8	61.0±6.5	82
Ulnar sensory d.l.	N.S. (.18)	58	2.63±0.33	2.73±0.45	67
Ulnar sensory v.	N.S. (.06)	58	50.2±5.3	48.1±6.6	67
Median sensory v.	N.S. (.30)	67	47.8±5.4	46.8±6.4	81
Peroneal d.l.	N.S. (.88)	67	3.99±0.90	3.97±0.92	81
Peroneal v.	N.S. (.68)	67	52.3±4.6	52.7±8.4**	81
Sural v.	N.S. (.77)	65	42.5±3.8	42.7±4.4	79
Hematocrit	N.S. (.48)	100	43.5±3.6	43.2±3.6	137

* These t-test significance levels were computed using log-transformed data due to non-normality of the variable.

** Indicates the standard deviation is significantly different between groups by 1% or less, using untransformed data.

*** d.l. = Distal latency for nerve.

**** v. = Velocity for nerve conduction.

Table 12. Chi square test, comparing incidence of symptoms in exposed and control communities.

Variable	Chi Square Sig. Level	Control Percent	& No.	Exposed Percent	& No.
Abnormal Dermatology	N.S. (P=.22)	2.9%	3	6.3%	9
Palmer/Plantar Keratosis	N.S. (P=.76)	1.0%	1	1.4%	2
Diffuse Palmer/Plantar Hyperkeratosis	N.S. (P=.46)	1.9%	2	3.5%	5
Any type Cancer	N.S. (P=.61)	4.8%	5	3.5%	5
Numbness & Tingling	N.S. (P=.80)	11.4%	12	10.4%	15
Touch & Temperature Sense	N.S. (P=.13)	1.0%	1	4.2%	6
Reduced Energy	N.S. (P=.70)	10.5%	11	9.0%	13
Nerve Conduction Slowing	N.S. (P=.51)	11.9%	8	15.7%	13
Anemia	N.S. (P=.47)	5.0%	5	7.3%	10
Sex = Female	N.S. (P=.82)	55.2%	58	53.8%	78

4. Discussion and Conclusions

The present drinking water standards, promulgated in the National Interim Primary Drinking Water Regulations, has a "Maximum Contaminant Level" of 0.05 mg/l for arsenic. This standard is based on the assumption that, at an average water intake of two liters per day, arsenic intake from water would not exceed 100 micro-grams per day (36.5 mg/year).

In this study the exposed population consumed an average of more than 150 mg of arsenic from well water per year. This indicates the amount of arsenic consumed from water by the exposed population was four times (150/36.5 = 4.1) the maximum allowed by the current standard.

The study showed a clear relationship between the amount of arsenic consumed in drinking water and the amount of arsenic in scalp hair (n=185, r=.47, P= <.0001) and urine (n=234, r=.70, P= <.0001). This relationship was expected since the ingestion of an element obviously will result in its excretion. Arsenic levels found in hair and urine samples were used as tools to evaluate arsenic exposure.

In the case of each health indicator studied, the exposed communities showed a slightly higher percentage of abnormalities. Dermatological signs were found in 6 percent of the exposed participants compared to 3 percent of the controls. Nerve conduction abnormalities occured in 16 percent of the exposed compared to 12 percent of the controls. Anemia was found in 7 percent of the exposed and 5 per cent of the controls. However as shown in table 12, these differences were not statistically significant. Furthermore there were factors that tended to discredit the differences.

The few dermatological signs observed were scattered singly among individuals of the whole study population, rather than occurring together on individuals with higher arsenic exposure. This implied that the signs may have been incidental and not related to arsenic. A hint of a familial tendancy toward anemia was found in one exposed community. In general, typical neurological signs and symptoms of arsenic intoxication were not present in any of the participants.

The hypothesis that arsenic levels found in Hinckley and Deseret drinking water would result in signs and symptoms of chronic arsenic poisoning, could not be confirmed by this study. Residents of Hinckley and Deseret appeared to be as healthy as Delta residents. When community of residence was ignored and health indicators were compared with measures of arsenic exposure, the participants with higher arsenic exposure did not show evidence of health problems any more than did participants with lower arsenic exposure.

This research was supported by a grant from the U.S. Environmental Protection Agency. EPA Project Officer was D. G. Greathouse, Health Effects Research Laboratory, Cincinnati, Ohio.

5. REFERENCES

Brown, M. and J. Benedetti. 1979. Two-Way Frequency Tables - Measures of Association. In: BMDP Biomedical Computer Programs P-Series 1979. Univ. Calif. Press. pp. 245-277.

Goodnight, J.H. 1979. TTEST Procedure. Pages 425-426. in SAS User's Guide. SAS Institute Inc. Raleigh N.C.

Heyman, A., J.B. Pfeiffer, Jr., R.W. Willett, and H.M. Taylor. 1956. Peripheral neuropathy caused by arsenical intoxication. New England J. of Med. 254(9):401-409.

Miyata, K., S. Kosho, and T. Nagai. 1970. Clinical observations on chronic arsenic poisoning. Sakai Iho 88:4 (not seen).

Mizuta, N., et al. 1956. An outbreak of acute arsenic poisoning caused by arsenic contaminated soy sauce; a clinical report of 200 cases. Japan. J. Int. Med. 45:867 (not seen).

Morton, W., G. Starr, D. Pohl, J. Stoner, S. Wagner, and P. Weswig. 1976. Skin cancer and water arsenic in Lane County, Oregon. Cancer 37:2523-2532.

National Academy of Science. 1977. Medical and biologic effects of environmental pollutants - Arsenic. National Academy of Sciences, Washington, D.C. 332 pp.

Sall, J.P. 1979. FREQ Procedure . Pages 215-220. in SAS User's Guide. SAS Institute Inc. Raleigh, N.C.

Southwick, J.W., A.E. Western, M.M. Beck, T. Whitley, R. Isaacs, J. Petajan and C.D. Hansen. 1980. Community Health Associated with Arsenic in Drinking Water in Millard County, Utah. U.S. Environmental Protection Agency Publication No. EPA-600/1-81-064. 63 pp. (NTIS PB 82-108374 =$8.00).

Tseng, W. P., M. H. Chu, J. M. Fong, C.S. Lin and S. Yeh. 1968. Prevalence of skin cancer in an endemic area of chronic arsenicism in Taiwan. J. National Cancer Inst. 40(3):453-463.

Vallee, B.L., D.D. Ulmer, and W.E.C. Wecker. 1960. Arsenic toxicology and biochemistry. A.M.A. Arch. of Ind. Health, 21:56/132-75/151.

Zaldivar, R. 1974. Arsenic contamination of drinking water and foodstuffs causing chronic poisoning. Beitr. Path. Bd. 151:384-400.

19

MORTALITY AMONG WORKERS EXPOSED TO ARSENIC AND OTHER SUBSTANCES IN A COPPER SMELTER

Philip E. Enterline and Gary M. Marsh
Department of Biostatistics
Graduate School of Public Health
University of Pittsburgh
Pittsburgh, PA 15261

A study of the mortality experience of 2802 men who worked a year or more during the period 1940-64 at a copper smelter where arsenic exposure occurred showed a 2 fold excess in respiratory cancer. Using a time weighted measure of exposure, a life table method for accumulating dose, and a 10 year lag period the excess ranged from 1.5 in the lowest exposure category to 2.5 in the highest. Neither duration of exposure nor time since first exposure contributed strongly to the respiratory cancer excess. The 2 fold excess in respiratory cancer deaths held for workers with relatively short exposures (< 10 years) and with a relatively short latent period (< 20 years) as well as for those with longer exposure and latent periods. This appeared to be because for workers terminated alive the respiratory cancer excess tended to disappear with time, and because workers in high exposure jobs tended to terminate more quickly than workers in low exposure jobs. A somewhat unorthodox analytic method showed a relationship between the respiratory cancer excess and exposure duration when exposure intensity was held constant. Here, arsenic exposure intensity was shown to make an independent contribution to the respiratory cancer excess.

1. Introduction

This is a report on the mortality experience of 2802 men who worked a year or more during the period 1940-64 at a copper smelter at Tacoma, Washington where exposures to arsenic and other substances occurred. It represents additional follow-up and editing of data reported in 1980 [1]. An earlier report has also been made on retirees from this smelter [2,3]. In that report each worker was classified by an index of exposure based on urinary arsenic levels that characterized men in the various departments in the smelter. This earlier study showed an overall three fold excess for respiratory cancer with the excess related to the arsenic exposure index--ranging from a relative risk of 1.1 for the lowest exposure group to 8.3 for the highest.

The copper smelter at Tacoma is one of the nation's largest, producing at times 10% of the refined copper in the United States. It started operations as a copper smelter in 1913 and for many years has employed approximately 1,000 people working in three shifts. The ore received has a high arsenic content and contains, in addition to copper and arsenic, other elements such as iron, sulfur, gold and silver. About 400,000 tons of ore are received annually. Mixed raw materials are ignited at the roasters to burn off some of the sulfur and to preheat the material prior to charging to a reverberatory furnace. Liquid metal is withdrawn from the reverberatory furnace and transferred to a converter where silica is added and air is blown through burning off the sulfur. Silicon combines with iron and forms slag. The melt (called blister copper) is electrically further purified.

The process by which arsenic is produced has been described by Pinto and McGill.[4] When the ore or concentrates are smelted arsenic goes off in the flue gases where along with other dusts it is caught by electrostatic precipitators or filters as a dust. These dusts are then roasted in Godfrey roasters and arsenic trioxide is driven off and collected in what are known as arsenic kitchens. These are long brick rooms in which the hot roaster gases bearing trioxide fume and vapor are circulated as they cool. Arsenic oxide settles to the floor as a coarse dust and also forms chrystalline deposits on the walls. It is removed from the kitchens and sold as a dust containing about 97% As_2O_3. The Tacoma copper smelter is currently the only facility in the United States commercially producing arsenic.

There are many points of potential emission of arsenic trioxide dust into the atmospheric environment. These occur at all points of transferring materials and when cleaning the arsenic kitchens. A considerable amount of dust is generated so that the plant itself is considered as an area source. In addition, arsenic dust is discharged through the company's 565 foot chimney.

2. Exposure levels

Levels of airborne arsenic trioxide have been measured at and around the arsenic plant at the smelter. Table 1 summarizes some reports of airborne arsenic concentrations for the period 1938-57.[5] Levels varied, but were very high particularly prior to 1951. All levels are very high in relation to the current PEL of .01 mg/m^3.

Men working in areas where arsenic concentrations were high generally wore respirators and protective clothing. In the early years respiratory protection was crude, often consisting only of a neckerchief tied around the face. In the 1940's a respirator was introduced which consisted of a hard rubber frame holding pieces of surgical woolsheeting over the nose and cheeks. This was tested and found to be 99% effective. For most workers, therefore, airborne arsenic is not a good measure of exposure. A more satisfactory measure is probably urinary arsenic excretion.[6]

Table 1. Reports of airborne arsenic trioxide concentrations, Tacoma smelter, 1938-57.

Year	Location	Concentration (mg/m^3)
1938	Godfrey roaster	6.6
	Charge bins	33.0
	Kitchens	2.3
	Ball mill	.8
	Flue (Godfrey roaster)	62.4
1947	Arsenic plant (9 samples)	.9-39.3
1948	Godfrey roaster	19.2
	Fire floor	14.0
		1.7
		9.2
	Dust from flue	259.0
1950	Arsenic plant (7 samples)	.8-41.4
	Godfrey #1 (3 samples)	1.5-17.9
	Godfrey #2	71.0
1951	Arsenic plant (8 samples)	1.4- 7.5
	Godfrey feed floor	1.2
1952	Arsenic plant (5 samples)	1.1- 7.7
1953	Arsenic plant (6 samples)	4.9-32.5
1955	Arsenic plant (3 samples)	.1- .9
1956	Arsenic kitchen (2 days)	.5- 2.5
1957	Arsenic kitchen (3 areas)	.1- .2

Urinary arsenic levels were determined for workers in the smelters starting in 1948. All departments were represented by at least one sample. Generally areas of the smelter where high airborne arsenic was likely to occur had the largest number of samples taken. The results of sampling during the period 1948-52 have already been described in an earlier publication.[4]

Table 2 shows mean urinary arsenic values by time period and department. These varied considerably but were generally highest in the Cottrell, Arsenic, Roaster and Boilerroom. These were in fact the high exposure areas selected by Pinto and Bennett for an earlier report.[7] Clearly urinary arsenic levels have been dropping, with levels during the period 1973-75 only about 50% of the levels observed in the period 1948-52. In the earlier report on retirees, the data for 1973 were used to characterize exposure by department, but it was noted that for workers in that study actual exposure must have been much higher since most of the retirees studied had been employed prior to 1950.

For the purpose of this study estimates of historic exposure have been made. These are simple linear interpolations and extrapolations of data shown in Table 2, except that for the period prior to 1950 urinary arsenic values for 1948-52 are assumed to apply. Using 1948-52 data to estimate earlier periods is probably an underestimate in light of data shown in Table 1. On the other hand linear extrapolations back in time lead to unrealistic values.

Table 2. Average urinary arsenic excretion by department, Tacoma smelter 1948-75.

	μg/ As/l Urine			
Department	1948-52	1973	1974	1975
Cottrell	553	526	524	470
Arsenic	804	516	777	522
Roaster-furnace	556	414	383	320
Boilerroom-waste heat-powerhouse	787	409	302	275
Janitor	176	289	35	181
Repair-handygang-maintenance	328	288	188	234
Steel-riggers-welder	624	272	458	293
Reverberatory-mudmill-slag shot	346	269	232	180
Carpenter-paint shop	219	255	167	168
Yard-engine-crane	506	226	98	273
Pipe	1258	218	267	144
Slimes-nickel plant	--	--	--	40
Acid	142	180	160	92
Electric	354	171	164	157
Lead burners	2144	166	151	59
Converter-flues-blast	161	160	196	152
Anode	103	98	82	65
Refinery-tank house-metals-cathode	103	98	114	75
Office	108	88	80	117
Warehouse-truck driver	80	82	115	134
Watchman	135	77	82	79
Sample-bucking room	215	81	107	79
Mason	270	260	--	216
Crushing plant	--	222	--	300
Laundry	115	201	--	334
SO_2	--	--	--	109
Machine shop-blacksmith-tool	120	102	--	98
Fire	--	140	--	138
Mobil equipment-vacuum-switchman-fork lift operator	--	119	--	169
Martin mill-mill operator	--	115	--	111
Track	--	112	--	121
Mobil equipment repair-construction-mechanical	--	101	--	87
Refined casting-casting house	67	58	54	--
Average of 22 departments where samples were taken in all four periods	447	231	220	191

A special study has been reported on a comparison of airborne concentrations of arsenic with urinary arsenic levels for 24 workers wearing personal air samples for 5 successive days.[6] None of these workers wore a respirator and all were asked to refrain from eating fish (an important source of arsenic) for 2 days preceeding and during the study. Data from this study indicate that airborne arsenic levels expressed as $\mu g/m^3$ can be estimated as roughly 1/3 urinary arsenic expressed as $\mu g/l$.

3. Methods

All males who worked a year or more during the period 1940-64 were identified from personnel records available at the Tacoma smelter. The completeness of these records was verified by matching names and social security numbers on social security tax returns filed with the U.S. Bureau of Internal Revenue by the plant owner with names and social security numbers of workers identified from personnel records. This verification method has been described elsewhere.[8] From this verification process it is believed that no workers were missed.

Arsenic exposure was estimated for each man for each departmental assignment on the basis of the estimated average urinary arsenic level for men in that department. This value was calculated for each year of employment. A total exposure estimate was made by summing values across all jobs and all years of employment.

Deaths were identified by a check against company files, social security files, and by clearing with drivers license bureaus in various states, the Veterans Administration, the U.S. Post Office, and by personal contact. Results are summarized in table 3. As of December 31, 1976, 1061 deaths are known to have occurred. One thousand six hundred ninety of the cohort were verified alive and the vital status of 51 was unknown. For the 1061 deaths, death certificates were obtained for all but 4.4%.

Table 3. Follow-up status as of 1976 for workers employed a year or more 1940-64, Tacoma smelter.

	Number	Percent
Total	2802	100.0
Known status	2751	98.2
Alive	1690	
Dead	1061	(100.0)
Death certificate	1014	(95.6)
No death certificate	47	(4.4)
Status unknown	51	1.8

4. Description of cohort

Table 4 provides some details on this cohort of 2802 workers. About 40% were hired at ages under 25 with a mean age at hire of 29.4. Since a year or more work experience in the period 1940-64 was required for eligibility, follow-up actually started on January 1, 1941. Many workers had considerable work experience prior to 1941, as table 4 shows. This is reflected as a mean age at entry into follow-up of 48.5 for workers hired prior to 1930 and 33.4 for workers hired during the period 1930-39.

Table 4. Year of hire by age at hire, Tacoma smelter workers.

Age at Hire	Year of Hire				
	Before 1930	1930-39	1940-49	1950-63	Total
Under 20	81	65	142	80	368
20-24	87	210	316	138	751
25-29	122	136	208	86	522
30-34	90	114	145	44	393
35-39	76	87	126	46	335
40-44	49	54	105	50	258
45-49	9	17	50	18	94
50-54	2	4	17	7	30
55-59	0	0	13	3	16
60-64	0	0	5	0	5
65-69	0	0	0	0	0
Total	516	687	1127	472	2802
Mean age at hire	29.4	29.0	29.8	28.8	29.4
Mean age on entry to follow-up	48.5	33.4	31.3	30.2	34.8

Table 4 shows that hiring practices with regard to age were fairly uniform over the years, with not much difference between workers hired in the earlier years and those hired later. A large number of workers were hired during the period 1940-49 reflecting in part increased effort associated with World War II.

5. Mortality experience

Table 5 shows the mortality experience of the cohort through 1976. Here deaths for which death certificates could not be located are included in the total, and under unknown causes of death. The 51 workers who could not be traced are counted up to the time of last observation (usually termination). Nearly all workers employed during 1940-64 were white. For table 5, SMR's were calculated using the mortality experience of white males for the entire U.S. to estimate expected deaths.

Overall mortality is slightly elevated. Cancer deaths are significantly elevated ($P<.01$) due mainly to cancer of the respiratory system. Deaths for other causes of death are unremarkable. Generally SMR's are lower than those reported for retired workers from the same smelter.[2,3]

Table 6 shows deaths and SMR's for cancer by type of cancer. Here SMR's are calculated both on the basis of U.S. white male mortality rates and white male mortality rates for the State of

Table 5. Observed deaths and SMR's for 2802 smelter workers who worked a year or more 1940-64 followed through 1976, by cause of death.

Cause of death (7th revision code)	Observed Deaths	SMR
All causes of death	1061	103.2
Tuberculosis (001-019)	4	27.6**
Malignant neoplasms (140-205)	231	123.6**
Buccal Cavity and Pharynx (140-148)	7	110.7
Digestive Organs & Peritoneum (150-159)	65	108.9
Esophagus (150)	3	66.2
Stomach (151)	17	122.1
Large Intestine(153)	21	120.4
Rectum (154)	9	122.4
Biliary Passages and Liver (155-156)	3	64.1
Pancreas (157)	11	106.0
All other digestive ograns (residual)	1	71.6
Respiratory System (160-164)	104	189.4**
Larynx (161)	2	67.7
Bronchus, Trachea, and Lung (162-163)	100	194.9**
All other respiratory system (residual)	2	305.0
Prostate (177)	11	79.0
Testis and other genital (178-179)	1	92.6
Kidney (180)	6	133.3
Bladder and other urinary organs (181)	4	63.0
Malignant melanoma of skin (190)	0	--
Eye (192)	1	492.7
Central Nervous System (193)	3	59.8
Thyroid Gland (194)	0	--
Bone (196)	2	175.0
Lymphatic & Haematopoietic (200-205)	17	93.8
Lymphosarcoma and Reticulosarcoma (200)	4	93.2
Hodgkins Disease (201)	2	83.9
Leukemia and Aleukemia (204)	6	78.7
Other lymphopoietic tissue (202, 203, 205)	5	130.4
Other malignant neoplasms (residual)	10	82.3
Benign neoplasms (210-239)	2	78.6
Diabetes mellitus (260)	12	84.8
Stroke (330-334)	91	111.4
Heart Disease (400-443)	412	92.5
Hypertension without Heart Disease (444-447)	1	18.8
Nonmalignant Respiratory Disease (470-527)	60	108.6
Influenza and Pneumonia (480-493)	24	92.9
All Other Respiratory Disease (Residual)	36	122.4
Ulcer of Stomach and Duodenum (540-541)	7	75.7
Cirrhosis of Liver (581)	22	101.9
Chronic Nephritis (592)	6	87.5
External Causes of Death (800-998)	81	94.2
Accidents (800-962)	61	100.6
Suicides (963, 970-979)	17	84.8
Other External Causes (Residual)	3	56.2
Other Causes of Death (Residual)	85	86.1
Unknown causes	47	--

*P< .05, **P< .01

Washington. For the latter, data are available only for the period 1950-76. For years prior to 1950 the 1950 rates are assumed to apply. This may slightly understate SMR's for the period 1941-49.

Table 6. Cancer mortality 1941-76 of 2802 workers who worked a year or more 1940-64, by type of cancer, Tacoma smelter workers.

Type of cancer	Observed Deaths	SMR†	SMR°
All Cancer	231	123.6**	129.9**
Buccal Cavity and Pharynx	7	110.7	142.9
Digestive Organs	65	108.9	117.4
Respiratory System	104	189.4**	198.1**
Prostate	11	79.0	75.8
Kidney	6	133.0	131.6
Lymphatic and Haematopoietic Tissue	17	93.8	93.8
All Other	21	61.9	63.6

† U.S. Expected
o Washington State Expected
**P < .01

There are some differences in SMR's using these two bases for calculating expected deaths due to lower mortality rates in the State of Washington for most types of cancer. In the tables which follow Washington State cancer death rates are used to calculate expected deaths and SMR's since it is felt these are probably most appropriate for this cohort. Pierce County, the county in which the smelter is located, has cancer mortality rates very much like those for the entire State of Washington.

Table 7 shows cancer deaths and SMR's in relation to period of hire. Given the working conditions and high airborne arsenic levels that undoubtedly prevailed many years ago and the long latent period represented, the 516 workers hired before 1930 but who worked a year or more after 1939 (table 4) are of particular interest. Four hundred and five or 78% of these workers died during the period 1941-76. Cancer was significantly elevated due mainly to cancer of the large intestine and of the respiratory system. Only cancer of the respiratory system was consistently high for each of the 4 cohorts shown, however, although for workers hired during the period 1930-39 the respiratory cancer excess was not very great. There is no clear evidence of a trend for respiratory cancer SMR's across successive cohorts.

6. Dose-response relationships

Table 8 shows the relationship between a time weighted estimate of arsenic exposure lagged 0 and 10 years and respiratory cancer. In this table, and in keeping with the generation life table method used here, workers are counted in their highest exposure category and in all lower exposure categories through which they progressed and in which they were followed for deaths.

Table 7. Year of hire by number of cancer deaths 1941-76, showing SMR's, Tacoma smelter workers.

	Year of Hire							
	Before 1930		1930-39		1940-49		1950-63	
Type of Cancer	Observed Deaths	SMR	Observed Deaths	SMR	Observed Deaths	SMR	Observed Deaths	SMR
All Cancer	90	140.9**	65	126.1	61	115.3	15	158.1
Buccal Cavity and Pharynx	2	125.4	1	66.0	3	197.4	1	371.6
Digestive Organs	30	132.6	22	143.8	13	86.7	0	--
Stomach	9	143.6	3	93.0	5	166.6	0	--
Large Intestine	12	208.5*	5	120.3	4	96.5	0	--
Rectum	3	107.7	5	286.3	1	59.4	0	--
Pancreas	3	74.6	6	186.4	2	61.5	0	--
Respiratory System	38	246.7**	22	132.3	34	196.6**	10	313.4**
Lung	37	256.3**	21	133.5	32	195.3**	10	330.2**
Prostate	5	85.1	4	118.0	2	65.3	0	--
Kidney	3	201.2	2	146.1	0	--	1	385.8
Lymphatic and Haematopoietic Tissue	4	66.7	8	149.5	3	51.3	2	167.8

*P<.05
**P<.01

Table 8. Respiratory cancer deaths and SMR's by cumulative arsenic exposure lagged 0 and 10 years, Tacoma smelter workers

Cumulative Exposure (μg/As/l) (urine-years)	Lag			
	0 Lag		10 Year Lag	
	Observed Deaths	SMR	Observed Deaths	SMR
<500 (302)	8	202.0	10	155.4
500-1500 (866)	18	158.4	22	176.6*
1500-3000 (2173)	21	203.2**	26	226.4**
3000-6000 (4543)	26	184.1**	22	177.6*
7000+ (13457)	31	243.4**	24	246.2**

* P<.05
**P<.01
()Mean of class interval

Table 8 shows some relationship between arsenic exposure and respiratory cancer, best perhaps where exposures are lagged 10 years. Whatever lag period is used, however, the relationship is much weaker than that previously reported for retired workers from this same smelter.[2,3] Data in table 8 are not strictly comparable with data from the earlier study for several reasons. For one thing, exposures were based on historic rather than 1973 urinary arsenic levels. Also, in the earlier report follow-up started after exposure stopped (at retirement) whereas in table 8 follow-up started at varying points in the work experience of the workers. For workers hired before 1930, for example, table 4 suggests that follow-up started after about 10 years of employment. On the other hand for workers hired after 1939 table 4 suggests that follow-up started after only a year of employment. This is not the method prescribed by the generation life table method, since it means that in compiling table 8 some workers in the highest exposure category did not move through all lower exposure categories shown in table 8, while others did. Thus, all workers were not given "credit" for surviving exposure duration categories below the one in which they ended.

It would, of course, be impossible to move workers through all exposure duration categories where workers were hired before the follow-up period started. In the earlier study of retirees this rule was not followed at all since here exposure categories were fixed when follow-up began. That is workers appeared only in a single exposure category, rather than several as in table 8.

Table 9 explores the effects of ignoring some exposure years for workers hired before 1940 on the dose-response relationship. This shows SMR's for respiratory cancer for each of the four cohorts described in table 4. There is no consistent difference in dose-response relationships for cohorts for whom all person years are counted (hired after 1940) and cohorts for whom some person years are missing (hired before 1941). The sharpest positive dose-response relationship appears for men hired during the years 1930-39. Curiously, as table 7 showed, this is the cohort with the smallest respiratory cancer excess. If arsenic trioxide exposure causes respiratory cancer, then the 1930-39 cohort must have been the least exposed--or the least likely to get respiratory cancer for some other reason. From table 9 it does not appear that dropping person years for workers hired before 1940 is the reason for the overall lack of a dose-response relationship in table 8.

Table 9. Respiratory cancer deaths and SMR's by exposure level and year of hire, Tacoma smelter workers.

Cumulative Exposure (μg As/l) (urine-years)	Year of hire: Before 1930 Observed Deaths	SMR	1930-39 Observed Deaths	SMR	1940-49 Observed Deaths	SMR	1950 & later Observed Deaths	SMR
Under 1500	1	279.2	4	78.0	15	182.8*	6	372.7*
1500-7000	18	248.9	11	127.6	15	210.5*	3	203.8
7000+	19	243.8	7	243.6	4	204.1	1	92.5

*P<.01

Table 10 shows the respiratory cancer mortality experience of 582 retired workers at ages 65 and over drawn from the entire cohort of 2802 workers in this study. This includes a few workers and deaths not in the earlier study since it covers retirements through 1976 instead of 1972 and deaths during the period 1941-76 instead of 1949-73. Also a few workers eligible for the earlier study but missed are included. Excluded are men who retired before 1941. Exposure was calculated using historic rather than 1973 urinary arsenic levels used in the earlier study. Exposure intervals are those used in the earlier report.

Table 10. Respiratory cancer deaths and SMR's by exposure level, 582 retired workers at ages 65 and over, Tacoma smelter.

Cumulative Exposure (μg As/l urine-years)	Number of Deaths	SMR
< 2000	1	142.3
2-3000	3	181.8
3-6000	6	136.8
6-9000	5	305.3
9-12000	6	345.5*
12-15000	6	393.2**
15000+	7	304.9*

* P< .05
**P< .01

Table 10 shows a somewhat stronger dose-response relationship than table 7, with almost a 4 fold excess in respiratory cancer in the next to the highest dose category. Of course, exposure categories differ from those in table 8 since table 10 deals with older workers with larger time weighted exposures, and this could make a difference. Findings are similar to those in the earlier study, however. As in the earlier study, exposure categories in table 10 were fixed at the time follow-up started (at time of retirement).

Table 11 shows respiratory cancer deaths and SMR's by duration of exposure and time since first exposure. Apparently one reason for the weak dose-response relationship in table 8 is that neither duration of exposure nor time since first exposure make strong contributions to excess respiratory cancer among these smelter workers. The reason for the weak contribution of time since first exposure and duration of exposure appears to be, in part, because SMR's are high shortly after temination of employment but decline thereafter. Note that for workers with less than 10 years of employment the SMR is highest 10-19 years after hire date; 10-19 years of employment, 20-29 years after hire date, etc. Of particular interest is the fact that SMR's are high in the decade immediately following termination of exposure--265.4, 278.0, and 265.3 and these are not related to exposure duration.

From table 11 it appears that if arsenic is indeed a carcinogen it has two notable characteristics: a fairly short latent period in some cases, and a disappearance of effects after termination of exposure. Only for persons exposed less than 10 years is there a long observation period following termination of exposure. Here the SMR approaches 100 after 30 years (137.9). If we consider that for persons exposed less than 10 years an effective dose might have been achieved on average in about 5 years, and that for 10-19 years latency deaths occurred on average in about 15 years then for these cases the latent period for arsenic induced cancers must have been around 10 years.

In an experimental study follow-up would ordinarily start only when dose had been completed. This, of course, eliminates the death free years needed to qualify for a particular dose category--years that contribute to expected deaths but not to observed deaths. This suggests that perhaps another way to look at data from this copper smelter is to divide workers into mutually exclusive groupings based either on attained dose at some point in time or on total exposure by the end of the study. The first method seemed useful in a study of nickel exposed workers where exposure in the first 20 years of employment was related to subsequent respiratory cancer mortality.[9] This ignores deaths in the first 20 years, however, and is probably not appropriate here in view of the considerable respiratory cancer excess 10 years or so after initial exposure. The second method is similar to studying retirees except that short exposures are often due to voluntary quits rather than to involuntary retirements.

Table 11. Respiratory cancer deaths and SMR's by duration of exposure and latency, Tacoma smelter workers.

	Latency (Years)									
	<10		10-19		20-29		30+		Total	
Duration (years)	Observed Deaths	SMR	Observed Deaths	SMR	Observed Deaths	SMR	Observed Deaths	SMR	Observed Deaths	SMR
< 10	1	55.6	10	265.4*	17	210.1**	12	137.9	40	178.9**
10-19	-	-	6	156.4	8	278.0*	4	137.9	18	187.2*
20-29	-	-	-	-	13	197.0*	13	265.3**	26**	226.1*
30+	-	-	-	-	-	-	20	221.3**	20	221.3**
Total	1	55.6	16	210.4*	38	216.3**	49	191.8**	104	198.1**

*P < .05
**P < .01

Table 12 presents data on the relationship of duration of exposure to respiratory cancer according to the second method. Here workers are followed only from their point of termination or retirement--in the way an experiment might have been conducted. Workers still employed on December 31, 1976, have been excluded since only terminations are included here. Also shown is the estimated urinary arsenic concentration during employment. Unlike tables 8 and 9, but like table 10, in this table workers contribute person years to only a single exposure category. Also, unlike tables 8 and 9, follow-up starts only when the exposure criteria is fulfilled. This is, therefore, the mortality experience from respiratory cancer for a group of workers after they leave employment and includes workers who terminated because of respiratory cancer.

Table 12. Respiratory cancer deaths and SMR's by duration and intensity of exposure, Tacoma smelter workers.

Duration of Exposure (Years)	Intensity Estimate: Low†			Intensity Estimate: High††		
	No. at Risk	Observed Deaths	SMR	No. at Risk	Observed Deaths	SMR
<10	687	15	169.9	824	25	203.9**
10-19	149	7	268.2*	168	11	321.6**
20-29	225	10	278.6**	171	16	577.6**
30+	179	9	302.0**	165	11	347.0**

* P<.05
**P<.01

† low - < 290 μg As/l urine (mean 163)

††high - ≥ 290 μg As/l urine (mean 477)

Of course, workers who terminated employment because of respiratory cancer may introduce a bias here since for these the criteria for being in their exposure category is to have respiratory cancer. That is, death determines the dose rather than dose the death. There is no way of knowing how many of these cases are present in table 12 and how they bias the dose-response relationship. For 21 workers who died of respiratory cancer the termination date and death date were the same. This is clearly a problem in any study that uses duration of employment as a surrogate for dose, but is most serious when follow-up and dose accumulation proceed concurrently.

Table 12 shows that both duration of exposure and intensity of exposure contribute to respiratory cancer mortality. It also shows that perhaps one of the reasons for the weak relationship between duration of exposure and respiratory cancer SMR's in table 11 is that workers in high exposure jobs tended to leave employment more quickly that workers in low exposure jobs. Note that of 1511 men in the cohort who worked less than 10 years 824 or 55% were in high exposure jobs whereas of 344 workers employed 30 years or more only 165 or 48% were in high exposure jobs. This phenomenon has been described previously by Fox and Collier in discussing the healthy worker effect.[10] Where workers enter employment in high exposure areas of a plant or factory, and where this exposure tends to discourage long employment, then duration of employment may be a poor surrogate for amount of exposure.

7. Interactions with SO_2

An important question is whether arsenic is the agent responsible for excess respiratory cancer in copper smelters or whether there is some interaction with SO_2, an important contaminant in all smelters where sulfide ore is used.[11] According to plant industrial hygiene measurements there were two departments at the smelter with high airborne arsenic levels but which differed in SO_2 levels. At the Cottrell arsenic exposures were considered very high (> 500 $\mu g/m^3$) according to measures taken during the years 1938-47 while SO_2 exposures were considered low to moderate (5-20 PPM). On the other hand, in the arsenic department arsenic exposures were considered very high according to measurements made 1938-68 (> 500 $\mu g/m^3$) but SO_2 exposures were considered nil. Table 13 compares the mortality experience of men who worked in the Cottrell department with those who worked in the arsenic department but never in the Cottrell department. Respiratory cancer SMR's were quite similar suggesting that SO_2 exposure did not play an important role in the respiratory cancer excess at this copper smelter. For this cohort the average exposure duration of respiratory cancer deaths from these two departments was relatively short (8.4 years in the Cottrell and 3.1 years in the Arsenic department) suggesting the important role of arsenic exposure in respiratory cancer occurring among men in these two departments as shown in table 13. Men who never worked in these two departments had a much smaller cancer excess. There were SO_2 exposures in some of these other departments, however, particularly in the roaster-furnace department.

Table 13. Respiratory cancer deaths and SMR's for selected departments, Tacoma smelter workers.

Department	Observed Deaths	Expected Deaths	SMR
Cottrell (ever)	11	2.97	370.4**
Arsenic (never Cottrell)	27	8.07	334.6**
All other	66	43.87	150.4**
Total	104	54.91	198.1**

*P<.05, **P<.01

8. Discussion

This study confirms previous reports and findings by others which demonstrate that exposure to arsenic is related to increases in respiratory cancer in man.[11-17] It does, however, add some new information regarding the relationship between arsenic and cancer. For one thing the carcinogenic response apparently can occur rather quickly--in this study apparently in about 10 years. For another, earlier observations that the effects of arsenic exposure tend to disappear with time are supported with these additional data.[2,3] Workers who worked 10 years or less showed large excesses in respiratory cancer in the decade following termination of employment and these excesses all but disappeared in the next 20 years. These observations strongly suggest that in this situation arsenic was acting as a cancer promoter rather than as an initiator. They also explain the lack of any relationship between latency and respiratory cancer.

Apparently the effects of arsenic exposure do not fit the model used here to measure dose. That is, for arsenic effective dose is not simply a multiplication of time times dose rate or intensity. Short exposures seem to have a disproportionately greater effect than longer exposures. Whether this is due to the epidemiologic nature of observations made here and the fact that some workers are simply more susceptible than others rather than something about arsenic *per se* could only be determined with an experimental study. It is also possible that the effects of early exposures fade while new exposure is added, so that it is not historic but recent exposure that is most important in any particular case, and the cumulative exposure to arsenic as a measure of dose has no overall meaning.

Perhaps the clearest association between arsenic exposure and respiratory cancer in this study was observed when a somewhat unconventional method of calculating dose was employed--a method that more closely approaches that used in experimental studies than in epidemiological studies. This method took advantage of the fact that natural exposure groupings are formed by employees terminating after varying employment durations. Grouping in this way also permits a meaningful subdivision by dose rate or intensity in a study where employment and follow-up proceed concurrently, and this seems to be a useful piece of information.

Cancer other than respiratory cancer also appeared to be slightly in excess in this study and this contributed to an excess in deaths for all causes combined. None of these excesses were statistically significant however. Considering that arsenic is excreted in the urine, it is of particular interest that urinary tract cancer did not appear to be unusually high. Numbers of deaths for most individual cancers were too small to have much meaning. Data lend no support, however, for an excess in lymphoma and leukemia, nonmalignant respiratory disease, cardiovascular disease, or liver cancer found in other studies.[13,14,15,17,18]

If arsenic is indeed a human carcinogen it behaves differently in humans than most other carcinogens have in other epidemiologic studies. Some of the usual analytic methods did not work and the dictum of a relationship between cancer and exposure duration and between cancer and time since first exposure did not hold.

This investigation was supported by Grant Number R01 CA 20540 awarded by the National Cancer Institute, DHEW. The authors gratefully acknowledge the assistance provided by Kenneth W. Nelson, Vice President, Environmental Affairs, ASARCO, Sherman Pinto, Medical Director at ASARCO's Tacoma copper smelter, and personnel at the Tacoma copper smelter. They also gratefully acknowledge the assistance of Frank D'Amico and the late Kenneth Wassam, research assistants at the University of Pittsburgh.

References

[1] Enterline, P.E. and Marsh, G.M., Mortality studies of smelter workers, Am. J. Ind. Med., 1: 251-259 (1980).

[2] Pinto, S.S., Enterline, P.E., Henderson, V. and Varner, M.O., Mortality experience in relation to a measured arsenic trioxide exposure, Env. Hlth. Persp., 19: 127-130 (1977).

[3] Pinto, S.S., Henderson, V., and Enterline, P.E., Mortality experience of arsenic-exposed workers, Arch. Env. Hlth, 325-331 (Nov.-Dec. 1978).

[4] Pinto, S.S. and McGill, C.M., Arsenic trioxide exposure in industry, Ind. Med. and Surgery, 22:7: 281-287 (July 1953).

[5] Memorandum from L.D. White to K.W. Nelson dated February 11, 1980.

[6] Pinto, S.S., Varner, M.O. and Nelson, K.W., Arsenic trioxide exposure and excretion in industry, J. Occup. Med. 18: 633-635 (1976).

[7] Pinto, S.S. and Bennett, B.M., Effect of arsenic trioxide exposure on mortality, Arch. Env. Hlth., 7: 583-591 (Nov. 1963).

[8] Marsh, G.M. and Enterline, P.E., A method for verifying the completeness of cohorts used in occupational mortality studies, J. Occup. Med., 21:10: 655-670 (1979).

[9] Enterline, P.E. and Marsh, G.M., Mortality among workers in a nickel refinery and alloy manufacturing plant in West Virginia, USA. JNCI, to be published.

[10] Fox, A.J. and Collier, P.F., Low mortality rates in industrial cohort studies due to selection for work and survival in the industry, Brit. J. prev. soc. Med. 30: 225-230 (1976).

[11] Lee, A.M. and Fraumeni, J.F., Arsenic and respiratory cancer in man: an occupational study, J. Nat. Cancer Inst., 42: 1045-1052 (1969).

[12] Feldstein, A.L., Arsenic and respiratory cancer in man: follow-up of an occupational study preliminary report. Presented at the Arsenic Symposium, November 3-6, 1981.

[13] Lubin, J.H., Pottern, L.M., Blot, W.J., Tokudome, S., Stone, B.J. and Fraumeni, J.F., Respiratory cancer among copper smelter workers: recent mortality statistics, J. Occup. Med., 7: 779-784 (1981).

[14] Axelson, O., Dahlgren, E., Jansson, C.D. and Rehnlund, S.O., Arsenic exposure and mortality: a case-referent study from a Swedish copper smelter, Br. J. Ind. Med., 35: 8-15 (1978).

[15] Ott, M.G., Holder, B.B. and Gordon, H.L., Respiratory cancer and occupational exposure to arsenicals, Arch. Env. Hlth., 29, (Nov. 1974)

[16] Rencher, A.C., Carter, M.W. and McKee, D.W., A retrospective epidemiological study of mortality at a large copper smelter, J. Occup. Med., 19: 754-758 (1977).

[17] Mabuchi, K., Lilienfeld, A.M. and Snell, L.A., Lung cancer among pesticide workers exposed to inorganic arsenicals, Arch. Env. Hlth., 34: 312-320 (1979).

[18] Roth, F., The sequelae of chronic arsenic poisoning in Moselle vintners, Ger. Med. Monthly 2: 211-217 (1957).

20

ARSENIC AND RESPIRATORY CANCER IN MAN: FOLLOW-UP OF AN OCCUPATIONAL STUDY

Anna Lee-Feldstein
Biostatistics Department
The University of Michigan
School of Public Health
Ann Arbor, MI 48109

The earlier report by Lee[1] and Fraumeni on the mortality experience of 8047 white male copper smelter workers exposed to arsenic trioxide in western United States followed the mortality of the group during the period 1938-63. The present study is a follow-up of the same group with mortality observed during 1938-77. The group of men were found to have an excess in total mortality when compared with white males in the same region, due largely to respiratory cancer, diseases of the heart, emphysema, and vascular lesions of the central nervous system. Three hundred and two men in this group had died of respiratory cancer by late 1977, nearly three times the expected number. The excess respiratory cancer mortality increased with length of employment and was positively related to degree of arsenic exposure, with more than five times as many deaths as expected found in the heavy arsenic-exposure group.

1. Introduction

In a study supported by the National Cancer Institute, Lee and Fraumeni [1] reported in 1969 on the mortality experience during 1938-63 of 8047 white male workers exposed to arsenic trioxide in a copper smelter in western United States. A comparison was made with the mortality of white males in States in the same area, and a threefold increase in respiratory cancer deaths was observed among the smelter employees. The excess in respiratory cancer mortality was as high as eight-fold among men working more than 15 years and heavily exposed to arsenic; further, it showed a gradient in proportion to level of arsenic exposure. Findings of the study were in support of the hypothesis that inhaled arsenic is a respiratory carcinogen in man, although the influence of sulfur dioxide and other chemicals present in the smelter could not be discounted.

Since the Lee and Fraumeni paper was published, several studies of copper smelter workers have reported excesses in respiratory cancer [2-7]. The most recent of these, by Lubin *et al.* [7], surveys the mortality experience of the Lee-Fraumeni study group *for the period 1964-77 only* and relates this mortality to smelter employment prior to 1964. The present study was designed as a broader follow-up of the Lee-Fraumeni

[1]Present author

study; it includes 1) mortality information from both the Lee-Fraumeni study and the Lubin study, 2) smelter employment experience prior to 1964 (from Lee and Fraumeni) and 3) recently collected smelter employment experience for the period 1964-77. The total period of mortality follow-up for the present study is 40 years (1938-77) and this mortality is related to the complete smelter employment experience of the study group through September 30, 1977.

2. Methods

Included in this study were 8045 white males in the Lee-Fraumeni study. These men were employed at the smelter for 12 or more months prior to December 31, 1956, and their mortality experience was observed from January 1, 1938, to September 30, 1977.

Smelter employment information for the period January 1, 1964, through September 30, 1977, was obtained for this study from company records. Mortality follow-up information was obtained from company records, Social Security Administration records, and death registers of various State health departments. Table 1 summarizes the status of the study group at the cut-off date of the present study. Compared with the Lee-Fraumeni study, nearly twice as many men were deceased by September 30, 1977.

Table 1 Status of study group, September 30, 1977

Known to be living		3679
Employed by smelters	442	
Other[a]	3237	
Known to be deceased		3550
Not known to be living or deceased		816
Total study group		8045[b]

[a]Includes persons receiving benefits or making claims to Bureau of Old Age and Survivors Insurance after September 30, 1977.

[b]Two persons in the original study group of 8047 who were females have been deleted.

Death certificates and vital statistics tapes were obtained for the smelter workers known to have died during the additional observation period, 1964-77. For this period, underlying causes of death were classified according to the seventh revision of the International Lists of Diseases and Causes of Death [8], which is quite comparable to the sixth revision coding used by Lee and Fraumeni for deaths occurring before 1964.

The life table method [9] as programmed by R. Monson [10] was used to analyze mortality of the smelter workers over the entire 40 year period of observation. Individual workers were entered into the life table at the end of one year of employment or in 1938, if employed at least 12 months prior to that date. Persons lost to follow-up were withdrawn from the study when lost. For each calendar period in the analysis, individuals were assigned to one of five groups (called cohorts by Lee and Fraumeni) on the basis of their total years of employment (table 2). These length of employment groups were selected to clarify the effect on mortality of various periods of exposure to environmental agents in the smelter. It should be noted that groups 1 and 2 are defined somewhat differently than Lee and Fraumeni's cohorts 1 and 2.

Table 2 Description of length of employment groups, with number of smelter workers, numbers of deaths, person years at risk, and duration of smelter employment (based on total work experience through Sept. 30, 1977).

Length of employment group[a]	Number of persons[b]	Number of deaths	Number of person years of follow up[c]
1 (25 or more years)	1899	1169	27,053
2 (15 to 24 years)	1138	586	26,556
3 (10 to 14 years)	678	328	19,734
4 (5 to 9 years)	1082	433	30,854
5 (1 to 4 years)	3248	1006	88,279
TOTAL	8045	3522	192,476

[a]Employees in all cohorts were living on Jan. 1, 1938.

[b]Group assignment of each person here was based on his status at the termination of employment or on September 30, 1977 (whichever date was earlier).

[c]Represents cumulative follow-up experience over the study period, 1938-77, with a total of 67,569 person years of follow-up in the period 1964-77. Individuals were initially counted at risk upon completing 1 year of employment or on Jan. 1, 1938, if employed at least a full year before that date. In each calendar year of the study period, employees were counted in the group reflecting their cumulative work experience to date.

The study group was compared with the combined white male population in Idaho, Wyoming and Montana by applying age- and cause-specific death rates for 5-year age groups and for the calendar intervals (1938-42, 1943-47, etc.) for these states to the study population at risk, thus obtaining expected deaths for our study group for each cause. Expected deaths were then summed over age groups and calendar periods to obtain total expected deaths by cause. Observed and expected deaths were compared by the standardized mortality ratio (SMR = (observed/expected)X100).

To determine the effect of arsenic on mortality, the study group was categorized by exposure to various levels of arsenic. From measurements made in the smelter, each work area in the smelter had been rated on a scale from 1 to 10, indicating the relative amount of arsenic trioxide (AS_2O_3) in the atmosphere [11]. Arsenic exposure associated with smelter operation is documented elsewhere [12-14]. As in the Lee and Fraumeni study, jobs in the areas known as the arsenic kitchen, Cottrell, and arsenic roaster were rated as "heavy" arsenic exposure areas. "Medium" arsenic exposure was associated with four work areas known as the converter, reverberatory furnace, ore roaster and acid plant, and casting. All other job areas had "light" arsenic exposure. Measurements in the work areas may have varied over time, but it is reasonable to assume that the 3 broad categories denoting relative exposure to arsenic remained fixed. Since most men worked in several different areas, to be conservative we classified each individual into the one of three arsenic categories denoting his heaviest (maximum) exposure to arsenic during his entire smelter work experience.

3. Results

We observed 3,522 deaths in the entire study group, compared with 2,729 deaths expected ($p < 0.01$). Of the thirteen specific causes of death considered, tuberculosis, digestive and respiratory cancer, vascular lesions of the central nervous system, diseases of the heart, emphysema, and cirrhosis of the liver showed a significant excess of observed over expected (table 3). Similar results were observed in the Lee and Fraumeni study for tuberculosis, respiratory cancer, diseases of the heart and cirrhosis of the liver. When the causes of death shown in table 3 were analyzed further by length of employment group, respiratory cancer was found to be highly significantly in excess of expectation in each group. Further, respiratory cancer SMR's increased monotonically with length of employment, agreeing with findings in the Lee-Fraumeni study (table 4).

Smelter workers in each of the three arsenic-exposure groups had significant excesses in respiratory cancer mortality (table 5). A definite gradient was associated with degree of arsenic exposure, with the ratio of observed to expected mortality from respiratory cancer being approximately 5.1, 4.5 and 2.3 in heavy, medium, and light arsenic-exposure groups, respectively. This result is in accord with earlier results of Lee and Fraumeni (except that their reported excess for men with heavy exposure to arsenic was 7 times expected). With the exception of length of employment group 1, the excess in respiratory cancer mortality increases with increased degree of exposure to arsenic

Table 3 Observed and expected deaths due to selected causes, with standarized mortality ratios (SMR's) among smelter workers, 1938-77.

Cause of Death	List No.[a]	Number of deaths: Observed	Number of deaths: Expected	SMR[b]
Tuberculosis	001-019	53	27.93	190[c]
Respiratory	001-008	47	25.51	184[c]
Other	010-019	6	2.42	248
Malignant neoplasms	140-199	609	370.74	164[c]
Digestive	150-159	167	133.58	125[c]
Respiratory	160-164[d]	302	105.81	285[c]
Other	140-148, 165-170 177-181, 190-199	140	131.35	107
Vascular lesions of central nervous system	330-334	262	211.56	124[c]
Diseases of heart	400-443	1366	1056.55	129[c]
Influenza and pneumonia	480-483, 490-493	88	76.05	116
Emphysema (1963-77 only)	527	90	34.58	260[c]
Cirrhosis of liver	581	76	36.53	208[c]
Accidents	800-962	288	280.27	103
Motor vehicle	810-825, 830-835	106	115.55	92
Other	800-802, 840-862	182	164.72	110
Suicide and homicide	963-964, 970-979 980-985	83	86.08	98
All other causes	Residual	606[e]	548.50	
Total		3522	2728.79	129[c]

[a]Seventh revision of International Lists of Diseases & Causes of Death.

[b]SMR = (observed/expected) x 100.

[c]Significant at 1% level. [15]

[d]Among the 302 deaths from respiratory cancer, the site was lung and bronchus (162,163) in 289 cases, larynx (163) in 9, mediastinum (164) in 3 and (160) in 1.

[e]Includes 19 emphysema deaths occuring before 1963, when emphysema death rates are not available for individual states.

Table 4 Observed and expected deaths for respiratory cancer, with standardized mortality ratios (SMR's), by length of employment group, 1938-77.

Length of employment group	Respiratory cancer deaths		
	Observed	Expected	SMR
1 (25+ yrs)	127	31.14	408[a]
2 (15-24 yrs)	44	17.00	259[a]
3 (10-14 yrs)	25	10.77	232[a]
4 (5-9 yrs)	29	12.73	228[a]
5 (1-4 yrs)	77	34.16	225[a]

[a]Significant at 1% level.

Table 5 Observed and expected deaths from respiratory cancer, with standardized mortality ratios (SMR's), by degree of arsenic exposure, 1938-77.

Respiratory cancer mortality	Maximum exposure to arsenic (12 or more months)[a]		
	Heavy	Medium	Light
Observed	33	93	136
Expected	6.45	20.85	58.86
SMR	512[b]	446[b]	231[b]
(Number of men in (arsenic category[a]	451	1585	4448))

[a]The remaining 1562 men in the study worked less than 12 months in their category of maximum arsenic exposure and had an SMR of 204 (40 observed respiratory cancer deaths and 19.64 expected).

[b]Significant at the 1% level.

in the various length of employment groups. (Group 1 contains a large group of men first employed prior to 1925 and, for this group, men having medium exposure to arsenic had the greatest observed excess in respiratory cancer mortality. It is possible that jobs identified as "medium" arsenic areas involved intermittant heavy exposure to arsenic in the earlier days of the smelter operation.)

The latent period for respiratory cancer in our three arsenic-exposure groups was estimated by the interval from first employment to death. Since age at first employment is an important covariate, we considered men who were over 30 years of age when first employed separately from men aged 30 and under at first employment. Among men aged 30 and under at first employment, the median latent period was 39.5, 42, and 39 years, respectively, for heavy, medium and light arsenic exposure groups, with maximum latent periods of 55, 62, and 62 years, respectively, in these groups. Men in the medium exposure group were on the average 2 years younger at first employment than the remainder of the group first employed at age 30 or under. Among men over thirty at first employment, the median latent period was the same (23 years) for all arsenic-exposure categories; however, the maximum observed latent period was 43, 53, and 55 years, respectively, for heavy, medium and light arsenic exposure groups.

4. Discussion

The excess of observed deaths among smelter workers in this study was due largely to respiratory cancer, diseases of the heart, emphysema, and vascular lesions of the central nervous system. No excess mortality was observed for pneumonia and influenza. Except for mortality from respiratory cancer, no significant cause of death showed a positive gradient with length of employment, which is an indication that mortality for these other causes is related to factors other than exposure to environmental agents in the smelter (including other work experience).

The smelter workers had nearly a threefold excess in observed mortality from respiratory cancer, with a positive gradient associated with degree of arsenic exposure. We also considered factors other than environmental agents in the smelter which might be related to respiratory cancer mortality. Adjustments for differences of age have been made in the life table calculations, thus removing the effect of increasing respiratory cancer rates associated with increase in age. Further, it hardly seems likely that the excess of respiratory cancer deaths and the gradients in mortality associated with exposure to arsenic in the smelter can be totally explained by factors affecting respiratory mortality, such as socioeconomic status, medical care availability, and genetic susceptibility.

Cigarette smoking and lung cancer have been linked in numerous reports [16]. Smoking histories for all persons in the present study were not available. Smoking histories have been collected for an independent study involving a sample of 1800 men from the present study group. The investigators in that study have indicated that their findings suggest that smoking alone does not account for the observed respiratory cancer excess seen among smelter workers in the present study [17].

The excess we are seeing here in respiratory cancer deaths among smelter workers is most likely related to exposure to carcinogenic agents in the workplace. Arsenic has been investigated in many non-mining occupational studies as a potential carcinogen in man, many of them published since the Lee and Fraumeni study [2-7, 18, 19]. The percentage of deaths due to respiratory cancer is quite similar in these studies, except for the large percentages reported among insecticide manufacturers by Ott _et al._ [18] (21%) and among copper smelter employees in Japan by Tokudome and Kuratsune [2] (18.5%). The collected evidence from the present study and others cited here supports the hypothesis that arsenic is a respiratory tract carcinogen among various occupationally exposed groups.

It is possible that other agents in the smelter which are correlated with levels of arsenic, such as SO_2, may enhance the hypothesized carcinogenic effect of arsenic. But, our findings continue to support the hypothesis of Lee and Fraumeni, stated more than 12 years ago: "...exposure to high levels of As_2O_3, perhaps in interaction with SO_2 or unidentified chemicals in the work environment, is responsible for the excessive number of respiratory cancer deaths among smelter workers."

The author wishes to thank the Environmental Epidemiology Branch of the National Cancer Institute for its support in collection of data used in this study; special recognition is due to Linda M. Pottern and B.J. Stone (NCI) for their diligent efforts in determining vital status of the study group and in the acquisition and coding of death certificates. Also to be thanked are those staff members of the Epidemiology Department of the University of Michigan who were responsible for the careful collection and coding of recent employment histories.

References

[1] Lee, A.M. and Fraumeni, J.F. Jr., Arsenic and respiratory cancer in man: an occupational study, J. Natl. Cancer Inst., 42: 1045-1052 (1969).

[2] Tokudome, S. and Kuratsune, M., A cohort study on mortality from cancer and other causes among workers at a metal refinery, Int. J. Cancer, 17: 310-317 (1976).

[3] Rencher, A.C., Carter, M.W., McKee, M.W., Mortality at a large western copper smelter, JOM, 19: 754-758 (1977).

[4] Pinto, S.S., Henderson, V., Enterline, P.E., Mortality experience of arsenic-exposed workers, Arch. Environ. Health, 33: 325-332 (1978).

[5] Wall, S., Survival and mortality pattern among Swedish smelter workers, Int. J. Epid., 9: 73-87 (1980).

[6] Enterline, P.E., and Marsh, G.M., Mortality studies of smelter workers, Am. J. Ind. Health, 1: 251-259 (1980).

[7] Lubin, J.H., Pottern, L.M., Blot, W.J., Tokudome, S., Stone, B.J., and Fraumeni, J.F., Jr., Respiratory cancer among copper smelter workers: recent mortality statistics, JOM, 23: 779-784 (1981).

[8] World Health Organization, Manual of the International Statistical Classification of Diseases, Injuries, and Causes of Death, (Geneva: World Health Organization, 1967).

[9] Cutler, S.J., and Ederer, F., Maximum utilization of the life table method in analyzing survival, J. Chronic Dis, 8: 699-712 (1958).

[10] Munson, R.R., Analysis of relative survival and proportional mortality, Computers and Biomedical Research, 7: 325-332 (1974).

[11] Hendricks, R.V., and Archer, V.E., Unpublished data.

[12] Rockstroh, H., Zur atiologie des bronchialkrebses in arsenverarbeitenden nickelhutten: beitrag zur syncarcinogenese des berufskrebses, Arch. Geschwulstforsch, 14: 151-162 (1959).

[13] Holmqvist, I., Occupational arsenical dermatitis; study among employees at copper ore smelting work including investigations of skin reactions to contact with arsenic compounds, Acta Dermatovener (Stockholm) 31 (suppl 26): 1-214 (1951).

[14] Dunlap, L.G., Perforations of the nasal septum due to inhalation of arsenous oxid, JAMA, 76: 568-569 (1921).

[15] Bailar, J.C. III, and Ederer, F., Significance factors for the ratio of a Poisson variable to its expectation, Biometrics, 20: 639-643 (1964).

[16] Report of the Surgeon General: The Health Consequences of Smoking, U.S. Department of Health, Education, and Welfare, Publication No. (HSM) 72-7516, Washington, D.C., U.S. Govt Printing Office (1972).

[17] Welch, K., Personal communication.

[18] Ott, M.G., Holder, B.B., Gordon, H.G., Respiratory cancer and occupational exposure to arsenicals, Arch. Environ, Health, 29: 250-255 (1974).

[19] Mabuchi, K., Lilienfeld, A.M., Snell, L.M., Lung cancer among pesticide workers exposed to inorganic arsenicals, Arch. Environ. Health, 34: 312-320 (1979).

DISCUSSION

E. P. Radford: I think you've seen some evidence from the human data that arsenic may not be a pure carcinogen, in the sense that we frequently think of it as an initiator. I'd like to point out something that was not emphasized today except in Dr. Lee-Feldstein's Table 5, which I think bears some repetition. It isn't only in smelters where a relationship of human cancer to arsenic has been shown. One of the most striking studies wasn't even on her slide. I'm referring to the pesticide plant in Baltimore, Maryland, which was studied by Baetjer, Levin and Lilienfeld, which has never been published in the open literature. These data were very impressive. Roughly half of a small group of the retired workers from that plant died of lung cancer, which is about as severe an epidemic as we've seen, except for the uranium miners in Central Europe. The point is that where arsenic exposures have occurred and where adequate checks on cases of lung cancer have been made, an excess has generally been found. This fact adds considerable strength to a causal relationship, but does not indicate whether arsenic compounds are carcinogens or promoters. One of the points I would like the panel to address is the question of the strength of evidence on arsenic, based on a number of different types of exposure, different chemical forms of the agent, and the presence or absence of other confounding factors which might modify the dose-response.

Another general comment I would like to make is that exposure conditions in these various populations, except possibly those from drinking water, are very difficult to obtain quantitatively. I submit, that for the field of arsenic carcinogenesis to be carried forward, it is going to be necessary to get much more quantitative information on actual exposure conditions. I think the efforts that have been made and the various studies are excellent. They are the best one can expect given that they are retrospective studies. Unfortunately, they still remain questionable. Thus, even though in the case of arsenic we have perhaps half a dozen really good quantitative studies of the effects, the dose-response relationship still remains somewhat obscure.

I would like to ask Dr. Enterline a question. Does he really think that the decrease in the relative risk of lung cancer as a function of time after arsenic exposure ceased, or the termination of the work experience, is a statistically significant change? Because if it is, it would be quite contrary to the situation that we see with other occupational carcinogens, such as asbestos or ionizing radiation.

P. E. Enterline: I think the answer is yes. I really believe that there is a decline with time in the relative risk among exposed workers. I say that for three reasons. First, in my study, if you were to look at the people exposed for five years, you'd see that the peak

SMR occurred roughly ten years after the midpoint of that five-year exposure and thereafter dropped. Secondly, this shows up in the earlier retirees' study. For men as they grew older and after exposure ceased, the SMR dropped rather sharply. And finally, there is a paper (Brown, C.C., and Chu, K.C. "Implications of the Multistage Theory of Carcinogenesis Applied to Occupational Arsenic Exposure." Submitted for publication) that analyzes the same data that Dr. Anna Lee-Feldstein presented that shows the same thing. Dr. Lee-Feldstein has not looked at it in that way. I think that this is unlike most of the carcinogens that I know about, but like radiation-induced leukemia, it does show a decline in risk as the time since last exposure increases.

E. P. Radford: That's true for radiation-induced leukemia, but not for lung cancer.

S. Lamm: Dr. Enterline, I'd like to follow up on your answer and to suggest another area where we ought to look for similarity. One area where we've had tremendous experience has been looking at the leukemias associated with benzene exposure. Benzene and arsenic are the two occupational carcinogenic exposures for which there is considerable multiple evidence of an effect in humans, but which has not been demonstrated in any other animals; therefore, it may be a species-specific effect.

We've found an interesting similar story in the benzene exposure/leukemia relationship--after exposure, the risk of leukemia seems to go down quite markedly. I think that if you look at almost any of the benzene/leukemia studies, you find that 80% of the leukemias occur within the two years after the final exposure. That is not true for most of the other types of cancer studies that we have looked at. This might suggest that with arsenic and benzene, we're dealing with a different pathological process and that possibly the types of toxicological studies we need for investigating such processes would be different from the standard carcinogenic rat-type studies.

P. E. Enterline: I think that's an interesting analogy.

S. A. Peoples: The differences I see in the studies between the people who drank the water in Utah and the exposure in a smelter is the nature of the arsenic. I've analyzed water found in Utah and in many places in Nevada where they have levels of 100 parts per billion or better of arsenic and it's all pentavalent. I'm pretty sure that if you analyze the arsenic in a smelter, you'll find it's probably trivalent.

J. W. Southwick: Dr. Irgolic will make a presentation tomorrow that will describe this in more detail. According to his data, approximately 86% of the arsenic in waters of our study communities was the arsenate (pentavalent).

M. L. Roy: I'd like to ask Dr. Southwick, do you have data on mortality in the two Utah towns?

J. W. Southwick: Yes, although I'm really not prepared to share the preliminary data that we have on mortality. I think it would be interesting to the group just to give a hint of what we've seen. We attempted to choose a number of communities in Utah that we could look at for mortality due to different diseases including cancer. In our arsenic-exposed community there was an apparent excess of cancer mortality. However, when you looked at the data closely, that excess was due to the fact that the people in the arsenic-exposed community tended to live longer than the others and most of the cancer was associated with the 70, 80, and 90-year-olds.

M. L. Roy: But did you notice an excess of a particular type of cancer?

J. W. Southwick: No, the cancers were of the whole garden variety.

L. J. Goldwater: I suspect that occupational exposure to arsenic can produce adverse effects other than cancer. I have heard practically nothing about either acute or chronic, systemic effects or even local effects due to occupational exposures. I wonder if anybody has studied these other effects.

E. P. Radford: I was going to introduce this session by pointing out that there are well-known effects of arsenic that occur at high levels; acute arsenic poisoning which we tend to forget about. Just a week ago, as a result of my name being on this symposium, I received a call from a man who lives not very far from here who told me about the sad case of his wife having had acute arsenic poisoning. The source of the arsenic is unknown to this day. It was a classic case with definitely elevated urinary and hair arsenic with residual sleeve and glove neuropathy; you can read about the whole syndrome in textbooks. Perhaps Leonard Goldwater's question could be amplified to say that if there are people who are aware of these isolated cases, it would be important, I think, to collect them, because apparently there are sources of arsenic not readily identified but sufficient to give serious effects.

P. E. Enterline: It should be noted that five years after termination of exposure for the men in that smelter, there was a very high death rate. Now I haven't looked at all the reasons for this high death rate, but this is one of the things that confuses the analytic procedure. Men were terminating from the smelter apparently for health reasons and, therefore, not working very long. In other words, these health problems seem to be showing up early in their worklives, after five and ten years. I think that it would be worth looking at why these terminations were taking place and why the death rate was so high in the period immediately after the termination. It could be simply the obvious--one way to terminate is to become ill or die. However, when looking at a set of data by length of exposure, you have to realize that one of the reasons exposure is short is that people are removed from the population because of poor health. This is an aspect of the data that I don't think has been thought about very carefully.

E. P. Radford: One of the things that impressed me about Dr. Lee-Feldstein's data is that the SMRs, particularly cardiovascular, were so high. Do you have any thoughts on that, Anna?

A. Lee-Feldstein: I really don't have anything to add to what I said before about heart disease. I did want to mention that there is also some indication of excess mortality from respiratory diseases, other than cancer; and there is some excess mortality due to mental disorders.

W. L. Marcus: There is an animal model published in 1979 by Maltoni and Scartino which showed that not only did benzene produce leukemia, but the animals got solid tumors of the zymbal gland as a result of the exposure. Maltoni has a history of doing very unusual experiments, but when they're repeated, they turn out to be correct. There are two other studies I'd like to mention--the Antafagusta Study in Chile and the Taiwan Study. The data that we have show, in fact, lesions called cancer seem to miraculously disappear within a year after the successful treatment of the water. The Taiwan Study, which included an enormous number of people, also showed a dose-related effect from a flourescent compound, as well as arsenic. We sent Dr. Chie Wu, a native Chinese, to talk to the person that developed this data. Dr. Wu was told that alkaloids produced peripheral vascular contraction. Arsenic also has that capacity, though to a very minor extent. This would, therefore, explain the so-called Black Foot Disease. I bring this up in relation to the Water Quality Criteria Document. I wrote numerous times to Dr. Albert explaining the flaws of the experiments he was using to determine those numbers. We pointed out the experimental problems item for item and they chose to ignore them. The Water Quality Criteria Document refers to about 22 nanograms/liter in drinking water. If the water quality criteria number were correct, Dr. Southwick's study would have produced a larger number of effects.

A. P. Borgias: Dr. Enterline, I was wondering how you felt about the validity of urinary arsenic concentrations as a measure of worker exposure and do you have any other suggestions for an alternative analytical technique?

P. E. Enterline: Initially, the reason we used the urinary arsenic as a measure was because of the protective clothing and the fact that workers were supposed to wear a mask in certain areas and in some areas I think they did. Therefore, atmospheric arsenic might not have been a good measure of the biologic intake of arsenic. There is a small study of 24 men who did not wear a mask but wore a personal sampler, and who had daily urinary arsenics. The airborne measure and the urinary measure had a correlation coefficient of 0.5. (It isn't large, but there must have been some errors in the two measurements, and at least it was in the right direction.)

A. Lee-Feldstein: An additional reference (19) is a study from northern Sweden of lung, liver and kidney tissue from deceased smelter workers. Copper smelter workers who had been exposed to arsenic were found to have seven times the amount of arsenic present in the lung tissue when compared with the lung tissue of men who lived some fifty kilometers away, where there was comparatively little arsenic in the environment.

I. Harding-Barlow: When you analyze lung tissue, you will find that practically all the pollutants are present in very high concentration. This is because the lungs try to filter out the pollutants that we're exposed to. Therefore, you have to compare concentration levels on the log-normal basis and I'm surprised the Swedish investigators didn't do this.

S. Lamm: Dr. Southwick, reference was made to the Water Quality Criteria Document and its risk analysis for skin cancer. While I won't try to go through a risk analysis here, one must consider the likelihood of an arsenic skin cancer death occurring. The Arsenic Water Quality Criteria Document concluded that three arsenic skin cancer deaths would occur annually in the U.S. from water containing 2 μg/l of arsenic. When the calculation was made, it was found that the medium time to tumor for these cancers was 2636 years. One of the problems that developed within their model was an absence of a variable to account for competing risks of death.

Another factor that goes into the calculation is that the average daily intake of water is two liters per day. You indicated that you had considered the water intake of the individuals in your area. What number did you use?

J. W. Southwick: It was not much different, but it was slightly more. Average daily water intake was about 2.2 liters. Using our water consumption data, we calculated that the exposed group averaged 150 milligrams of arsenic from well water per person per year. The Criteria Document assumed that arsenic at 0.05 milligrams per liter of water would not exceed 36.5 milligrams of arsenic per year. Thus, we had an arsenic exposure of 4.1 times the maximum allowed by the standard adopted in the Criteria Document. And, by the way, folks out there may live a long time, but not 300 years!

S. L. Malish: In the semiconductor industry, many kilos of gallium-arsenide are now being used. The gallium-arsenide is made from arsine. Would it be worthwhile, to biologically monitor the people handling arsine, which is obviously very toxic, and the gallium-arsenide? Frequently the manufacture of gallium-arsenide is done under hoods and under vacuum chambers and things of that nature, but the scientists where I work are somewhat concerned about the health aspects.

I. Harding-Barlow: Some of the semiconductor companies have tried to monitor arsenic in urine of workers. The only times they measured high concentrations of total arsenic were when the workers had eaten seafood.

E. P. Radford: These were arsine-exposed workers?

I. Harding-Barlow: This was for the type of exposure that is seen in the electronics industry.

E. P. Radford: That might imply then that arsine is so toxic in its own right that you're not likely to see any arsenic effects as such.

S. A. Peoples: Arsine has a completely different type of toxicity affecting the red cells and causing anemia. Of course, one could measure the arsenic concentration in the urine, but if there are any reasonable amounts of arsine absorbed by the workers, they would be quite ill and the poisoning would be more easily detected than poisoning with trivalent or pentavalent arsenic.

A. V. Colucci: This question is for both Dr. Enterline and Dr. Lee-Feldstein. Even in the closely circumscribed atmosphere of the occupational scenario, the smelter environment is loaded with a myriad of compounds, many of which could be extremely carcinogenic. To what extent did you, could you, or would you be able to disentangle those competing factors?

A. Lee-Feldstein: We looked at sulfur dioxide separately as an exposure variable in the original paper, and we found that we didn't see the nice neat gradient in respiratory cancer mortality relative to level of SO_2 exposure that we did see in relating respiratory cancer mortality to arsenic exposure. It is difficult to separate the effects of arsenic and SO_2 in this study group, since many work areas with heavy or medium arsenic exposure also afford medium or heavy exposure to SO_2. We concluded in the original paper that arsenic, perhaps in interaction with SO_2 and other unidentified chemicals in the environment, is responsible for the excess in respiratory cancer deaths. My present information does not contradict that conclusion.

A. V. Colucci: The arsenic is associated with particulate matter and SO_2 is a gaseous component. Since the particulate matter has a number of chemicals in it, have you been able to analytically separate what these materials are and run similar correlations on other chemicals which should follow similar gradings?

E. P. Radford: Well, if they're intimately associated with the arsenic, that is, there is always a one-to-one correspondence between the arsenic level and the particulates, you're bound to get the same correlation. There are new techniques available by which if you have differential exposures to elements in the workplace, you can sort them out statistically. But if there's a one-to-one correspondence, there's nothing you can do. This was one reason why I stressed that in the pesticide plants and in the smelters an effect on lung cancer can be seen. I think the case for arsenic can be strengthened if we can eventually get quantitative dose response data and show that they are the same regardless of all the other things that might or might not be in the same environment.

P. E. Enterline: I just wanted to add that other substances such as nickel subsulfide (Ni_2S) cause cancer. Smelters generally have higher death rates than you might anticipate, and it's hard to look at all the different kinds of smelters and connect them with a single substance like arsenic. There must be a lot of things going on in smelters that we haven't even tried to measure.

F. H. Nielsen: The studies by Lee-Feldstein show that many workers exhibited respiratory disorders. This indicates that lung physiology has been altered, probably by materials other than arsenic. Perhaps the elevation in lung and respiratory disorders can be more closely correlated with something like particulate matter. Furthermore, if particulate matter is present, the particulates may contain some unidentified carcinogenic material, for example, nickel subsulfide. Perhaps some workers are also exposed to other carcinogens like nickel carbonyl. Thus, one must be careful not to relate any increase in cancer appearance to just one agent such as arsenic. Also, I think this is another example of trying to incriminate arsenic in a variety of types of cancer without any consideration to its form or route of administration.

P. E. Enterline: I thought one interesting thing about Anna Lee-Feldstein's presentation was the fact that emphysema seemed to be increased in people with short exposures. If it's something in the general environment, you would expect duration of exposure would somehow be related to this effect.

E. P. Radford: In a symposium we had in Pittsburgh about two years ago, Dr. Marvin Schneiderman pointed out that the short-timers may be the heavily exposed ones who have done the dirty jobs. That may be one reason they leave that particular work early.

P. Mastradone: My comment deals with Dr. Lee-Feldstein's data. I was thinking that the data on emphysema tends to show--at least it has in the past--an association with exposure to aerial particulate matter. I was wondering just how much you thought subtracting that kind of data out of your lung cancer assessment situation might show arsenic to be considered a carcinogen?

A. Lee-Feldstein: I have not yet looked at that for this follow-up study. We originally had some data on other agents such as silica, lead fumes, and ferromanganese dust, and I really feel that the other agents that had been measured in the atmosphere (other than SO_2) were not strongly implicated. However, we didn't make a formal attempt to "subtract them out." We had the same kind of problem that we have with sulfur dioxide: the exposure to these various agents was simultaneous.

P. Mastradone: My idea is that something similar to silicosis might be occurring where you get constant irritation of the lung lining. It seems to me that this would have a tendency to promote cancer of the lung.

A. Lee-Feldstein: In fact, Lee and Fraumeni found that excess respiratory cancer mortality was inversely related to degree of exposure to silica, which seems to contradict your theory, at least as far as silica is concerned.

E. P. Radford: Apropos to that comment, there is no evidence of a synergistic effect of silicosis or chronic lung disease and radiation-induced lung cancer.

J. W. Southwick: I'd like to respond to the oft-stated comment of "why worry about the Drinking Water standard for arsenic?" The reason is that excess arsenic in water violates regulations. That's why you worry about it. And, to give you a follow-up on our study community, as soon as we had submitted our report to the research branch of EPA showing no observed health effects, the enforcement branch of EPA descended on the community of Hinkley and said, "You can't have water with arsenic at this level; it's in violation of the standard." And they said, "Well, we don't have any money to do differently." EPA essentially said, "That doesn't count. You still have to correct the problem. Our regulations don't allow you to have a public drinking water system the way it is." So the town applied for a federal grant to correct the problem, and the last I checked on it, I think they were in the mill for almost a million dollar grant to correct the problem EPA said they had.

C. Gordon: Dr. Enterline, in your 1977 update of the Tacoma Study, you said that pre-1950 exposures were five to ten times the 1973 exposures. However, on the chart (Table 7) which showed exposure pre-1950, '50s to '60s, '60s to '70s, and post-'70, which I presume is the '73 data, it looked like the general differential between pre-1950 and 1973 was by a factor of about two, not five to ten. I was wondering if you had any comment on that.

P. E. Enterline: I think that was a statement that Sherman Pinto put in the paper, and I assumed that he knew something more than is on the chart there. I do know that the air data we have that goes back to 1937 shows extremely high levels in the 1930s and so perhaps he was referring to air concentrations and made some kind of estimate. But you're absolutely right. The chart shows about a doubling and the paper does have a statement in it as you say.

I. Harding-Barlow: Your Table 2 showed urine levels but without a 45° slope.

P. E. Enterline: That's right. I think the urinary arsenic levels were about double around 1950, compared with 1973.

I. Harding-Barlow: But this doesn't mean anything with regard to air because you do not have a 45° slope. Therefore, you do not have a one-to-two correlation between urine and the air.

P. E. Enterline: Yes, that's right.

C. Gordon: In May, 1980, EPA published a risk assessment for airborne exposure to arsenic making use of Dr. Enterline's original study, Dr. Lee-Feldstein's original study and the Baltimore study, which was at a chemical plant. I was wondering whether any of the panelists had any comments on the EPA risk assessment.

P. E. Enterline: I've seen it. The amazing thing is that even with crude wild methods, it came out with about the same kind of answer for all three studies. I should add it's pretty ingenious.

I. Harding-Barlow: I would like to point out that using these three studies and the EPA extrapolation techniques, one comes up with a p value of ten to the minus two, for the air equivalency of our normal air, food, and water exposure to arsenic. As explained in my paper, if one assesses normal intake of arsenic from food, air, and water and converts it to an air equivalency by making the proper assumptions as to differences in percentages absorbed, et cetera, a value for p can be assessed and it is approximately 10^{-2}.

E. P. Radford: How can you equate that to inhaled arsenic?

I. Harding-Barlow: The details are given in my paper both for air and water equivalencies and lung and skin cancer probability risks.

M. Schneiderman: What are the confidence limits on these estimates?

I. Harding-Barlow: Even at plus or minus 50%, this is still an extremely high lifetime probability risk.

E. P. Radford: I for one would feel that it would be necessary to see the calculations by which you go from food levels to airborne concentrations before one can assess their equivalence.

E. A. Woolson: Just to make three quick points. Using urinalysis as a measure of industrial exposure is fine, as long as you speciate the arsenic. If you're exposed industrially, it'll be present in the urine as either arsenate, arsenite, or cacodylic acid. You don't want to confuse the arsenic compounds you get from eating seafood and probably most of your other food. Most of the arsenic in vegetable crops grown on high arsenic soils, for instance, is not there as inorganic arsenic. It's as an organic complex: It will hydrolyze. This is discussed in greater detail in my presentation. Secondly, the Baltimore pesticide plant which utilized lead and calcium arsenates, to improve efficacy, frequently used lime sulfur as an adjuvant. I've done some studies with lime sulfur and it causes a reduction of some of the arsenate to arsenite. Therefore, they had exposure to copper, a wide variety of sulfur, oxisulfur compounds and sulfides, as well as lead and calcium arsenate. In terms of arsenic essentiality, we've seen it many times in plant studies where we grow plants on low arsenic containing soils and add, as a treatment, a variety of arsenic levels. We do get a stimulation in growth and yield at low arsenic levels. Again, maybe this is an indication of essentiality for plants. We've not analyzed it on a statistical basis, so we can't really say anything about it definitely, but it is an indication.

W. L. Marcus: I'd like to talk more about a study of the ASARCO plant. As part of a thesis, Dr. Genevieve Matanoski did a study funded by EPA. She used census tracts and interviews with the people in those areas to determine mortality. Dr. Matanoski, in looking at mortality records, was able to show that as people lived farther away from the plant there was a decrease in the number of lung cancers. As I recall, at our request she obtained soil samples near the homes and was trying to correlate the levels of arsenic in the soil samples as a measure of exposure in relationship to distance from the plant.

E. P. Radford: I'm familiar with Dr. Matanoski's study, because I was involved in the inception of it. It is true that that census tract had a high lung cancer rate which was one of the reasons why the arsenic connection was investigated. But I think the recent report published by Dr. Matanoski ("Cancer Mortality in an Industrial Area of Baltimore," Matanoski, G., Landau, E., Tonascia, J., et al., in Environmental Research, Vol. XXV, pp. 8-28. Academic Press, 1981) indicates that there is not a clear relationship between the proximity of the plant and the high incidence of lung cancer. There was a possibility that the freight cars bringing in arsenic trioxide might have dribbled small amounts near the plant. But I don't think that the connection between arsenic and lung cancer is clear cut at all.

W. L. Marcus: I want to make a second point to Dr. Southwick since my office sets regulations for drinking water. We are looking very seriously at raising the allowable arsenic levels. Those levels were originally set, as were most of the MCLs, in 1964 by the United States Public Health Service. For many of them there was not a clear reason for the numbers they chose. There are towns in the United States with very high levels of arsenic, including Fairbanks, Alaska, where the drinking water has levels similar to those presented by Dr. Southwick or a little higher, in which there was no evidence of increased mortality or cancer. So we're very seriously considering raising the level although I don't know by how much. The reason the enforcement people became involved is because that is the way the law reads. If you exceed the level, they have very little choice. However, under the Variance and Exception Provisions, you could have applied for a variance and received it for at least two years.

J. W. Southwick: An exemption was applied for and granted to the community of Hinkley for its public water system. The unincorporated community of Deseret had no mechanism for getting a variance or an exemption because they did not have a public water system (each home had its own private well). As a consequence, the Farmers' Home Administration, and others who loan money for home construction, refused to loan money in Deseret because there was no variance or exemption granted. This has a tremendous economic impact because a large power plant is going in and the area anticipates a big development boom, so it's a very painful thing for them. Not only is our government possibly going to spend a million dollars to "improve" that water system, but its arsenic standard is having a dramatic economic impact on these small communities.

SESSION V - ENVIRONMENTAL PERSPECTIVES

Session Chairman: Edwin A. Woolson
U. S. Department of Agriculture
Agricultural Environmental Quality Institute
Beltsville, MD 20705

INTRODUCTION

We've already seen the wide variety of sources from which we get arsenic as a by-product, whether intentionally produced or not. This by-product is used in a wide variety of applications. The fact that man has changed the form of arsenic from an insoluble form present in sulfidic minerals to a more soluble, volatile, and biologically active form has caused some problems to man, plant, and animals.

In the following presentations, we examine what happens to this soluble, bioavailable arsenic, arising from man's activity or as a result of processing by Mother Nature.

21

COMMERCIAL SCALE REMOVAL OF ARSENITE, ARSENATE, AND METHANE ARSONATE FROM GROUND AND SURFACE WATER

Norman E. Krapf
Vineland Chemical Company, Inc.
West Wheat Road
Vineland, N.J. 08360

A commercial waste water treatment plant designed for the removal of methane arsonates and inorganic arsenic from ground and surface waters as well as industrial waste waters is described.

This project was undertaken primarily to treat groundwater which had been contaminated by leachate from a saltpile. This by-product of sodium methane arsonate manufacture contained a small percent of arsenicals, chiefly methane arsonates. The saltpile accumulation was inevitable because removal was prohibited by law and approved landfills did not exist.

This saltpile was removed when an approved disposal site became available. Thereupon further entry of arsenical waste water into the environment was curtailed.

A thorough literature search failed to reveal any prior efforts to remove organic arsenic from industrial wastewaters. All of the reports were based on laboratory or pilot plant experiments with the exception of a Taiwanese plant designed to render water potable by removal of inorganic arsenic.

Hundreds of laboratory experiments with many reagents and combinations led to the conclusion that ferric hydroxide floc was the most cost effective and efficient method of arsenic removal.

Over a period of three years 5,000,000 gallons of waste salt leachate contaminated water were treated by a batch process whereby over 90% of the arsenic was removed from the leachate.

Subsequently a plan for a continuous treatment works was submitted to and approved by the N.J. Dept. of Environmental Protection. A plant was constructed which currently treats water at the rate of 36,000 gallons per day, reducing 30 ppm arsenic to 0.7 ppm or less. The arsenic in the water before and after treatment was speciated. Inorganic arsenic in treated water was reduced to potable water levels and the methane arsonate content reduced 97%. The separated ferric hydroxide sludge is dewatered and removed to an approved landfill. Further research towards improvement of the technology is in progress.

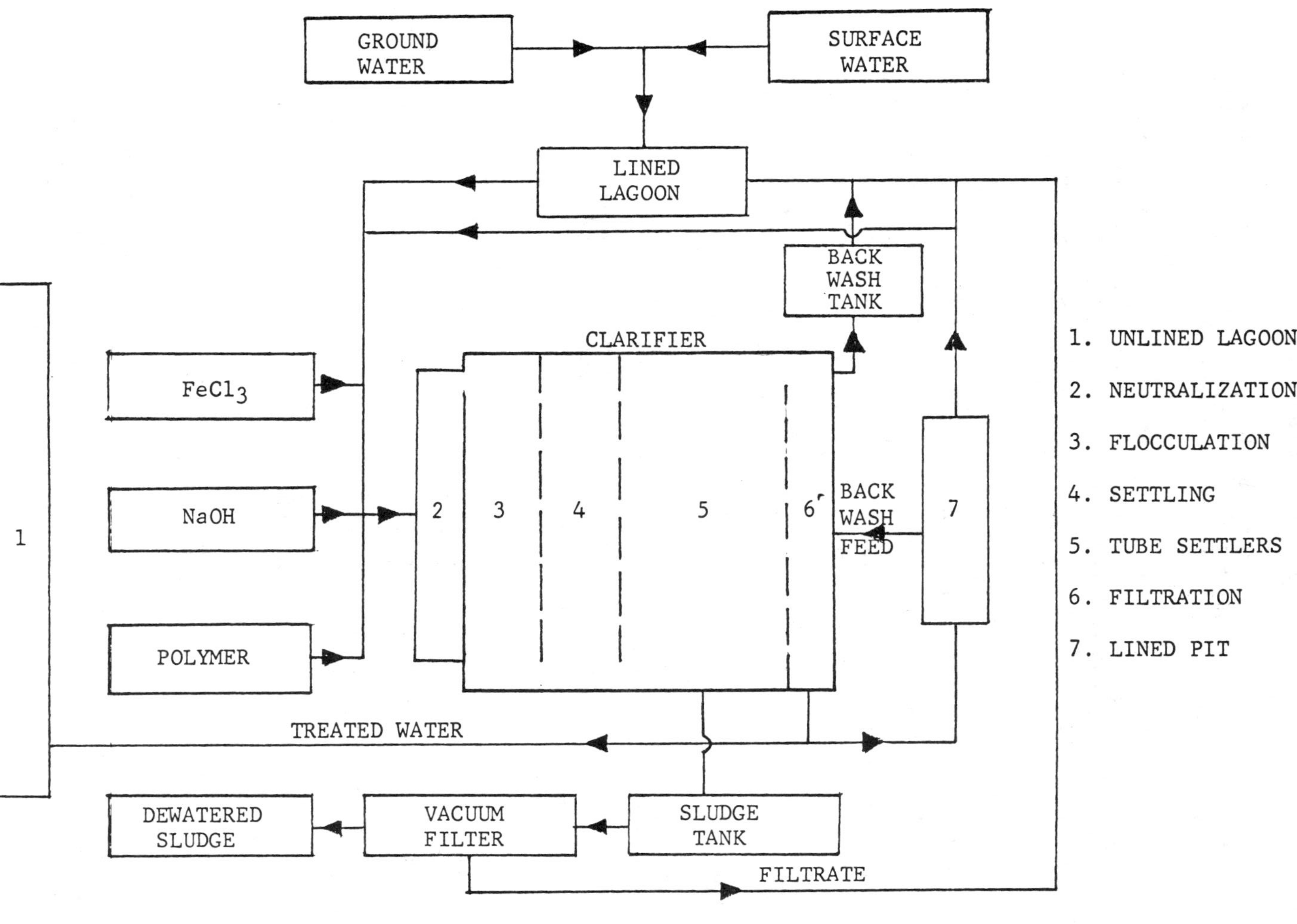

FIGURE 1. FLOW DIAGRAM - WASTE WATER TREATMENT PLANT VINELAND CHEMICAL CO., INC.

1. Introduction

Arsenic has a long history of beneficient uses. These uses have been overshadowed by its occasional misuse to create beneficiaries.

The enormous compilation of research in the literature attests to the recognition of the unique nature of arsenic compounds, many of which have highly desirable biological properties, enabling them to be used as medicines, wood preservatives and herbicides (1,2)[1].

More recently arsenic and its derivatives have found valuable applications in the electronics industry. Many older uses such as in glass manufacture are also well known. Ongoing research, creating a legion of new and often useful compounds, indicates that arsenic will continue to require smelting and processing industries, not only for present compounds but for those as yet to be discovered and utilized.

Apropos of the foregoing, dimethyl arsinic acid and methane arsonic acid were well known for their medicinal properties in the treatment of chronic eczema, anemia, leukemia and psoriasis (2).

Agronomic research with the above compounds originated at Vineland Chemical Company over a quarter century ago (3) and, with the research and collaboration of many scientists in the field, has resulted in their worldwide use as herbicides (4). More recently, new uses in biological preparations have come to the fore.

Inevitably, in the manufacture of any composition there is an escape potential to the environment. In some industries this is not particularly grievous, but in the case of biologically active chemicals such as arsenicals, control of emission is mandatory because of the relatively low tolerance for inorganic arsenic considered safe in potable water (0.05 ppm) in U.S.A. (5) or even the 0.20 ppm European drinking water standard (6). Everyone familiar with arsenic knows we are exposed to it in our everyday environment and, that indeed, it is a small part of everyone of us.

Arsenic occurs naturally in many foods, especially in seafood reported to contain as much as 80 ppm (7). Relatively recent evidence indicates that arsenic as a trace element is essential, at least to some animals (8-11). As with some other micro essential elements, higher concentrations are extremely toxic. This is particularly true of selenium and to some extent other metals. The essentiality of selenium,

---

[1]Figures in parentheses indicate the literature references at the end of this paper.

poisonous in large quantities, has only recently been demonstrated (11,12). Arsenic, therefore, although probably essential in trace amounts because of its toxicity at higher levels, must be limited in potable water. Hence, in any large scale process involving arsenic, prudence and law dictate effluent control to tolerable limits.

Control must be exercised in some regions of the world where naturally occurring arsenic and other minerals in ambient water would otherwise preclude potability.

2. Prior Literature

A complete search of the literature by visual reference to abstracts and articles, followed by computer search, revealed that aqueous arsenic removal techniques fell into three major categories: ion exchange (13-16), activated alumina adsorption (17,18) and precipitation or adsorption by metal hydroxides--prominently ferric hydroxide (19-29).

All of the literature dealt with "arsenic" without specifying the valence state or whether organic or inorganic. Standard reference solutions of sodium arsenite or sodium arsenate were mentioned in some articles but no reference dealt with the methane arsonates or any other organo arsenicals.

All the literature described only laboratory or pilot plant procedures with one exception (24). A water treatment plant with a capacity of 34,000 gallons per day was built in Taiwan to remove about 1 ppm of naturally occurring arsenic from well water to render it potable. In the Taiwan plant ferric hydroxide precipitation was preceded by chlorination, indicating either that the arsenic was in an arsenite form or that interfering substances required oxidation. Private communication (30) revealed that speciation of the Taiwan water proved it entirely inorganic.

3. Analytical Methods Employed

The Taiwanese plant was reported to have reduced the well water arsenic content to potable levels utilizing silver diethyldithiocarbamate as the analytic method. We have found this method unreliable for methane arsonate analysis. However, graphite furnace atomic absorption spectroscopy, using deuterium background correction, has proven highly reliable at ten to 100 ppb concentrations--the range with which we were concerned.

4. Investigations of Various Methods of Arsenic Removal

In our efforts to remove arsenic from water we first experimented with an alumina column. This was quite effective, but regeneration posed a problem because each regeneration was followed by loss of adsorption capacity. Also, considerable alumina loss occurred due to sodium aluminate formation in the regeneration process (31). The method was considered impractical and abandoned.

We also investigated the use of the ion exchange resins IR-120 and IR-400 manufactured by Rohm & Haas. Totally arsenic retentive at first, in a few hours the resin no longer removed arsenic. The resin bed, not being anion selective, was also adsorbing other anions in the water and lost its capacity to retain arsenic. This limited capacity ruled out commercial feasibility, even as a final polishing or trace removal step.

After hundreds of experiments, we found that ferric hydroxide as a complexing precipitant was the most efficient and cost effective method of arsenic removal whether inorganic or as methane arsonate. Pre-treatment with oxidants such as hydrogen peroxide, chlorine, sodium hypochlorite, potassium permanganate and others did not significantly improve arsenic removal. The carbon-arsenic bond in methane arsonic acid is highly resistant to oxidation in aqueous systems at ambient temperatures. The use of reducing agents to generate arsines or arsine-oxides, followed by ferric hydroxide precipitation while an improbable commercial method, nevertheless was also investigated and found ineffective. Many other experiments involving retention on clay, ferrous minerals, sulfur compounds, etc. likewise did not equal ferric floc precipitation. Several years ago semi industrial plant scale trials with various arsenic sequestering compounds including ferric hydroxide were initiated by H.P. Loomis (32) who also concluded that the arsenic content of water could best be reduced by ferric hydroxide.

In the mid seventies, with the informal consent of the New Jersey Department of Environmental Protection (DEP), Vineland Chemical Company initiated commercial efforts utilizing ferric hydroxide to treat 160,000 gallons of water by a batch process. The water subjected to treatment was subsurface leachate from the immediate vicinity of an erstwhile pile by-product waste salt containing residual arsonates. The objective of course was to remove arsenic in the groundwater and arrest the potential formation of a plume. The State consent was the first ever granted for experimentation on this scale. The process consisted of adding ferric chloride to the water with air agitation at an iron-arsenic ratio of 7 to 1. The pH was then adjusted to 5.5 by addition of sodium hydroxide and the mixture allowed to stand and settle the ferric floc overnight. The following day the clear supernatant water was discharged. We were able to achieve a 90-95% reduction of the arsenic in the groundwater by this method. The information gathered during the several years of treatment in this fashion proved invaluable in the design of our continuous treatment plant. The groundwater contained arsenic in the form of arsenite, arsenate and methane arsonate ions. Five million gallons of water were treated over a period of 3 years by this method which succeeded in reducing the arsenic content of the groundwater to 30-40 ppm at the area where the saltpile had been, and to less than 10 ppm in more distant areas. The separated ferric hydroxide sludge containing the removed arsenic from this process was dewatered and sent to an approved landfill.

In April of 1980, we began a 24-hour-a-day continuous treatment of groundwater from the same contaminated area. Leachate from a waste saltpile containing some residual arsenic had been stored outdoors and was the source of arsenic. The saltpile had repeatedly been covered to prevent weather exposure. Occasional high winds made it impossible to keep a cover intact despite many improvisations. Accumulation of this wastepile was inevitable at the time because no approved hazardous waste disposal sites were available. Some had been temporarily approved but then closed.

With the approval of the New Jersey DEP, we installed an automated continuous treatment works with a water treatment design capacity of 85,000 gallons per day. However the capacity of any plant utilizing flocculation is determined by the residence time required for floc settlement. The rate of settling for proper clarification limited the operating maximum of our plant to 40,000 gallons per day.

In order to halt any further arsenical groundwater intrusion, the entire area around the plant was paved with an impervious asphalt and concomitantly graded so that all rain-water was collected in a 500,000 gallon EPDM-lined lagoon. This was designed to contain any inadvertent arsenic release due to spills, leaks, etc. during manufacture and transport. The treatment plant is now being used to treat any contaminated plant effluent as well as contaminated groundwater.

All groundwater and plant runoff are piped into the 500,000 gallon lined storage lagoon. Water to be treated is pumped from the lagoon at 25 gallons per minute into the first of 4 sections of the treatment plant processing equipment Simultaneously ferric chloride solution is added by a metering pump; thereafter sodium hydroxide solution is added in a neutralization section to automatically adjust the pH to 5.5. Hercofloc 847, a polyelectrolyte, is also added to this section at 5 ppm to enhance flocculation and settling. The water, now containing ferric hydroxide and adsorbed arsenicals as a floc, enters the bottom of the mixing section of a clarifier. The supernatant water then flows into a floc settling and separation section. The clear water rises through slant tubes, and overflows into and through a dual media filter (see figure 1 for flow diagram).

The treated water is analyzed twice daily by means of a graphite furnace atomic absorption spectrophotometer with deuterium background correction. At the present time the discharge limit is set at 0.7 ppm arsenic. Water above this standard is recycled. It would appear from simple stoichiometry that increasing the ferric content would proportionately reduce the aqueous arsenic concentration. However it was found that the reduction of arsenic becomes asymptotic with increased iron to arsenic ratio thus limiting the amount removable.

Our current operation reduces groundwater now containing a maximum of 30 ppm of arsenic to 0.7 ppm.

The settled floc or sludge is periodically pumped from the bottom of the clarifier to a holding tank and then dewatered by a continuous vacuum filter. The water extracted from the sludge is recycled.

5. Monitoring Plant Water

The manufacturing plant cooling water effluent is monitored continuously by a conductivity meter which normally reads 30 to 40 micromhos indicating a very low level of electrolytes. If arsenic salts enter the cooling water due to equipment failure, an alarm sounds and the water is automatically diverted to the lined lagoon until the problem is corrected. Individual sources of cooling water are provided with conductivity probes in order to rapidly identify any problem.

6. Conclusion

The Vineland treatment plant is the first continuously operating plant treating water to remove both organic and inorganic arsenic. As of this writing only methyl arsonates have been removed utilizing this technology along with small amounts of inorganic arsenates and arsenites. However, there is reason to believe that other organic arsonates could be removed in the same fashion and perhaps, with pretreatment, other organoarsenicals.

We are continuing research designed to further reduce arsenic in the treatment plant effluent. The methanearsonates are many times less toxic than the inorganic arsenicals (33), but also appear to be more difficult to remove. Samples of our arsenic containing water have been speciated before and after treatment (34). The treatment does reduce the concentration of <u>inorganic</u> arsenic to potable level. However, organic arsenic continues to persist at about 0.7 ppm. Our current research is directed toward reducing this level.

Grateful acknowledgment is expressed for the engineering concepts advanced by Kenneth Mann of M & M Engineering, the legal interfacing with the Department of Environmental Protection by Franklin Riesenburger and George Pazianos, the speciation testing by Dr. Woolson and the dedicated cooperation of the Vineland Chemical Company staff including Dr. Mookerjee, Dr. Narayanan, Raj Ravi, Robin Rayborn, Arthur Schwerdtle and Porter Loomis as well as the conscientious efforts of the many technicians also involved.

References

(1) Moore, L.O., Ehman, P.J., "Arsenic and its Compounds", Encycl. Chem. Process Des., Eds. McKetta, J.J., Cunningham, W.A. Dekker, New York, N.Y. 3, 396 (1977).

(2) Stecher, P.G. et al "The Merck Index" An Encyclopedia of Chemicals and Drugs, 8th Edn., Pub. by Merck & Co., Inc., Rahway, N.J., 185, 670, 956 (1968).

(3) Schwerdtle, A., U.S. Patent 2,678, 265 (1954); 3,056,821 (1962); 3,030,199 (1962); 3,068,088 (1962); 3,338,780 (1967); and 3,410,885 (1968).

(4) U.S. Environmental Protection Agency "Substitute Chemical Program, Initial Scientific Review of MSMA/DSMA", EPA-540/1-75-020 (1975).

(5) U.S. Dept. of Health, Education and Welfare, "Occupational Exposure to Inorganic Arsenic, New Criteria" (1975). New Publication No. (NIOSH) 75-149.

(6) Goldsmith, J.R., Deane, M., Thom, J., and Gentry, G., Water Research, Pergamon Press, 6, 1133 (1972).

(7) Uthus, E.O., and Nielsen, F.H. "Nutrients and Growth Regulations" Pub. C.R.C. Handbook of Nutrition & Foo (1979) Tables 14 and 18 and references therein.

(8) Anke, M., Grun, M., Partschefeld, M., "The Essentiality of Arsenic for Animals", Proceedings, University of Missouri Annual Conference, Vol. 10, pp 403-9 (1976).

(9) Nielsen, S.H., "Evidence of the Essentiality of Arsenic, Nickel & Vanadium and their Possible Nutritional Significance". Advances in Nutritional Research, Vol. 3, pp 157-172 (1980).

(10) Nielsen, S.H. and Schuler, T.R., "Arsenic Deprivation Studies in Chicks". Federation Proceedings, Vol. 36, p. 893 (1978).

(11) Mertz, W., "The Essential Trace Elements". Science, 213, 1332 (1981).

(12) Schrauzer, G.N., "Selenium & Cancer: A Review". Bioinorganic Chemistry, Vol. 5, 275-81 (1976).

(13) Slovak, Z., and Docekal, B., Anal Chem Acta, 117, 293 (1980).

(14) Ambro, I.B., J. Chromatography, 102, 454 (1974).

(15) Kennedy, D.C., Chem. Eng. June 16, 106 (1980).

(16) Sandhu, S.S., Envir. Sc. & Tec., 13, 477 (1979).

(17) Gupta, S.K., and Chen, K.Y., J. Water Pol. Cont. Fed., 493 (1978).

(18) Bellack, E., J. Am. Water Works Assoc., 454 (1971).

(19) Patterson, J.W., Waste Water Treatment Technology, Ann Arbor, Mich. Ann Arbor Science, Inc., (1975) Ch. 1; Christensen, D.C., and McNeese, J.A., Proc. Ind., Waste Conf., (1977) Pub. vol. 32, 242 (1978).

(20) "Waste Water Treatment Technology", Second Ed., National Technical Information Service, U.S. Dept. of Commerce, (1973) PB-216 162.

(21) Hau, Ton Jung, Kogyo Yosui, 123, 21 (1968).

(22) Halloo, J., Szebenyi, I., Toth, J., Vermes, E., and Illenyi, T., Period. Polytec. Chem Engg., 12, 283 (1968).

(23) Speas, V.E., and Mnookin, N.M., U.S. Patent 2,080,582 May 18 (1937).

(24) Shen, Y.S., J. Am. Water Works Assoc., 65, 543 (1973).

(25) Gulledge, J.H., and O'Connor, J.T., J. Am. Water Works Assoc., 548 (1973).

(26) Kunin, R., Rohm and Haas Co., Philadelphia, Private Communication.

(27) (a) Kimihiko, S., Kunihiko, T., Nobuaki, K., Jpn Kokai Tokyo Koho 79,04,874 (cl co2c5/02 13 Jan. 1979; (b) Eduardo, F.A., Saneamiento 54, 148 (1970).

(28) Cohen, J.M., Municipal Envir. Lab., Cincinnati, Ohio "A Review on Trace Metal Removal by Waste Water Treatment" (year of pub. not available).

(29) DeCarlo, E.H., and Zeitliu, H., Fernando, Q., Anal. Chem., Vol. 53, 1104 (1981).

(30) Woolson, E.A., USDA, Beltsville, Maryland, Private Communication (1981).

(31) Mookerjee, P., Vineland Chemical Co. Staff, Unpublished Work (1977).

(32) Loomis, H.P., Vineland Chemical Co. Staff, Unpublished Work (1963).

(33) Von Endt, D.W., Kearney, P.C., and Kaufman, D.D., Agr. and Food Chem., 16, 17 (1968). Frost, D.V., "Arsenicals in Biology--Retrospect and Prospect". Fed. Proc. 26, 194-208 (1967); Elaborate studies have been made on the

effects of methanearsonates on several animal species and data have been compiled on "The Risk Assessment of Methanearsonates" Prepared by The NACA Industry Task Force on Methanearsonates, (1976); see also ref. 5.

(34) The speciation of arsenic in the water samples was performed by the courtesy of Dr. E.A. Woolson, USDA, Beltsville, Maryland.

DISCUSSION

Name not given: What does it cost to treat water at the New Jersey wastewater treatment plant and what did it cost to build the plant?

A. Schwerdtle: I was afraid somebody was going to ask that. On good days we think it's about a penny a gallon, and on bad days, ten. On most days we think it's about two cents a gallon treated. We don't have an exact accounting of everything that was put into it, but I would say that it's well over half a million dollars but under one million. It depends on the amortization rate.

R. D. Wauchope: Is there a particular reason you adjust to pH 5.5?

A. Schwerdtle: We've tried fractional pHs to a tenth from four to nine. It is the optimum.

J. Andelman: Is this aquifer used by anyone as a water supply and, if so, what arsenic levels are found in the waters?

A. Schwerdtle: It is just the top soil that has the higher concentration of arsenic and there are clay barriers just beneath the soil. We have a number of wells in the area and the water in the aquifer has zero arsenic according to our tests.

R. T. Price: I was wondering in all the work that you've done in your lab studies, what is the lowest concentration of arsenic that you've ever found?

A. Schwerdtle: Are you referring to the methanearsonates?

R. T. Price: Yes, sir.

A. Schwerdtle: We've found that the practical lowest level with our present knowledge to be 0.7 parts per million.

R. T. Price: Is that more of an average or is that a low value?

A. Schwerdtle: That's a little on the high average side. We've gone below it, but as I said, there is not a natural stoichiometric relationship between the ferric hydroxide and the concentrations, so there is a point beyond which further addition of ferric hydroxide has no effect.

R. T. Price: Have you had any trouble with any of your floc rising after it's settled down?

A. Schwerdtle: Yes, and if you want to overcome that, increase the temperature.

R. T. Price: Can you give me a range of temperature that has been best for you?

A. Schwerdtle: I would say probably 35 degrees centigrade would be the maximum required. We do not know why that happens; we just haven't had the problem lately.

NOTE: A. Schwerdtle responded for N. E. Krapf who was not present.

22

DETERMINATION OF ARSENIC AND ARSENIC COMPOUNDS IN WATER SUPPLIES

Kurt J. Irgolic, R. A. Stockton and D. Chakraborti[1]
Department of Chemistry, Texas A&M University
College Station, Texas 77843

During the past few years it became apparent that accurate and precise values for trace element concentrations in environmental samples give only part of the information required for the assessment of the impacts of trace elements on biological systems. The concentrations of trace element compounds, which exert beneficial or detrimental effects on the basis of their distinct chemical and/or physical properties, must also be known for complete characterization of a sample. To determine the potential impact of arsenic-containing water on biological systems, knowledge of the type of arsenic compounds present is of utmost importance.

Water samples devoid of organisms will very likely contain only inorganic arsenic in form of arsenite and/or arsenate. Biologically mediated reactions can transform inorganic arsenic compounds into methylated arsenic derivatives, arsenocholine, arsenobetaine, arsenolipids and arsenocarbohydrates. Methods for the determination of total arsenic concentrations including inductively coupled argon plasma emission, flame atomic absorption and graphite furnace atomic absorption spectrometry, polarography and colorimetry have detection limits in the low ppb range. Arsenic compounds can be determined by gas chromatography; reduction of arsenic compounds to volatile hydrides followed by separation of the arsines and detection by flame AA, microwave emission or GC detectors; separation by ion chromatography or high pressure liquid chromatography with element-specific detectors; and differential pulse polarography. Detection limits in the ppt range have been reported for methylarsenic compounds.

Since water samples can rarely be analyzed immediately after collection, they must be preserved

---

[1]On leave from the Department of Chemistry, Jadavpur University, Calcutta-32, India.

to prevent losses and transformations of arsenic compounds. The preservation methods found in the literature are summarized. Water samples from Hinckley and Delta, Utah; Antofagasta, Chile; Alaska; Taiwan; Nova Scotia; Fallon, Nevada; Michigan and New Hampshire were analyzed by several independent methods. Only arsenite and arsenate were found in varying ratios. To assure that losses of arsenic and chemical changes of arsenic compounds would be detected unpreserved samples, samples preserved with nitric acid and samples preserved with ascorbic acid were analyzed. In most cases the preserved and unpreserved samples gave the same results.

1. Introduction

Analytical chemistry has made phenomenal progress in developing sensitive methods for the determination of trace elements in water and other matrices. Not too long ago the limits of detectability were at the part per million level. Now, determinations of trace elements are routinely carried out at low part per billion concentrations, and several methods with detection limits in the part per trillion range are available. As analytical methods for the determination of trace elements are extended to ever lower concentrations, increasing attention must be paid to matrix interferences. Interlaboratory calibration exercises are no guarantee that such interferences are detected, unless several independent analytical techniques are employed. Although the precision of such results may be adequate to excellent, their accuracy may be rather poor.

During the past few years it became apparent that accurate and precise values for trace element concentrations in environmental samples give only part of the information required for the assessment of impacts of trace elements on biological systems. Trace elements occur in nature as chemical compounds. These compounds interact with biologically important molecules and exert their beneficial or detrimental effects on the basis of their distinct chemical and/or physical properties. The well-known difference in toxicity between arsenite, arsenate and various organic arsenic derivatives may serve as an example of the compound-specific activity of a trace element.[1][2] To completely characterize, for instance, a water sample, the determination of the "total trace element" concentration must be followed by the determination of the concentrations of the compounds, which contain this trace element. Whereas many methods exist for the determination of "total trace element" concentrations, the choice of techniques for the determination of trace element compounds is generally much more limited. A sample does not contain only one trace element and its compounds. Other trace elements and their compounds are usually present as cationic, anionic or neutral

[2]Figures in brackets indicate the literature references at the end of this paper.

species. Although the determination of all of these species is still too time consuming to be practical, the total trace element concentrations can now be easily determined by sequential or simultaneous inductively coupled argon plasma emission spectrometry. The availability of such data makes it possible to look for the presence of trace elements and trace element compounds, which could enhance or reduce their detrimental effects through chemical interactions. In this paper the analytical techniques available for the determination of arsenic and arsenic compounds are reviewed and the experiences gained during the analyses of water supplies famous for their arsenic content are discussed.

2. Arsenic Compounds in Water Samples

Arsenic is associated with igneous and sedimentary rocks in the form of inorganic arsenic compounds. Arsenic is found frequently as a component of sulfidic ores in form of arsenides of nickel, cobalt, copper and iron. The arsenic concentration in igneous rocks varies from 0.06 ppm to 113 ppm [2,3]. Weathering of arsenic-containing rocks produces arsenic trioxide, arsenites and arsenates which might be retained in soils, dissolved in water, transported into lakes and into the ocean and deposited in sediments. Sedimentary rocks, including coal, have been found to contain 0.1 ppm to 2900 ppm of arsenic [2,3]. The arsenic concentrations in fresh surface waters and ground waters are generally below 50 ppb, the maximal permissible concentration in drinking water [4]. Seawater contains arsenic in the range of 1 ppb to 8 ppb [5]. Sediments are generally richer in arsenic than the waters with which they are in contact [4]. In the absence of living matter and without any input of man-made arsenic compounds water most likely will contain only arsenite and/or arsenate.

Many reactions are known which produce organic arsenic compounds from inorganic arsenic derivatives [6]. Some of these reactions occur in aqueous systems. The conversion of inorganic to methylated arsenic compounds mediated by various organisms is well established [7,8,9]. Methylarsines, methylarsonic acid and dimethylarsinic acid may thus be formed. Microorganisms can also demethylate methylarsenic derivatives [10]. These methylation and demethylation reactions, the reduction of arsenate to arsenite, the oxidation of arsenite, the reduction of pentavalent methylarsenic compounds to methylarsines, and the oxidation of these arsines set up an arsenic cycle [10]. Arsenic compounds are distributed among the various compartments of the environment including air, water, soil, sediments, rocks, plants and animals. Living matter seems to be a rich source of many organic arsenic compounds. Arsenic is known to be present in marine and freshwater organisms in the form of water- and lipid-soluble organic arsenic compounds [11,12]. The suggestion was made [11] that arsenic is incorporated into lipids by replacing nitrogen in lecithins.[3] A phospholipid containing 0.5 percent arsenic was isolated from the alga

[3]Formulas for these compounds are given in the paper "Biochemical Reactions in Environmental Situations."

Tetraselmis chui [13]. Hydrolysis of arsenic-containing lipids would yield water-soluble arsenocholine.[3] Arsenobetaine,[3] an oxidation product of arsenocholine, was isolated from the Australian rock lobster [14]. *Chaetoceros concavicornis* grown in arsenate-containing, axenic media is reported to have synthesized 2-carboxyarsenocholine[3] [15]. An arsenic containing sugar[3] was isolated from *Ecklonia radiata* [16]. Most of these investigations employed marine organisms. It is very likely, that freshwater organisms are capable of synthesizing the same or similar arsenic compounds. The isolation and identification of organic arsenic compounds has just begun. Many more, yet unknown arsenic compounds will be discovered and their properties determined as analytical techniques are improved for the speciation of such derivatives.

In natural freshwater not polluted with arsenic compounds through human activities, arsenite and arsenate are expected to be found. Based on the current knowledge of the arsenic cycle in nature, methylarsenic compounds, arsenobetaine and arsenocholine-type compounds might also be present.

Man-made arsenic compounds are used as pesticides, herbicides, fungicides, cotton desiccants, wood preservatives, feed additives and drugs [17,18]. Most of these compounds are either inorganic, belong to the class of methylated arsenic derivatives or substituted benzenearsonic acids and will be detected and quantitatively determined using the techniques discussed in the next section.

3. Analytical Techniques for the Determination of Arsenic and Arsenic Compounds

Several methods are available, which can be relied upon to give "total arsenic" concentrations irrespective of the nature of the chemical compounds containing arsenic. These methods are neutron activation analysis [19,20], X-ray fluorescence spectroscopy [20], plasma emission spectrometry [20] and perhaps flame and flameless atomic absorption spectrometry [20]. The detection limits for these methods are listed in table 1. Most of the other techniques, which are generally said to give "total arsenic" concentrations, provide only values for "total inorganic arsenic" as the sum of arsenite and arsenate.

The choice of methods for the determination of specific arsenic compounds is rather limited. These methods are briefly reviewed below and are summarized in tables 1-4. The tables do not contain all of the variants of the techniques reported in the literature for the determination of arsenic and arsenic compounds, but only selected procedures, most of which we have used in our laboratories.

---

[3]Formulas for these compounds are given in the paper "Biochemical Reactions in Environmental Situations."

TABLE 1

THE DETERMINATION OF ARSENIC AND ARSENIC COMPOUNDS AT TRACE LEVELS

Method	Total Arsenic	Arsenite	Arsenate	Alkylarsenic Compounds	Other Organic Arsenic Compounds	References
Neutron activation analysis	<1[b]	-	-	-	-	[19,20]
Flame atomic absorption spectrometry	500[b]	-	-	-	-	[21]
Graphite furnace atomic absorption spectrometry	0.2[b]	-	-	-	-	[22]
X-ray fluorescence	0.05[b]	-	-	-	-	[20]
Plasma emission spectrometry	20[b]	-	-	-	-	[20,23]
Silver diethyldithiocarbamate (colorimetry)	20[a]	20[a]	20[a]	$CH_3AsO_3H_2$, 20[a]	-	[24]
	∿10[b,d]	∿10[b]	∿10[b]	-	-	[25,26]
	100[a,d]	100[a,c]	100[a,c]	-	-	[27]
Molybdenum blue (colorimetry)	∿1[b,e]	∿1[b,e]	∿1[b]	-	-	[30]

[a]Detection limits are given in ppb of arsenic compound. [b]Detection limits are given in ppb of arsenic. [c]Separation of arsenite from arsenate is achieved by pH dependent reduction with $NaBH_4$. [d]Total arsenic refers to arsenite + arsenate. [e]After oxidation of arsenite to arsenate.

3.1. Silver diethyldithiocarbamate method

The determination of inorganic arsenic (arsenite and arsenate) by reduction of these compounds by zinc/HCl to arsine and the colorimetric measurement at 540 nm of the complex formed between silver diethyldithiocarbamate and arsine in pyridine requires only inexpensive instrumentation and has a detection limit of approximately 10 ppb (table 1) [24,25,26]. Methylarsonic acid can be determined in the presence of arsenate and arsenite after reduction to methylarsine at 440 nm [24]. Employing the pH-dependent reduction of arsenite and arsenate by sodium borohydride arsenite and arsenate have been determined separately [27]. Interferences have been studied [28,29].

3.2. Molybdenum blue method

Arsenate (but not arsenite) reacts with ammonium molybdate to form blue arsenomolybdate, the absorbance of which is measured at 865 nm. Arsenite can be oxidized to arsenate for the determination of total inorganic arsenic [30] with a detection limit of approximately 1 ppb.

3.3 Polarography

Arsenite can be determined in sulfuric acid medium by single sweep polarography (detection limit 5 ppb) [3] or in aqueous hydrochloric acid by differential pulse polarography (detection limit 0.3 ppb) [32]. Arsenate is polarographically inactive under these conditions. Dimethylarsinic acid [35,38], methylarsonic acid [35,36,38], and other aliphatic [36,39] and aromatic [39] arsonic and arsinic acids are reducible on the dropping mercury electrode (table 2). Arsenate becomes polarographically active in the presence of polyhydroxy compounds and can be quantitatively determined in 2.0 M aqueous perchloric acid in the presence of D-mannitol with a detection limit of approximately 6 ppb [33]. Arsenite, arsenate and total inorganic arsenic may be determined under these conditions.

3.4 Hydride generation technique

Arsenic compounds reducible to volatile arsines by sodium borohydride or other reducing agents can be easily identified and quantitatively determined with detection limits in the low part per trillion range (table 3). The hydride generation technique is most often used for the determination of arsenite, arsenate, total inorganic arsenic, methylarsonic acid, dimethylarsinic acid [40-46,47] and trimethylarsine oxide [42]. Ethyl-, propyl- and butylarsonic acid can also be determined by this method [46]. Speciation of these compounds is achieved by the pH-dependent reduction of arsenite and arsenate to arsine [40,43], separation of arsine and organic arsines by selective volatilization from a cold trap [40,42,46], or gas chromatographic separation of the arsines [42,46]. Helium-DC discharge, flame atomic absorption, microwave emission, flameless atomic absorption, flame ionization and electron capture were employed for the detection of the

TABLE 2

DETERMINATIONS OF ARSENIC COMPOUNDS BY POLAROGRAPHY

Method	Total Arsenic	Arsenite	Arsenate	Alkylarsenic Compounds		Other Organic Arsenic Compounds		References
Single sweep polarography	-	5[a]	-	-		-		[31]
Differential pulse polarography	-	0.3[a]	-	-		-		[32]
Differential pulse polarography	∿6[a]	∿6[a]	∿6[a]	-		-		[33,37]
Differential pulse polarography	2[a]	2[a]	-	-		-		[34]
Differential pulse polarography	-	-	-	$CH_3AsO_3H_2$	2,000[b]	-		[35,36]
	-	-	-	$CH_3AsO_3H_2$	18[b,c]	-		[38]
	-	-	-	$C_2H_5AsO_3H_2$	5,000[b]	-		[36]
	-	-	-	$C_3H_7AsO_3H_2$	1,320[b]	-		[36]
	-	-	-	$C_4H_9AsO_3H_2$	610[b]	-		[36]
	-	-	-	$C_5H_{11}AsO_3H_2$	270[b]	-		[36]
	-	-	-	$C_6H_{13}AsO_3H_2$	170[b]	-		[36]
	-	-	-	$C_7H_{15}AsO_3H_2$	140[b]	-		[36]
	-	-	-	$C_8H_{17}AsO_3H_2$	110[b]	-		[36]
	-	-	-	$C_6H_5CH_2AsO_3H_2$	250[b]	$C_6H_5AsO_3H_2$	520[b]	[39]
	-	-	-	$C_6H_5CH_2CH_2AsO_3H_2$	330[b]	-		[39]
Differential pulse polarography	-	-	-	$(CH_3)_2AsOOH$	2,000[b]	-		[35]
	-	-	-	$(CH_3)_2AsOOH$	8[b,c]	-		[38]
	-	-	-	$(C_2H_5)_2AsOOH$	500[b]	$(C_6H_5)_2AsOOH$	110[b]	[39]
	-	-	-	$(C_3H_7)_2AsOOH$	350[b]	$(C_6H_5CH_2)_2AsOOH$	130[b]	[39]
	-	-	-	$(C_4H_9)_2AsOOH$	210[b]	$(C_6H_5CH_2CH_2)_2AsOOH$	140[b]	[39]
	-	-	-	$(c\text{-}C_6H_{11})_2AsOOH$	150[b]	-		[39]

[a]Detection limits are given in ppb of arsenic. [b]Detection limits are given in ppb of arsenic compound. [c]After digestion with perchloric acid and reduction to arsenite with sulfur dioxide.

arsines (table 3). Figure 1 summarizes the various instrumental systems developed for the determination of arsenic compounds by the hydride generation technique.

Arsenic compounds can be separated by ion chromatography [44] or by high pressure liquid chromatography before their reduction to arsines. Atomic absorption spectrometers [44] or inductively coupled argon plasma spectrometers [55] may serve as detectors in these systems. Care should be exercised when arsenic compounds are reduced with sodium borohydride. Some arsenate may be reduced to arsine under conditions under which only arsenite should react, and cleavage of methyl groups from methylarsenic compounds has been observed during their reaction with sodium borohydride [43]. Improvements concerning the reduction conditions [43] and modifications of the instrumental system [43-45] made the hydride generation technique a more reliable method for the determination of arsenic compounds.

The disadvantages of the hydride generation method include the requirement that the arsenic compounds must be reducible to arsines, and the necessity that the arsines possess sufficient volatility to allow their transfer to the detectors. Many arsenic compounds exist which do not fulfill these requirements.

3.5. Gas chromatography

Volatile arsenic compounds such as methylarsines can be identified and determined directly by gas chromatography. Methylarsine, dimethylarsine and trimethylarsine were volatilized from water samples by stripping with helium and separated by gas chromatography on a column packed with 16.5% silicone oil DC-550 on 80-100 mesh Chromosorb WAW DMCS [42] employing flame ionization, electron capture or a flame atomic absorption detector.

Methylarsine, dimethylarsine, trimethylarsine, ethylarsine, propylarsine and butylarsine, obtained by reduction of the pertinent arsonic acids, arsinic acids or arsine oxide, were separated by gas chromatography with a mass spectrometer as an element-specific and compound-specific detector, which can also provide the elemental compositions of the compounds [46].

Trimethylarsine was separated from dimethyl selenide and tetramethyl tin by gas chromatography with a graphite furnace atomic absorption spectrometer as an element-specific detector [56].

Most of the arsenic compounds are not sufficiently volatile for gas chromatography. Such compounds must be chemically modified to increase their volatility. Methylarsonic acid combines with ethylene glycol [57], dimethylarsinic acid reacts with hydriodic acid [58], and both acids interact with allylthiourea [59] to form arsenic derivatives which can be chromatographed. The reactions of arsenic compounds with trimethylsilyl chloride produce volatile arsenic derivatives. The method is, unfortunately, non-selective, because several arsenic compounds gave the same product [60].

TABLE 3

DETERMINATION OF ARSENIC AND ARSENIC COMPOUNDS BY THE HYDRIDE GENERATION TECHNIQUE

Method	Total Arsenic	Arsenite	Arsenate	Alkylarsenic Compounds	Other Organic Arsenic Compounds	References
Hydride generation/DC discharge	0.02[a,b]	0.02[a]	0.02[a]	$CH_3AsO_3H_2$	-	[40,43]
				$(CH_3)_2AsOOH$	-	[40]
Hydride generation/flame AA	500 ng[b]	-	-	$CH_3AsO_3H_2$ 1 μg	-	[41]
				$(CH_3)_2AsOOH$ 1 μg	-	[41]
Hydride generation/GC-FID	0.0005[a,b]	0.0005[a]	0.0005[a]	$CH_3AsO_3H_2$ 0.0005[a]	-	[42]
GC-ECD	0.0005[a,b]	0.0005[a]	0.0005[a]	$(CH_3)_2AsOOH$ 0.0005[a]	-	[42]
GC-flame AA	0.0005[a,b]	0.0005[a]	0.0005[a]	$(CH_3)_3AsO$ 0.0005[a]	-	[42]
Hydride generation/heated quartz cell-AA preceded by separation by ion chromatography	-	10[a]	10[a]	$CH_3AsO_3H_2$ 10[a]	4-$H_2NC_6H_4AsO_3H_2$	[44]
				$(CH_3)_2AsOOH$ 10[a]		
Hydride generation/extraction with benzene or toluene/collection in cold toluene/or liquid N_2 trap-GC-microwave emission spectrometry	-	-	0.25[a]	$CH_3AsO_3H_2$ 0.25[a]	-	[46]
	-	-	-	$(CH_3)_2AsOOH$ 0.25[a]	-	[46]
	-	-	-	$C_2H_5AsO_3H_2$ 0.25[a]	-	[46]
	-	-	-	$C_3H_7AsO_3H_2$ 0.25[a]	-	[46]
	-	-	-	$C_4H_9AsO_3H_2$ 0.25[a]	-	[46]
Hydride generation/flameless AA	1.0[a]	-	-	-	-	[47]

[a]Detection limits are given in ppb of arsenic. [b]Total arsenic refers to arsenite + arsenate.

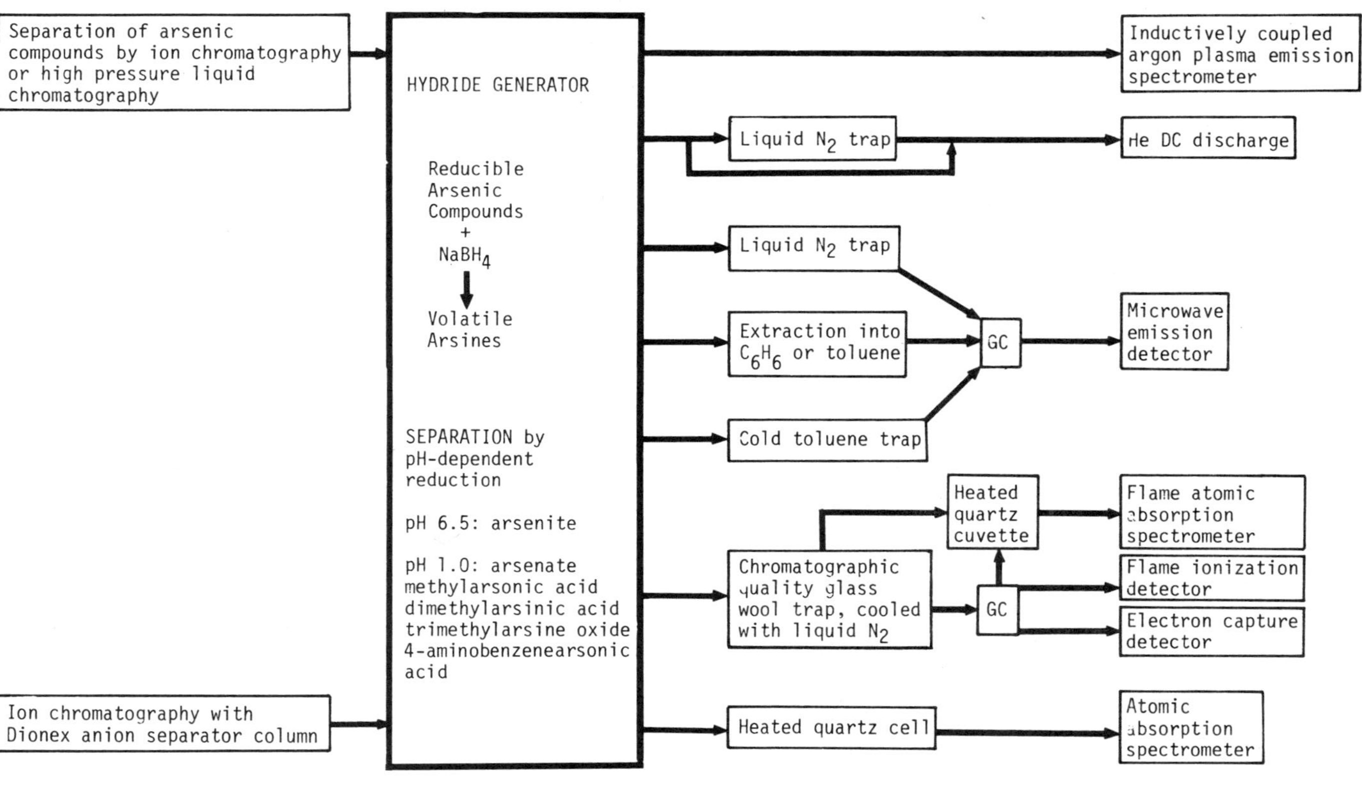

Figure 1. Instrument systems for the determination of arsenic and arsenic compounds by the hydride generation technique

Methods which digest samples to convert arsenic compounds to inorganic derivatives [61] and subsequently transform these into organic derivatives suitable for gas chromatography [20,62] cannot be employed for speciation but used only for the determination of total arsenic concentrations.

3.6. Flameless atomic absorption spectrometry

Flameless atomic absorption spectrometry is one of the most widely used methods for the determination of arsenic. Since the arsenic compounds are decomposed and atomized in graphite cups, graphite tubes, flames or specially designed cells at temperatures sometimes approaching 3000°C, only total arsenic can be determined. The signals generated by amounts of arsenic compounds containing the same amount of arsenic are dependent on the volatility of the arsenic compounds [52]. A truly quantitative determination of arsenic in a mixture of arsenic compounds seems to be possible only when the calibration curves have been obtained with standards of the same composition as the mixture to be analyzed.

Graphite furnace atomic absorption spectrometers have a detection limit for arsenic of approximately 0.5 ppb [48]. Solutions with much lower arsenic concentrations can be analyzed, when arsenic compounds are extracted by appropriate reagents (diethyldithiocarbamate [49], dibutyldithiophosphate [50]) from large volumes of water and enriched samples are injected into the graphite cup.

Arsenic compounds can be speciated either by selective extraction [49,50] or by separation prior to analysis by atomic absorption spectrometry.

High pressure liquid chromatography (HPLC) is potentially the best technique for the separation of non-volatile arsenic compounds. The column effluent can then be routed into the graphite cup or tube of a graphite furnace atomic absorption spectrometer (GFAA), which serves as an arsenic-specific, sensitive detector. Two automated systems combining HPLC with GFAA are now available. The first system uses a Perkin-Elmer GFAA with a specially adapted autosampler [51]. The second system employs a Hitachi-Zeeman GFAA with a sample valve, an injector and associated electronics to control the analysis sequence [63]. A computerized version of the HPLC-GFAA interface applicable to other graphite furnace atomic absorption spectrometers is now available [64]. These HPLC-GFAA systems have thus far been used to separate and determine arsenite, arsenate, methylarsonic acid, dimethylarsinic acid, triphenylarsine, *p*-aminophenylarsonic acid, arsenobetaine and arsenocholine. References to these investigations are given in table 4.

GFAA as an element-specific detector equipped with an arsenic lamp has the advantage of responding only to arsenic compounds. Water samples may contain substances in addition to arsenic compounds. Upon passing through an HPLC column these compounds may be separated and bands of arsenic compounds may overlap with bands of other substances.

TABLE 4

THE DETERMINATION OF ARSENIC AND ARSENIC COMPOUNDS BY FLAMELESS ATOMIC ABSORPTION SPECTROMETRY AND BY GAS CHROMATOGRAPHY

Method	Total Arsenic	Arsenite	Arsenate	Alkylarsenic Compounds		Other Organic Arsenic Compounds		References
GFAA	0.5[a]	-	-	-		-		[48]
Extraction by diethylammonium diethyldithiocarbamate/CCl_4-GFAA	1.0[a]	1.0[a]	1.0[a]	1.0[a,b]		1.0[a,b]		[49]
Extraction with dibutyldithio-phosphate/hexane-GFAA	-	0.006[a]	0.006[a]	-		-		[50]
HPLC-GFAA	-	-	-	-		$(C_6H_5)_3As$		[51]
HPLC-GFAA	-	~100[a]	~100[a]	$CH_3AsO_3H_2$	~100[a]	$4\text{-}H_2NC_6H_4AsO_3H_2$		[52]
	-	-	-	$(CH_3)_2AsOOH$	~100[a]			
HPLC-GFAA	-	~250[a]	~250[a]	$CH_3AsO_3H_2$	250[a]	-		[53]
				$(CH_3)_2AsOOH$	250[a]	-		[53]
HPLC-GFAA	~					$[(CH_3)_3AsCH_2CH_2OH]^+X^-$		[54]
						$(CH_3)_3AsCH_2COO$	~100[a]	[54]
GC-flame AA	-	-	-	CH_3AsH_2	0.0005[a]	-		[42]
				$(CH_3)_2As$	0.0005[a]			[42]
				$(CH_3)_3As$	0.0005[a]			[42]

[a]Detection limits are given in ppb of arsenic. [b]Determinable if compounds can be extracted into CCl_4. [c]Total arsenic refers to arsenite + arsenate.

The common detectors (refractive index, uv-visible spectrometer) will not respond specifically to arsenic compounds, whereas the GFAA detector will respond only to arsenic compounds. The identification of arsenic-containing fractions is, therefore, much easier with the HPLC-GFAA system.

The high sensitivity of GFAA for arsenic is, of course, retained in the HPLC-GFAA system for each injection. Upon migration through the chromatographic column the arsenic compounds are separated and spread into bands. Aliquots of 40-100 μL are withdrawn from the effluent. The aliquots taken from the center of a band, where the arsenic concentration is highest, must contain arsenic at a concentration of at least 0.5 ppb. The detection limit of the HPLC-GFAA system is, therefore, strongly dependent on the degree of band spreading. Under optimized conditions several nanograms of an arsenic compound placed on the column should produce discernable GFAA signals. The sensitivity of the HPLC-GFAA system can be increased by optimization of the chromatographic parameters, by using multiple injection - drying cycles before atomization or by preconcentration of arsenic compounds.

3.7. Inductively coupled argon plasma emission spectrometry

Inductively coupled argon plasma emission spectrometry (ICP) can determine many elements in a small volume of sample, is almost interference-free and is capable of analyzing many samples in a short time. ICP, unfortunately, does not have very good sensitivity for arsenic. Detection limits for arsenic vary from 10 to 80 ppb. A hydride generation attachment has recently become available, which promises to allow the determination of reducible arsenic compounds at sub-ppb levels.

ICP may serve not only as an arsenic-specific detector, but also as a multi element-specific detector for HPLC. Arsenite, arsenate, methylarsonic acid, dimethylarsinic acid and arsenobetaine were separated by HPLC and determined with an ICP with a detection limit of 2.6 ng/s at a flow rate of 1 mL [65]. Arsenite, arsenate, methylarsonic acid, dimethylarsinic acid, selenite and selenate were also separated by HPLC and determined by ICP. Selenium-containing bands and arsenic-containing bands overlapped. ICP as multi element-specific detector allowed the identification and quantitative determination of the overlapping bands containing compounds of different elements [66].

4. Preservation of Arsenic Compounds in Aqueous Solutions

In contrast to most metal ions, which do not change valence easily and are not transformed into organic compounds in aqueous solution, arsenic possesses a complicated chemistry characterized by equilibria, steady state conditions or situations approaching such conditions between trivalent and pentavalent inorganic arsenic compounds and various derivatives containing arsenic-carbon bonds. The analysis for arsenic compounds in water samples will correctly give the composition of the sample at the time of collection only if no arsenic compounds were lost, interconversions of arsenic compounds present in the samples did not occur and no new arsenic compounds were formed during the time between collection and analysis. The determination of total arsenic concentrations in samples requires only that no arsenic shall be lost and that the analytical techniques to be employed shall be applicable to all the arsenic compounds present in the sample.

Water samples can generally not be analyzed for trace elements immediately after collection. Several hours or more often several days elapse between collection and analysis. During this time the chemical nature of a particular trace element can change, trace elements can be lost by volatilization and/or can be adsorbed on container walls. The adsorption of many metal ions [67,68] and of phosphate ions [69] were studied but arsenic was rarely included.

Disagreements exist in the literature as to the extent of loss of arsenic from solution stored in various containers. These disagreements are probably caused by differences in composition of the samples and sample containers, the types of arsenic compounds present and the sensitivity of analytical methods.

Acidification of samples is usually considered necessary when trace elements are to be determined [67,68]. Several acids (nitric, perchloric, hydrochloric and acetic acid) were suggested as preservatives. Guimont and co-workers [70] mention the use of acetic acid as a preserving agent for arsenic solutions but give no details. Ray and Johnson [19] report losses of arsenic up to 70 percent within one week from seawater acidified with 9 mL concentrated hydrochloric acid per liter of sample. Talmi and Bostick [20] observed no losses of arsenate from distilled-water solutions containing 15 percent nitric acid or 5 percent perchloric acid upon storage for three weeks in polyethylene or soft glass containers. Whitnack and Brophy [31] reported no loss of arsenite from samples of well water kept for one week in polystyrene vials with snap caps of polyethylene. Al-Sibaai and Fogg [71] found that the total arsenic concentrations in dilute standard solutions containing only arsenite or arsenate (20 μg mL^{-1}) remained unchanged for at least 50 days when stored in borosilicate glass, soda glass or polyethylene in light or in the dark. Experiments with seawater containing 2 μg L^{-1} of arsenic gave different results [72]. When such filtered seawater samples equilibrated with 5 μCi of arsenic-74 were placed in high density polyethylene, soda glass or Pyrex bottles and stored in the dark over a period of five weeks, losses of arsenic occurred with all types of bottles [72]. With soda glass bottles sixteen percent had been adsorbed by the 15th day when equilibrium was attained. The uptake by polyethylene and Pyrex bottles was complete within ten days and amounted in both cases to approximately six percent [72].

Storage of water samples in the frozen state has been recommended [67,68,72,73]. Harrison and co-workers [74] found that frozen solutions (1 ppm As) did not lose arsenic upon freeze-drying. The average retention was 96.3±1.8 percent. Natural water samples stored below -15° or under dry ice did not lose arsenite [42]. Seawater samples kept in a refrigerator at 4° showed no measureable change in the arsenite/arsenate ratio within about ten days [75]. For longer storage the samples had to be frozen quickly with crushed dry ice within 15 minutes. They could then be stored frozen. In samples frozen in a freezer, spurious losses of arsenite were observed [76].

Andreae [42] stated that mono-, di- and trimethylarsine were stable in aqueous solutions for a few days when stored in air-tight containers. Methylarsonic acid and dimethylarsinic acid were lost measurably after a period of approximately three days from untreated water samples. Acidification with hydrochloric acid to make the sample 0.05 M prevented these losses.

Arsenite and arsenate are rather easily changed into each other. Arsenite in natural water samples is slowly oxidized to arsenate. Acidification of the sample increases the oxidation rate [42]. Fifty percent of the arsenite in an aqueous standard solution (20 μg mL^{-1}) was oxidized to arsenate after storage in a glass container for 33 days [71]. Arsines were oxidized by traces of air [42]. An arsenic cycle operates in nature [10] connecting inorganic with organic arsenic compounds. The chemical nature of many of the organic arsenic compounds participating in this cycle are still unknown.

In experiments carried out in our laboratories to check whether Pyrex or soft polyethylene bottles (Cubitainers manufactured by Kimberly) remove arsenic compounds from solution, Cubitainers were found not to change the concentration of aqueous 1 ppm solutions of arsenite, arsenate or dimethylarsinic acid. Pyrex flasks of 25 mL volume adsorbed approximately one percent of the arsenic [63].

Among the chemical changes to which arsenic compounds may be subjected in aqueous solutions, the oxidation of arsenite to arsenate has been reported repeatedly in the literature [10,71,42,43].

The most convenient and experimentally tested procedure for the preservation of arsenite was reported by Feldman [43]. Ascorbic acid at a concentration of 1 mg mL^{-1} prevented the oxidation of arsenite to arsenate in aqueous standard solutions at room temperature. Arsenate was not reduced to arsenite under these conditions [43]. However, special care must be taken not to expose arsenite solutions preserved by ascorbic acid to elevated temperatures. Ascorbic acid has been used to reduce arsenate to arsenite in boiling seawater [72]. Whether ascorbic acid scavenges oxygen from the solution, combines with arsenite producing a species less reactive toward oxidation than arsenite or acts both ways in preventing oxidation of arsenite, is not known. Natural water samples contain many other metal ions. Comprehensive experiments to elucidate the influence of such metal ions on the ability of ascorbic acid to preserve arsenite have not yet been carried out. Metal ions do catalyze the oxidation of ascorbic acid by oxygen [77].

It is not possible to recommend at this time a method for the preservation of water samples applicable to all the various arsenic compounds which may occur in water. Not enough is known, for instance, about the adsorption behavior of inorganic and organic arsenic derivatives in the presence of other compounds generally found in water. Even less knowledge exists about the conditions under which arsenic compounds are converted to other species in the presence or absence of other trace elements, which might serve as catalysts for the transformation of arsenic compounds.

5. Analysis of Water Samples for Arsenic Compounds

During the past two years we analyzed approximately fifteen water samples for arsenic compounds. These samples came from locations within the United States, from Taiwan, Nova Scotia and Antofagasta. The samples had total arsenic concentrations in the range 18 ppb to 8 ppm. The arsenite/arsenate ratios were in the range 0.007 to 3.4. No indications of the presence of methylated arsenic compounds reducible to volatile arsines have been found. Experiments with the HPLC-GFAA system, which would provide information about the presence of organic arsenic compounds not reducible to volatile arsines, detected only arsenite and arsenate. Comparison of total arsenic concentrations with the sum of arsenite and arsenate concentrations place an upper limit of a few ppb on the concentrations of any other arsenic compounds which might be present.

Other trace elements present in these water samples were determined by ICP. No particularly offensive elements were found in these water supplies with exception of beryllium (7.4 ppm) in the untreated Antofagasta samples.

The various physiological effects observed in populations exposed to these arsenic-containing drinking water supplies could have been caused by the presence of varying amounts of arsenite and arsenate.

It is also conceivable that one or more of the trace elements present in the water supplies act in concert with arsenic to cause the observed effects. Many more samples need to be analyzed and the results of these analyses correlated with epidemiological studies before a definite statement can be made about the interactions of trace elements with arsenite or arsenate.

On the basis of the experience gained with these samples the following procedure is recommended for the analysis of arsenite and arsenate in water samples:

- Collect three samples for each water supply to be analyzed in one-quart polyethylene Cubitainers.
- Preserve the first sample by addition of 0.1 percent by weight of ascorbic acid.
- Preserve the second sample by addition of 7 mL ultrapure, concentrated nitric acid to one quart of sample.

- Leave the third sample unpreserved.
- Fill all the containers to the brim to avoid air spaces.
- Analyze the samples as soon as possible after collection.
- The samples should be kept at or below room temperature during shipment and storage.

If the concentrations of arsenate and arsenite in the three samples determined by at least two independent methods agree, arsenite and arsenate were not chemically changed during the period between collection and analysis. Among the fifteen different water samples analyzed by this procedure thus far, the arsenite and arsenate values obtained from a set of preserved and unpreserved samples did not agree in only one case.

6. Interferences in the Determination of Arsenic

A completely interference-free technique for the determination of arsenic and arsenic compounds is not available, although ICP might come closest to this ideal. Several papers identifying interferences with the diethyldithiocarbamate method [28,29] and with atomic absorption spectrometry [78-90] of arsenic and other elements have appeared in the literature. Ubiquitous anions such as phosphate [78, 89] and sulfate [79,83] suppress the GFAA arsenic signal. Arsenic may interact with the carbon of the graphite cup or tube [80,84]; arsenic compounds may be volatilized during the ashing cycle [52,85,86]; and particulate matter in natural waters may enhance the arsenic signal [87]. The addition of nickel salts does not always counteract these interferences [88].

As an example of the potential problems associated with the analysis of arsenic our experience with the analysis of a water sample for arsenic is cited. Direct injection of this drinking water sample into the GFAA under normal operating conditions (atomization at 1100°) gave arsenic concentrations approximately 80 percent lower than those found by ICP and the hydride generation technique. The precision of the GFAA values was excellent. GFAA analysis of the sample at an atomization temperature of 900° gave correct results. The initial arsenic determinations by GFAA would have been in serious error, if other methods would not have been used also. The incorrect values would have been reported with great confidence. Experiences of this nature suggest that arsenic determinations in drinking water samples and samples with more severe matrix problems shall be carried out by at least two independent methods to increase the reliability of the results.

7. Conclusions

Total inorganic arsenic, arsenite, arsenate and organic arsenic compounds (mainly methylarsonic acid, dimethylarsinic acid, trimethylarsine oxide) reducible to volatile arsines can be easily determined by a number of techniques with detection limits as low as a few ppt.

Organic arsenic compounds not reducible to arsines can be identified and quantitated by high pressure liquid chromatography-graphite furnace atomic absorption spectrometry (GFAA) or inductively coupled argon plasma emission spectrometry (ICP) systems, in which the spectrometers serve as element-specific detectors. Graphite furnace atomic absorption spectrometry is probably the most widely used technique for the determination of arsenic and arsenic compounds. When water samples are analyzed for arsenic, one should always be mindful of potential interferences, and should use at least two independent methods for the quantitative determination of arsenic to increase the reliability of the results. The availability of ICP for simultaneous trace element analysis and of ion chromatography for anion analysis makes possible a complete, chemical characterization of a water sample. Whenever possible such a complete analysis should be carried out in addition to the determination of total arsenic and arsenic compounds. These chemical data can be used to check whether other trace elements are present which could interact with arsenic compounds and reduce or enhance their effects on biological systems.

The analyses of fifteen water samples collected at localities famous for their arsenic-rich waters showed that only arsenite and arsenate were present at concentrations ranging from 18 ppb to 8 ppm. In freshwaters a variety of organic arsenic compounds formed by organism from inorganic arsenic could be present. A complete characterization of a water sample containing arsenic requires that analysis techniques, which will detect and quantitate these organic arsenic compounds, are used.

The support of the investigations, which were carried out at Texas A&M University and which are reported in this paper, by the U.S. Environmental Protection Agency, Grant R 8047 740 10, by the Robert A. Welch Foundation of Houston, Texas, and by Texas A&M University's Center for Energy and Mineral Resources is gratefully acknowledged.

References

[1] Arsenic, National Academy of Sciences, Washington, D.C., 1977, p. 130.

[2] Boyle, R. W., and Jonasson, I. R., The geochemistry of arsenic and its use as an indicator element in geochemical prospecting, J. Geochem. Explor., 2, 251-296 (1973).

[3] Onishi, H., Arsenic, Chapter 33 in Handbook of Geochemistry, K. H. Wedepohl, ed. (Springer Verlag, Berlin, 1969).

[4] Arsenic, National Academy of Sciences, Washington, D.C., 1977. pp. 20-23.

[5] Woolson, E. A., Bioaccumulation of arsenicals, ACS Symposium Series, 7, 97-107 (1975).

[6] Doak, G. O., and Freedman, L. D., Organometallic compounds of arsenic, antimony and bismuth, Wiley Interscience, New York, 1970, pp. 17-26, 63-81, 120-235.

[7] Cox, D. P., Microbiological methylation of arsenic, ACS Symposium Series, 7, 81-96 (1975).

[8] Braman, R. S., Arsenic in the environment, ACS Symposium Series, 7, 108-123 (1975).

[9] Challenger, F., Biosynthesis of organometallic and organometalloidal compounds, ACS Symposium Series, 82, 1-22 (1978).

[10] Woolson, E. A., Fate of arsenicals in different environmental substrates, Environ. Health Perspectives, 19, 73-81 (1977).

[11] Irgolic, K. J., Woolson, E. A., Stockton, R. A., Newman, R. D., Bottino, N. R., Zingaro, R. A., Kearney, P. C., Pyles, R. A., Maeda, S., McShane, W. J., and Cox, E. R., Characterization of arsenic compounds formed by Daphnia magna and Tetraselmis chui from inorganic arsenate, Environ. Health Perspectives, 19, 61-66 (1977); and references cited there.

[12] Lunde, G., Occurrence and transformation of arsenic in the marine environment, Environ. Health Perspectives, 19, 47-52 (1977); and references cited there.

[13] Bottino, N. R., Cox, E. R., Irgolic, K. J., Maeda, S., McShane, W. J., Stockton, R. A., and Zingaro, R. A., Arsenic uptake and metabolism by the alga Tetraselmis chui, ACS Symposium Series, 82, 116-129 (1978).

[14] Edmonds, J. S., Francesconi, K. A., Cannon, J. R., Raston, C. L., Shelton, B. W., and White, A. H., Isolation, crystal structure and synthesis of arsenobetaine, the arsenical constituent of the western rock lobster Panulirus Longipes George, Tetrahedron Letters, 1543-1546 (1977).

[15] Cooney, R. V., Mumma, R. O., and Benson, A. A., Arsoniumphospholipid in algae, Proc. Natl. Acad. Sci. USA, 75, 4262-4264 (1978).

[16] Edmonds, J. S., and Francesconi, K. A., Arseno-sugars from brown kelp (Ecklonia radiata) as intermediates in cycling of arsenic in a marine ecosystem, Nature, 289, 602-604 (1981).

[17] Arsenic, National Academy of Sciences, Washington, D.C., 1977, pp. 26-43.

[18] Burrus, R. P., Jr., and Sargent, D. H., Technical and micro-economic analysis of arsenic and its compounds, EPA Report No. 560/6-76-016; NTIS PB-253 980, April 1976.

[19] Ray, B. J., and Johnson, D. L., A method for the neutron activation analysis of natural waters for arsenic, Anal. Chim. Acta, 62, 196-199 (1972).

[20] Talmi, Y., and Bostick, D. T., The determination of arsenic and arsenicals, J. Chromatogr. Sci., 13, 231-237 (1975) and references cited there.

[21] Kirkbright, G. F. and Ranson, L., Use of nitrous oxide - acetylene flame for determination of arsenic and selenium by atomic absorption spectrometry, Anal. Chem., 43, 1238-1241 (1971).

[22] Koizumi, H., Yasuda, K., and Katayama, M., Atomic absorption spectrometry based on the polarization characteristics of the Zeeman effect, Anal. Chem., 49, 1106-1112 (1977).

[23] Irgolic, K.J. and Stockton, R. A., unpublished results obtained with an ARL 34000 ICP.

[24] Peoples, S. A., Lakso, J., and Lais, T., The simultaneous determination of methylarsonic acid and inorganic arsenic in urine, Proc. West. Pharmacol. Soc., 14, 178-182 (1971).

[25] Standard Methods for the Examination of Water and Wastewater, 15th ed., American Public Health Association, Washington, D.C., 1980, pp. 174-175.

[26] Stratton, G., and Whitehead, H. C., Colorimetric determination of arsenic in water with silver diethyldithiocarbamate, J. Amer. Water Works Association, 54, 861-864 (1962).

[27] Howard, A. G., and Arbab-Zawar, M. H., Sequential spectrophotometric determination of inorganic arsenic(III) and arsenic(V) species, Analyst (London), 105, 338-343 (1980).

[28] Sandhu, S. S., and Nelson, P., Ionic interferences in the determination of arsenic in water by the silver diethyldithiocarbamate method, Anal. Chem., 50, 322-325 (1978).

[29] Jueneman, F., As you like it - or, that old Sb, Chemist-Analyst (J. T. Baker Chemical Co.), 64(3), 1-4 (1975).

[30] Johnson, D. L., and Pilson, E. Q., Spectrophotometric determination of arsenite, arsenate and phosphate in natural waters, Anal. Chim. Acta, 58, 289-299 (1972).

[31] Whitnack, G.C., and Brophy, R. G., A rapid and highly sensitive single-sweep polarographic method of analysis for arsenic(III) in drinking water, Anal. Chim. Acta, 48, 123-127 (1969).

[32] Myers, D. J., and Osteryoung, J., Determination of arsenic(III) at the parts-per-billion level by differential pulse polarography, Anal. Chem., 45, 267-271 (1973).

[33] Nichols, R. L., Polarographic determination of arsenate in the presence of polyhydroxy compounds, MS thesis, Texas A&M University, December 1978.

[34] Howe, L. H., Trace Analysis of arsenic by colorimetry, atomic, absorption, and polarography, EPA Report No. 600/7-77-036; NTIS PB-269 652 (1977).

[35] Elton, R., and Geiger, W. E., Jr., Electroactivity of cacodylic acid in aqueous and nonaqueous media, Anal. Lett., 9, 665-670 (1976).

[36] Bess, R. C., Irgolic, K. J., Flannery, J. E., and Ridgway, T. H., Polarographic reduction of alkylarsonic and dialkylarsinic acids, Anal. Lett., 9, 1091-1097 (1976).

[37] Henry, F. T., Kirch, T. O., and Thorpe, T. M., Determination of trace level arsenic(III), arsenic(V), and total arsenic by differential pulse polarography, Anal. Chem., 51, 215-218 (1979).

[38] Henry, F. T., and Thorpe, T. M., Determination of arsenic(III), arsenic(V), monomethylarsonate and dimethylarsinate by differential pulse polarography after separation by ion exchange chromatography, Anal. Chem., 52, 80-83 (1980).

[39] Bess, R. C., Irgolic, K. J., Flannery, J. E., and Ridgway, T. H., Polarographic reduction of aromatic arsonic and arsinic acids, Anal. Lett., 10, 415-421 (1977).

[40] Braman, R. S., Johnson, D. L., Foreback, C. C., Ammons, J. M., and Bricker, J. L., Separation and determination of nanogram amounts of inorganic arsenic and methylarsinic compounds, Anal. Chem., 49, 621-625 (1977) and references cited there.

[41] Edmonds, J. S., and Francesconi, K. A., Estimation of methylated arsenicals by vapor generation atomic absorption spectrometry, Anal. Chem., 48, 2019-2020 (1976).

[42] Andreae, M. O., Determination of arsenic species in natural waters, Anal. Chem., 49, 820-823 (1977).

[43] Feldman, C., Improvements in the arsine accumulation-helium flow detector procedure for determining traces of arsenic, Anal. Chem., 51, 664-669 (1979).

[44] Rica, G. R., Shepard, L. S., Colovos, G., and Hester, E. N., Ion chromatography with atomic absorption spectrometric detection for determination of organic and inorganic arsenic species, Anal. Chem., 53, 610-613 (1981).

[45] Crecelius, E. A., Modification of the arsenic speciation technique using hydride generation, Anal. Chem., 50, 826-827 (1978).

[46] Talmi, Y., and Bostick, D. T., Determination of alkylarsenic acids in pesticide and environmental samples by gas chromatography with a microwave emission spectrometric detection system, Anal. Chem., 47, 2145-2150 (1975).

[47] Fishman, M., and Spencer, R., Automated atomic absorption spectrometric determination of total arsenic in water and streambed sediments, Anal. Chem., 49, 1599-1602 (1977).

[48] Owens, J. W., and Gladney, E. S., The determination of arsenic in natural waters by flameless atomic absorption, Atomic Absorption Newsletter, 15, 47-48 (1976).

[49] Tam, K. C., Arsenic in water by flameless atomic absorption spectrometry, Environ. Sci. Technol., 8, 734-736 (1974).

[50] Chakraborti, D., DeJonghe, W., and Adams, F., The determination of arsenic by electrochemical atomic absorption spectrometry with a graphite furnace. Part 2. Determination of arsenic(III) and arsenic(V) after extraction, Anal. Chim. Acta, 120, 121-127 (1980), and references cited there.

[51] Brinckman, F. E., Blair, W. R., Jewett, K. L., and Iverson, W. P., Application of a liquid chromatograph coupled with a flameless atomic absorption detector for speciation of trace organometallic compounds, J. Chromatogr. Sci., 15, 493-503 (1977).

[52] Brinckman, F. E., Jewett, K. L., Iverson, W. P., Irgolic, K. J., Ehrhardt, K. C., and Stockton, R. A., Graphite furnace atomic absorption spectrometers as automated element-specific detectors for high pressure liquid chromatography: the determination of arsenite, arsenate, methylarsonic acid and dimethylarsinic acid, J. Chromatogr., 191, 31-46 (1980).

[53] Woolson, E. A., and Aharonson, N., Separation and detection of arsenical pesticide residues and some of their metabolites by high pressure liquid chromatography-graphite furnace atomic absorption spectrometry, J. Assoc. Off. Anal. Chem., 63, 523-528 (1980).

[54] Stockton, R. A., and Irgolic, K. J., The Hitachi graphite furnace-Zeeman atomic absorption spectrometer as an automated, element-specific detector for high pressure liquid chromatography: the separation of arsenobetaine, arsenocholine and arsenite/arsenate, Intern. J. Environ. Anal. Chem., 6, 313-319 (1979).

[55] Irgolic, K. J., Stockton, R. A., and Chakraborti, D., unpublished results.

[56] Parris, G. E., Blair, W. R., and Brinckman, F. E., Chemical and physical considerations in the use of atomic absorption detectors coupled with a gas chromatograph for determination of trace organometallic gases, Anal. Chem., 49, 378-386 (1977).

[57] Salmi, E. J., Merivuori, K., and Laaksonen, E., Condensation products of alkyl and aryl arsonic acids with alcohols, glycols, and α-hydroxycarboxylic acids, Suomen Kemistilehti, 19B, 102-108 (1946).

[58] Soderquist, C. J., Crosby, D. G., and Bowers, J. B., Determination of cacodylic acid (hydroxydimethylarsine oxide) by gas chromatography, Anal. Chem., 46, 155-157 (1974).

[59] Lodmell, J. D., The development and utilization of a wavelength selective multi-element flame spectrometric detector for the gas chromatograph, Diss. Abstr. Int. B, 34, 5359 (1974).

[60] Johnson, L. D., Gerhardt, K. O., and Aue, W. A., Determination of methanearsonic acid by gas-liquid chromatography, Sci. Total Environ., 1, 108-113 (1972).

[61] Stringer, C. E., and Attrep, M., Jr., Comparison of digestion methods for the determination of organoarsenicals in wastewater, Anal. Chem., 51, 731-734 (1979).

[62] Talmi, Y., and Norvell, V. E., Determination of arsenic and antimony in environmental samples using gas chromatography with a microwave emission spectrometric system, Anal. Chem., 47, 1510-1516 (1975).

[63] Irgolic, K. J., Speciation of arsenic compounds in water supplies, Final Report for EPA Project R 8047 740 10 (1981), pp. 14-32.

[64] Stockton, R. A., Bancroft, K. C. C., and Irgolic, K. J., publication in preparation.

[65] Morita, M., Vehiro, T., and Fuwa, K., Determination of arsenic compounds in biological samples by liquid chromatography with inductively coupled argon plasma-atomic emission spectrometric detection, Anal. Chem., 53, 1806-1808 (1981).

[66] Irgolic, K. J., Stockton, R. A., Chakraborti, D., ICP as a multi-element-specific detector for high pressure liquid chromatography: the determination of arsenic and selenium compounds, Proc., 1982 Winter Conference on Plasma Spectrochemistry, Orlando, Florida, January 4-9, 1982, in preparation.

[67] Zief, M. and Mitchell, J. W., Contamination control in trace element analysis, John Wiley & Sons, New York, 1976, pp. 27-33.

[68] Robertson, D. E., The adsorption of trace elements in seawater on various container surfaces, Anal. Chem. Acta, 42, 533-536 (1968).

[69] Ryden, J. C., Syers, J. K., and Harris, R. F., Sorption of inorganic phosphate by laboratory ware. Implications in environmental phosphorus techniques, Analyst, 97, 903-908 (1972).

[70] Guimont, J., Pichette, M., and Rheaume, N., Determination of arsenic in water, rocks and sediments by atomic absorption spectrometry, Atomic Absorption Newsletter, 16, 53-56 (1977).

[71] Al-Sibaai, A. A., and Fogg, A. G., Stability of dilute standard solutions of antimony, arsenic, iron and rhenium used in colorimetry, Analyst, 98, 732-738 (1973).

[72] Portmann, J. E., and Riley, J. P., Determination of arsenic in seawater, marine plants and silicate and carbonate sediments, Anal. Chim. Acta, 31, 509-519 (1964).

[73] Wagner, R., Sampling and sample preparation, Z. Anal. Chem., 282, 315-321 (1976).

[74] Harrison, S. H., LaFleur, P. D., and Zoller, W. H., Evaluation of lyophilization for the preconcentration of natural water samples prior to neutron activation analysis, Anal. Chem., 47, 1685-1688 (1975).

[75] Andreae, M. O., Distribution and speciation of arsenic in natural waters and some marine algae, Deep-Sea Res., 25, 391-402 (1978).

[76] Andreae, M. O., Arsenic speciation in seawater and interstitial waters: the influence of biological-chemical interactions on the chemistry of a trace element, Limnol. Oceanogr., 24, 440-452 (1979).

[77] Taqui Khan, M. M., and Martell, A. E., Metal ion and metal chelate catalyzed oxidation of ascorbic acid by molecular oxygen. I. Cupric and ferric ion catalyzed oxidation, J. Am. Chem. Soc., 89, 4176-4185 (1967).

[78] Pierce, F. D., and Brown, H. R., Comparison of inorganic interferences in atomic absorption spectrometric determination of arsenic and selenium, Anal. Chem., 49, 1417-1422 (1977).

[79] Walsh, P. R., and Fasching, J. L., Matrix effects and their control during the flameless atomic absorption determination of arsenic, Anal. Chem., 48, 1014-1016 (1976).

[80] Robinson, J. W., Garcia, R., Hindman, G. and Slevin, P., Difficulties in the determination of arsenic by atomic absorption spectrometry, Anal. Chim. Acta, 69, 203-206 (1974).

[81] Maessen, F. J. M. J., Balke, J., and Massee, R., Non-spectral interferences in flameless atomic absorption spectrometry using graphite mini-tube furnaces, Spectrochim. Acta, Part B, 33B, 311-324 (1978).

[82] Julshamn, K., Inhibition of response by perchloric acid in flameless atomic absorption, Atomic Absorption Newletter, 16, 149-150 (1977).

[83] Chakraborti, D., DeJonghe, W., and Adams, F., The determination of arsenic by electrothermal atomic absorption spectrometry with a graphite furnace. Part 1. Difficulties in the direct determination, Anal. Chim. Acta, 119, 331-340 (1980).

[84] Baird, R. B. and Gabrielian, S. M., A tantalum foil-lined graphite tube for the analysis of arsenic and selenium by atomic absorption spectrometry, Appl. Spectroscopy, 28, 273-274 (1974).

[85] Hocquellet, P., Application de l'atomisation electrothermique a la determination de As, Sb, Se et Hg par spectrometrie d'absorption atomique, Analusis, 6, 426-432 (1978).

[86] Alder, J. F., and Hickman, D. A., The influence of mineral acid and hydrogen peroxide matrices on elemental sensitivity in graphite furnace atomic absorption spectrometry, Atomic Absorption Newsletter, 16, 110-111 (1977).

[87] Poldoski, J. E., Molecular spectral interferences in the determination of arsenic by furnace atomic absorption, Atomic Absorption Newsletter, 16, 70-73 (1977).

[88] Chakraborti, D., Adams, F., and Irgolic, K. J., in preparation.

[89] Saeed, K., and Thomassen, Y., Spectral interferences from phosphate matrices in the determination of arsenic, antimony, selenium and tellurium by electrothermal atomic absorption spectrometry, Anal. Chim. Acta, 130, 281-287 (1981).

[90] Koreckova, J., Frech, W., Lundberg, E., Persson, J.-A., and Cedergren, A., Investigations of reactions involved in electrothermal atomic absorption procedures. Part 10. Factors influencing the determination of arsenic, Anal. Chim. Acta, 130, 267-280 (1981).

DISCUSSION

A. Schwerdtle: With regards to preservation of samples, I've worried a good deal about how much arsenic could get out of the glass bottles in which we kept our samples. We have found the arsenic compounds under alkaline conditions will react with glass and become part of it. I wonder if you have any special method for sample preservation. We've considered, of course, polyethylene. Would you care to comment on that?

K. J. Irgolic: We have looked into the problem of adsorption of arsenic compounds on glass. This situation, of course, becomes worse as the concentration decreases. When the parts per billion range is reached, it is tremendously important that you have the right container. We found that soft polyethylene hardly adsorbed any arsenic compounds. In addition, to make sure that we are not mislead by adsorption processes, samples were collected in polyethylene, as well as in glass bottles. We performed analyses using inductively coupled argon plasma emission spectrometry to see whatever else happens. Samples stored in certain glass bottles had boron and barium concentrations drastically higher than samples stored in polyethylene bottles. These elements were leached from the glass. Generally, at concentrations as low as 20 ppb adsorption did not create any problems. This statement is based on the fact that the samples were found to have the same arsenic concentrations upon storage in glass and polyethylene. One should mention the unlikely possibility that glass and polyethylene adsorb inorganic arsenic compounds to the same extent. Since our samples contained only arsenite and arsenate, I cannot make a definitive statement about the adsorption of organic arsenic compounds. Polyethylene might not be the material to use with organic arsenic compounds. No systematic study has been performed with the goal to find the most suitable container and the best conditions for the storage of samples containing organic arsenicals.

E. A. Woolson: If you're using glass and not the polyethylene, does the addition of acid in a glass container inhibit or enhance either the adsorption or leaching of arsenic from the glass surface?

K. J. Irgolic: We haven't observed any leaching of arsenic from the glass. As we heard from Corning, there generally is not very much arsenic in the type of glass used for containers. Leaching of arsenic was not a problem because the arsenic values in polyethylene and in glass-stored samples were the same within experimental error.

E. A. Woolson: Does it affect the adsorption?

K. J. Irgolic: Yes it does. Glass adsorbs arsenic compounds much better than polyethylene does.

E. A. Woolson: No, I mean a preservative, either nitric or ascorbic acid.

K. J. Irgolic: Nitric acid, of course, is added to prevent adsorption. We have not found any differences in arsenic values between unpreserved samples and samples preserved with either nitric or ascorbic acid. The best recommendation I can give is to collect three water samples: one unpreserved, one preserved with nitric acid and one preserved with ascorbic acid. If the concentration of arsenic is found to be the same, it will be difficult to question the analytical results.

I. Harding-Barlow: I wish that your institute would repeat the work that Thiers published in 1957. He published a comparison of trace analysis in containers and what was obtained from them. (Thiers, R. E., "Separation, Concentration and Contamination" in _Trace Analysis_, Editors: Yoe, J. H., and Koch, H. J., published by John Wiley & Sons, p. 637, 1957. Thiers, R. E., "Contamination in Trace Element Analysis," in _Methods of Biochemical Analysis_, Vol. V, p. 273, 1957.)

K. J. Irgolic: We very much would like to do this. We planned work along these lines, but, of course, such investigations are complex and time-consuming. We have plenty of containers and plenty of compounds, but not enough money and people.

23

SPECIATION OF ARSENIC IN FOSSIL FUELS AND THEIR CONVERSION PROCESS FLUIDS

C. S. Weiss[1], E. J. Parks, and F. E. Brinckman
Chemical and Biodegradation Processes Group
National Bureau of Standards
Washington, DC 20234

The increased use of coal and oil shale, as alternatives to petroleum, dictates the need to understand the chemistry and environmental fate of many toxic elements which are found at higher concentrations in these fossil fuels and their processing products. These elements include As, Be, Cd, Cr, Ge, Hg, and Se, and their ultimate environmental fate will depend not only upon the molecular form originally present, but also on the chemical transformations occurring during processing to convert these materials to conventional fuels, and during their final use.

The concentration and nature of one principal bioactive element, As, in the source materials (petroleum, coal, and oil shale) are reviewed in the context of the other elements present. The distributions of As species among products, process waters, and residues during typical coal conversion and oil shale retorting processes are discussed as are the methods developed at NBS to speciate As and Fe in the product oils generated during oil shale retorting.

1. Introduction

The presence of metals and metalloids, at trace and ultra-trace level concentrations, in fossilized organic material is due to the abiotic and biotic reactions occurring during the biogeochemical history of these materials. The physiochemical factors governing the depositional environment and subsequent fossilization (and migration) processes of petroleum, coal, and oil shale vary considerably; consequently the concentrations and chemical forms of trace elements in these matrices

[1]National Research Council--National Bureau of Standards, Postdoctoral Research Associate, 1980-1982.

exhibit wide variations. Table 1 compares the concentrations of a number of elements in petroleum, coal, and oil shale. In general, petroleum has the lowest concentrations of toxic metals and metalloids, including As, although large variations exist depending upon the particular deposit. For example, Hitchon et al. [1][2] analyzed 88 crude oils from Alberta, Canada. They found an average concentration of As of 0.111 ppm, but the concentration in a number of samples fell below the detection limit of 0.002 ppm and one sample had a concentration approaching 2.0 ppm.

Table 1.
Comparison of Selected Elemental Concentrations[a] in Petroleum, Coal, and Oil Shale

Element	Petroleum[b]	Petroleum[c]	Coal[d]	Oil Shale[e]
As	0.111	0.263	15	44.3
Be	--	--	2.0	--
Cd	--	--	1.3	0.64
Cr	0.093	0.008	15	34.2
Fe	10.8	40.7	1.6%	2.07%
Ge	--	--	0.71	--
Hg	0.051	3.236	0.18	0.089
Ni	9.38	165.8	15	27.5
S	0.83%	1.31%	2.0%	0.573%
Se	0.052	0.530	4.1	2.03
Si	--	--	2.6%	15%
U	--	0.060	1.6	4.5
V	13.6	87.7	20	94.2

[a]Concentrations in ppm except as noted.
[b]From Ref. [1], the average of 88 crude oils.
[c]From Ref. [2,3], the average of 10 crude oils.
[d]From Ref. [4], the average of 799 coals of varying ranks.
[e]From Ref. [5a].

The high concentrations of V and Ni in petroleum have been the focal point of metal speciation efforts because of their biogeochemical origin as well as the practical importance of V acting as a catalyst poison in petrochemical processing. The isolation and identification of Ni and V porphyrins has been accomplished by a combination of liquid chromatographic separations, UV visible spectroscopy, in the Soret region (380-500 nm), and

[2]Figures in brackets indicate the literature references at the end of this paper.

mass spectrometry [6]. Investigations of these biogeochemical markers as indicators of age, maturation, and migration have been performed. The presence of non-porphyrin vanadium containing compounds has been reported by Yen [7], who attributes their occurrence to complexes formed with mixed O, N, and S ligands.

Reports on the chemical nature of other elements, including As, are quite limited. In general most elements are associated with the asphaltene fraction (soluble in benzene, insoluble in pentane) obtained from petroleum. Filby [8] determined the distribution of a number of elements in the gel permeation chromatographic fractions of a filtered and solvent separated California tertiary crude oil. He found that As is associated with discrete low molecular weight compounds as well as associated with higher molecular asphaltene compounds, as presented in table 2. In the chromatographic fractions obtained from the asphaltene material the distribution of As is quite different from the distributions of Ni and Sb, but very similar to that of Fe, with a correlation coefficient of 0.99.

Table 2

Selected Elemental Concentrations in Molecular Weight Fractions in Resins and Asphaltenes from Petroleum[a]

	Molecular Weight Fraction						
	Resins			Asphaltenes			
Element (ppm)	300-1000	1000-4000	4000-8000	300-1000	1000-4000	4000-8000	8000-22,000
As	0.407	0.200	0.200	0.850	0.620	1.900	6.600
Cr	0.310	0.800	2.960	0.77	4.80	9.12	19.6
Fe	30.1	24.0	236.0	480.0	368.0	867.0	1934.0
Hg	22.0	44.0	72.0	72.0	20.9	90.0	350.0
Ni	206.0	110.0	80.2	1327.0	189.0	984.0	1060.0
Sb	0.0430	0.0026	0.0054	11.0	0.910	0.350	0.1040

[a]From Ref. [8]

The speciation of trace metals and metalloids in coal has been mainly limited to the characterization of the mineral constituents and organic associations utilizing: the analysis of float sink fractions, x-ray diffraction, scanning electron microscopy with energy dispersive x-ray analysis, low temperature and high temperature ashing, and solvent extraction [9,10,11]. The findings indicate that As is present in association with sulfide minerals, particularly pyrite, in contrast to Be, Ge, and Ga which appear to be organically associated.

The nature of As in oil shale is a particularly interesting problem because of the extremely high concentrations of As found in these remnants of ancient algal mats, with respect to both petroleum and coal. It is tempting to interpret these high As concentrations in terms of the biologically mediated reactions of As known to occur in modern algal systems [12]. Weiss et al. [13] analyzed methanol extracts of oil shale via high performance liquid chromatography-graphite furnace atomic absorption spectrometry (HPLC-GFAA). They found the presence of arsenate, a neutral As-containing compound, and phenylarsonic acid. However, because of experimental interference, alternate extraction procedures are necessary for quantitation.

2. The Chemistry of Trace Metal and Metalloids during the Conversion of Fossil Fuels

The development of processes to convert coal and oil shale to conventional liquid fuels, as alternatives to the use of petroleum, must include consideration of the chemistry of trace metals and metalloids, which are present at concentrations considerably higher than those found in petroleum. Two main areas of concern are their effect on processing, including catalyst poisoning, particulate agglomeration, and precipitation; and the ultimate environmental redistribution, and possible mobilization of toxic metals and metalloids.

The preliminary evaluation of elements which may be problematic during conversion processes, and therefore necessitate chemical speciation, must be based on sound quantitative measurements of trace elements in the source matrices, conversion products, process waters and gases, by-products and residues.

During the SRC I process the coal is first ground and dried, and then slurried with a donor solvent which has been derived from the process itself. The coal is reacted at temperatures greater than 400 °C and H_2 pressures of approximately 1500 psi. These conditions are used to fragment and hydrogenate the polymeric coal structures. The differences between the SRC I process and the SRC II process are increased hydrogen pressure for the latter, longer residence time in the reaction zone, facilitated by recycling a portion of the material through the process and the distillation of the product from the residue during the SRC II process as compared to mechanical filtration during the SRC I process.

The process conditions used to obtain conventional fuels from coal may generate a number of chemical reactions which can transform toxic metals and metalloids from their geologically stable forms. Filby et al. [14] have proposed the possible formation of volatile hydrides or elemental compounds of As, Hg, and Se in the reducing atmosphere present during coal liquefaction. For As two possible reactions are

$$As_2S_3(s) + 3H_2(g) \longrightarrow 1/2As_4^0(g) + 3H_2S(g)$$

and

$$As_2S_3(s) + 6H_2(g) \longrightarrow 2AsH_3(g) + 3H_2S(g).$$

The probability that these volatile and reactive forms survive the entire liquefaction process, including the process atmosphere's water content, is not high. This is evidenced by the retention of the major portion of the As in the residue during the SRC-II process. However, high concentrations of As, as well as Hg, Se and Sb, have been found in solids in liquid-liquid separators used during the pressure let-down portion of the process gases. This indicates volatile transport of these elements and subsequent precipitation within the separator tank. But the amounts of As actually transported in this manner are negligible as evidenced by the relatively low concentrations of As detected in the various oils produced during the SRC-II process [15].

The concentrations of many elements, including As are substantially reduced during the production of SRC I and II [15] as presented, in part, in table 3. The As concentraton is ∿ 2 ppm. The SRC I product does contain substantial concentrations of Cr, V, and Ti, which have been found to be

Table 3

Comparison of Selected Elemental Concentrations in Source, Products, and Process Effluents from the Solvent Refining of Coal and the Retorting of Oil Shale.

	SRC-I Process[a]					SRC-II Process[a]				Parahoe Retorting Process[b]			
Element (ppm)	Feed Coal	Product	Wet Filter Cake	PI	Process Water	Feed Coal	Total Oil	Vacuum Bottoms	Process Water	Raw Shale	Product Oil	Retorted Shale	Product Water
As	20.1	2.27	3.24	388	0.01	6.00	0.023	15.1	0.029	49	31	59.4	5.7
Cr	14.8	5.50	141	200	0.011	15.1	0.032	35.8	0.051	34	--	44.2	--
Fe	2.38%	0.026%	9.37%	13.7%	1.34	1.49%	3.40	3.50%	0.85	2.1%	49	2.4%	1.5
Hg	0.114	0.047	0.034	0.497	0.021	0.170	0.023	0.094	0.096	0.089	0.30	0.035	0.0023
Ni	12.4	< 2	142	170	0.014	8.62	< 0.02	19.6	0.031	28	2.74	32.1	0.54
S	--	--	--	--	--	--	--	--	--	0.6%	0.74%	0.68%	3.9%
Se	3.03	0.080	8.64	14.1	0.914	1.35	0.039	3.04	0.47	2.0	0.91	2.3	9.8
Si	--	--	--	--	--	--	--	--	--	15%	55	18%	41
U	--	--	--	--	--	--	--	--	--	4.5	--	5.10	--
V	29.2	13.7	103	226	< 0.001	21.3	0.021	39.8	< 0.02	94	0.25	129	0.045

[a]From Ref. [15].

[b]From Ref. [5a].

associated with catalyst deactivation during attempts to upgrade SRC. The residues (the wet filter cake in the SRC I process and the vacuum bottoms in the SRC II process) are enhanced in the concentrations of many elements. The benefication, including bioleaching, of these residues to provide sources of metals and metalloids is an area that requires further study. The disposal of these materials also necessitates the investigation of both the chemical and microbial leaching characteristics of toxic metals and metalloids.

The oil shale retorting process involves the initiation of pyrolysis with subsequent heating accomplished by the introduction of heated gases, although several modifications exist. Complex chemical reactions which may involve As have also been suggested for hydrothermal processes occurring in oil shale retorting [16]. Though biogeochemical origins imply that some organoarsenicals, such as methylarsonates may exist in kerogen [17], a plethora of abiotic reactions can also give rise to relatively volatile organoarsenicals during retorting. Among these, a group of arsenation reactions, originally described by Meyer in 1883 [18], may prevail since the necessary reactants, inorganic arsenic, polar organic molecules, water and heat all are present:

$$RX + AsO_4^{3-} \xrightarrow[H_2O]{\Delta} RAsO_3^{2-} ,$$

and

$$RAsO_3^{2-} + RX \xrightarrow{etc.} R_2AsO_2^{-} .$$

During the oil shale retorting process the majority of the trace elements are retained in the spent shale. Fruchter et al. [5a] have analyzed the distribution of trace elements during the retorting process and their results are presented in table 3. Considerable concentrations of As and Se are partitioned into the product oil and the process water, possibly due to the anionic nature of their acid forms. The concentrations of As, Se, and Fe in shale oil are considerably higher than in the oil produced during the SRC II process. Fruchter has also identified arsines in both coal liquefaction and oil shale retorting process gases at low concentrations [5b,5c].

Fish et al. [17] have chemically speciated As in process waters from a number of different retorting processes. They employed a high performance liquid chromatograph coupled to a graphite furnace atomic absorption (HPLC-GFAA) detector selective for As. They found the presence of a neutral arsenic containing compound, methanearsonic acid, phenylarsonic acid, and arsenate. The "fingerprints" they obtained from the different process waters indicated variations in the respective concentrations of As species depending on the retorting process employed.

3. Speciation of As in Shale Oil

Even though the presence of As in shale oil has necessitated the development of dearsenation processes to protect hydrotreating catalysts during shale oil processing, the molecular speciation of As in shale oil has not been reported. Curtin et al. [19] found that As was present throughout various boiling range fractions including a fraction obtained below 205 °C. The development of methods for the molecular speciation of As in shale oil would be valuable to further optimize dearsenation processes as well as process monitoring in support of dearsenation catalysts.

3.1. Experimental

The application of element selective chromatographic detectors to the molecular identification of numerous trace metal and metalloid species, in the presence of numerous interferring compounds, has provided valuable information in the areas of aquatic chemistry [20], biological and microbiological transformations of trace elements [21,22], and controlled release biocidal organotin polymers [23]. The HPLC-GFAA procedure involves the liquid chromatographic separation of the materials under investigation by ion-exchange, reversed bonded phases, or size-exclusion chromatography. The chromatographic effluent is then periodically sampled, via a specially designed sampling cup, by the graphite furnace atomic absorption spectrometer which is sensitive to the selected element. The details of the procedure have been published elsewhere [24]. Two modifications have been implemented during the course of this study and will be described in detail elsewhere [25,26]. The first utilizes a PTFE flow cup modified to provide both the interface between the HPLC and GFAA and entry of a second regulated flow stream containing a dilute solution of transition metal ion to "fix" or promote more favorable atomization of the analyte emerging from the HPLC column [25]. The post-column addition of 0.01 percent (wt./Vol.) Ni [as $Ni(NO_3)_2$ in tetrahydrofuran (THF)], in this manner allows the use of elevated charring temperatures with the retention of normally volatile As [27]. The second modification is the splitting of the chromatographic effluent into two streams, one of which is directed to a GFAA unit selective for As, and the other stream is directed toward a second GFAA unit for Fe, as depicted in figure 1 [26]. The selection of As and Fe was based on their presence at high concentrations in shale oil. The GFAA conditions used for As and Fe are presented in table 4.

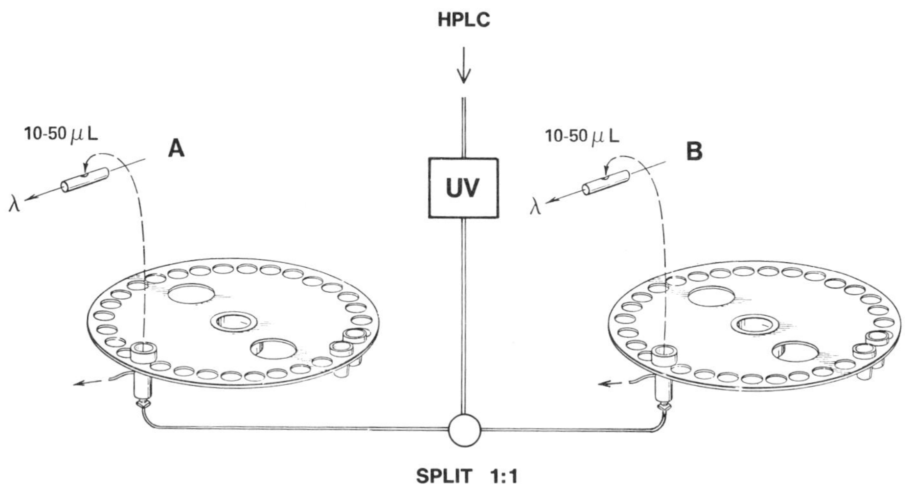

Figure 1. Schematic illustrates coupling of HPLC effluent through a 1:1 flow splitter to the PTFE flow cups which provide an automatic interface between the HPLC and the GFAA [24]. Either or both of the GFAA cups can be readily modified to permit prescribed amounts of a second solution into the mixing well for purposes of enhancing AA signals [25].

The size exclusion chromatographic separations were performed on both rigid and non-rigid polystyrene divinyl benzene packings using THF as a mobile phase. The calibration curve for the rigid μ-Styragel columns [2(1000 Å) and 1(100 Å) columns] is presented in figure 2. The columns are capable of separating compounds of differing molecular size, by comparison with polystyrene calibration standards, from less than 600 to greater than 19,000 daltons.

Table 4

The Graphite Furnace Atomic Absorption Conditions Used for the Detection of As and Fe

Element	As		Fe	
Wavelength (nm)	193.7		248.3	
Slit width (nm)	0.7		0.2	
Lamp	electrodeless discharge		hollow cathode	
Dry	85°	15 s	85°	15 s
Char Ramp	1100°	25 s	--	--
Char	1100°	5 s	1100°	30 s
Atomize	2700°	7 s	2700°	7 s

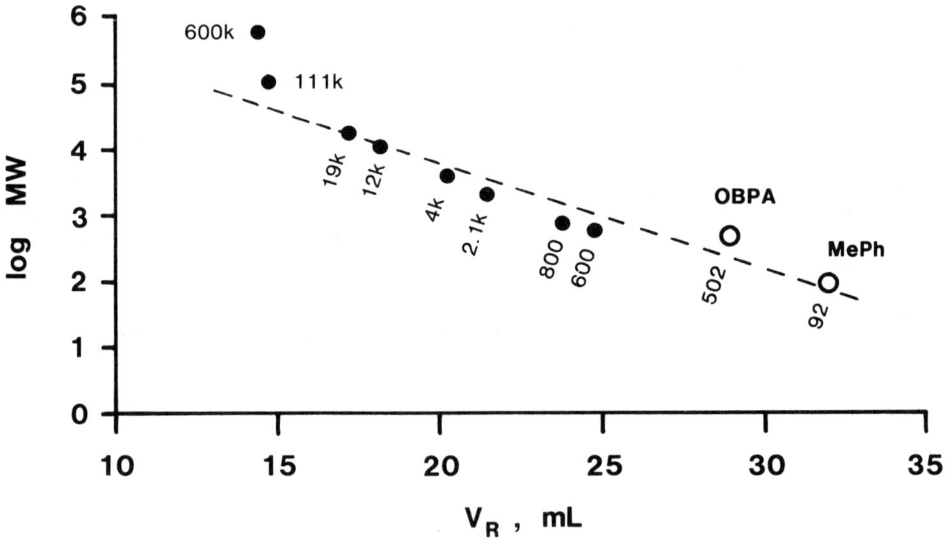

Figure 2. Molecular weight calibration curve derived from polystyrene MW standards, 10,10-ortho-bis-phenoxyarsine (OBPA), and toluene (MePh) at MW values indicated. The combination of two 10^3 Å μ-styragel and one 100 Å columns provided complete MW exclusion at 16 mL (19,000) and nominal linearity down to MW 100 using THF as the mobile phase.

The shale oils investigated during this study were: Paraho shale oil (obtained from Dr. R. H. Fish, Lawrence Berkeley Laboratory), National Bureau of Standards SRM 1580, and Crude Shale Oil A and Centrifuged Shale Oil number 4101 (Fossil Fuels Research Materials Facility, Oak Ridge National Laboratory). The oils were diluted with THF to concentrations near ten percent, centrifuged where necessary and filtered through a 0.45 μm filter prior to injection into the HPLC-GFAA system.

4. Results and Discussions

The effect of the post-column addition of Ni^{2+} to the chromatographic effluent is demonstrated in figure 3. One of the shale oils obtained from Oak Ridge National Laboratory was separated on the non-rigid Bio-Beads gel permeation columns. The enhancement in signal due to the addition of Ni is about two fold. The As elutes with macromolecular compounds of a molecular weight greater than 2,000. Throughout the course of this work Ni^{2+} was continually added to the chromatographic effluent, via a specially designed sample cup, before GFAA analysis.

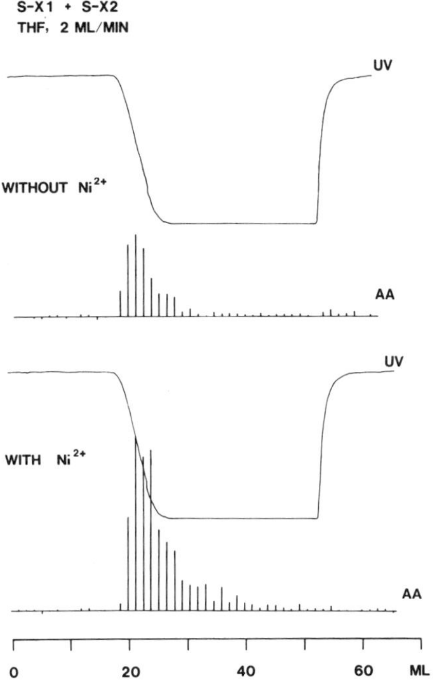

Figure 3. The size exclusion chromatographic separation of ORNL Shale Oil with and without the addition of Ni with both UV and As-selective GFAA detection.

The analysis of NBS shale oil SRM 1580, certified for organics only, is presented in figure 4 and exemplifies the dual element selective chromatographic detection. The As is eluted with a retention volume corresponding to a molecular species of close to 2,000 daltons, as is the Fe. The As also appears to be distributed among species of various molecular weights. Chromatograms of a Paraho shale oil and another oil obtained from ORNL are also presented in figure 4. The concentration of As in the Paraho oil is approximately 10 times the concentration present in the NBS shale oil and in the ORNL Centrifuged oil. In the Paraho oil the As and Fe coelute with a molecular species having an apparent molecular weight greater than 2,000 daltons. In the ORNL centrifuged oil, the As is distributed towards lower molecular weights and the Fe has a bimodal distribution with elution volumes corresponding to about 12,000 and 2,000 daltons.

The primary difficulty in trying to speciate metals in oils is the difficulty in finding appropriate standards for macromolecular compounds. This is evidenced by the reporting of the distributions of metals without the necessary standards for chemical speciation [28,29]. Our experimental spiking of the Paraho shale oil with methanearsonic acid is presented in figure 5. The methanearsonic acid is ligated by the molecular weight compounds of 2,000 daltons, again coeluting with Fe. We postulate that this association of As with Fe in macromolecules is analogous to the association of orthophosphate and metals such as Cr and Fe via oxygen bonding in polymers viz., [30],

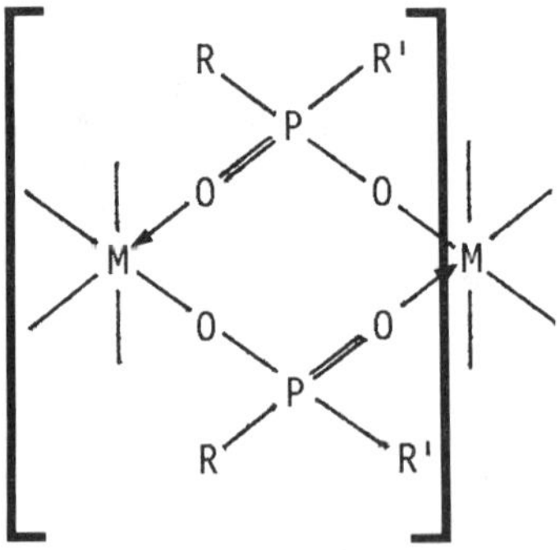

We envision Fe complexes of humic acids, the precursors of many fossil fuels, ligating the anionic forms of arsenic acids via Fe-O-As bonding by the reaction,

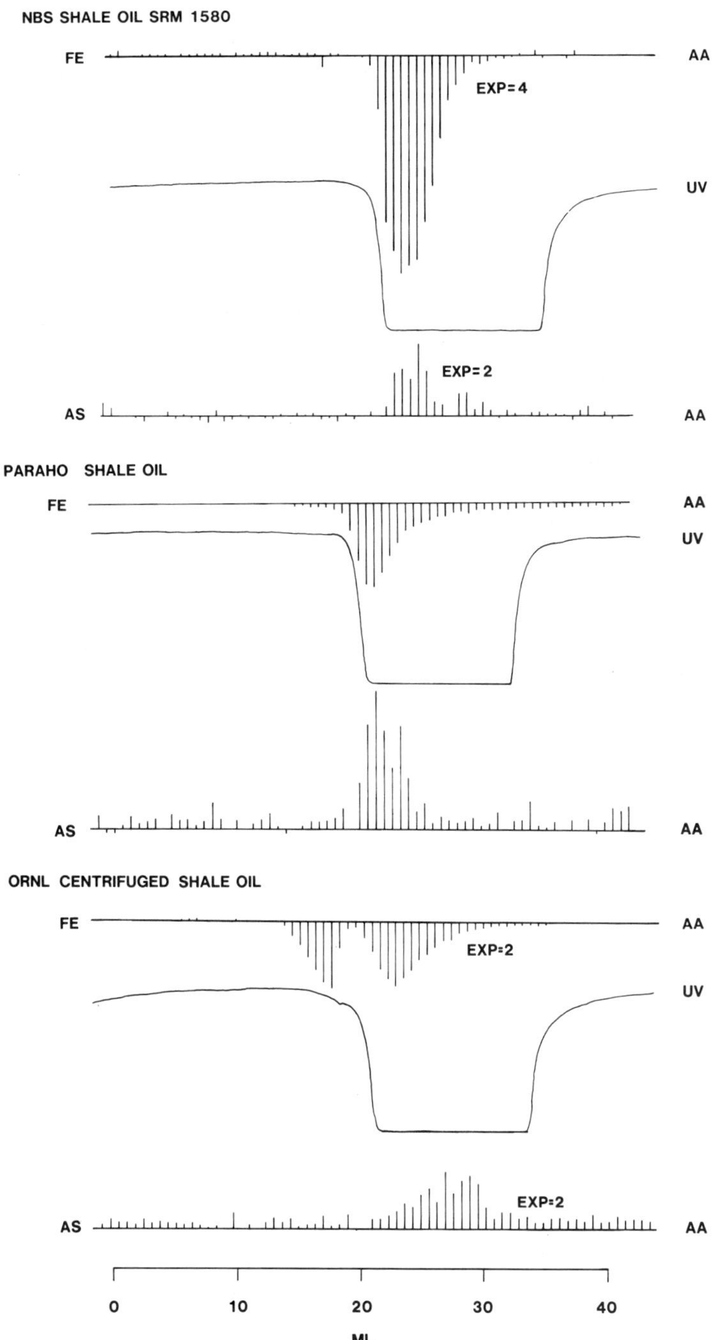

Figure 4. Dual element selective detection, for Fe and As, of the size exclusion chromatograms of various shale oils, as indicated.

$$MeAsO_3H^- + Humic \sim Fe \rightarrow Humic \sim Fe\,[O{-}As(Me)(OH){=}O]_2\,Fe \sim Humic$$

This may indicate the competitive chemistry occurring in iron oxide guard beds used for the commercial dearsenation of shale oil. Our continuing speciation studies on the chemical removal of the methanearsonic acid added to the shale oil, as well as the native As found in the shale oil, can provide information for the further optimization of dearsenation processes.

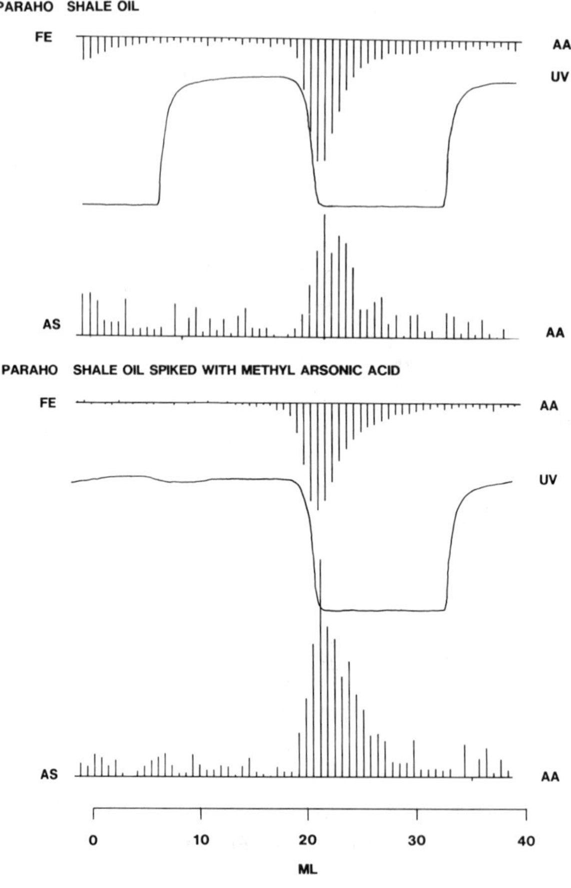

Figure 5. The addition of methyl arsonic acid to a solution of Paraho shale oil.

5. Conclusions

In summary, the discrete molecular speciation of As is possible in some cases. However, macromolecular As-containing species, evident in size exclusion chromatograms of various shale oils, must be evaluated in terms of: heteroatom content, correlations among a variety of elements, the extent and nature of the complexing capacity of the matrix for As, and the stability of these As containing compounds. The conclusions drawn during this study are: (1) HPLC-GFAA permits the chemical speciation of As in process fluids from oil shale retorting; (2) the association of As with Fe provides a preliminary chemical basis for understanding the presence of metals and metalloids in the macromolecular components of fossil fuels; (3) the anionic nature of As may explain its unique behavior with respect to other elements' cationic behavior; and (4) studies based on the uptake of methanearsonic acid by the macromolecular compounds in shale oil may provide beneficial insights for optimizing the removal of As from shale oil.

6. Acknowledgments

We acknowledge the partial financial support of the Office of Standard Reference Materials, NBS. We thank Mr. V. Jain for experimental assistance and Dr. K. L. Jewett for valuable suggestions. We are indebted to Dr. R. H. Fish (Lawrence Berkeley Laboratories) and Dr. R. Alvarez (Office of Standard Reference Materials, NBS) for samples of shale oil. Samples obtained from the Oak Ridge National Laboratory Fossil Fuels Research Matrix Program were provided by Dr. W. H. Griest.

References

[1] Hitchon, B.; Filby, R. H.; Shah, K. R. Geochemistry of trace elements in crude oils, Alberta, Canada in The Role of Trace Metals in Petroleum. Yen, T. F. ed. (Ann Arbor Science Publishers, Inc., Ann Arbor, 1975).

[2] Shah, K. R.; Filby, R. H.; Haller, W. A. Determination of trace elements in petroleum by neutron activation analysis Part I, J. Radioanal. Chem., 6:185-192; 1970.

[3] Shah, K. R.; Filby, R. H.; Haller, W. A. Determination of trace elements in petroleum by neutron activation analysis Part II, J. Radioanal. Chem. 6: 413-422; 1970.

[4] U.S. National Committee for Geochemistry. Trace-element geochemistry of coal resource development related to environmental quality and health, National Academy Press, Washington, DC, 1980.

[5a] Fruchter, J. S.; Wilkerson, C. L.; Evans, J. C.; Sanders, R. W. Elemental partitioning in an aboveground oil shale retort pilot plant, Environ. Sci. & Tech. 14: 1374-1381; 1980.

[5b] Fruchter, J. S.; Petersen, M. R. Environmental characterization of products and effluents from coal conversion process, in Analytical Methods for Coal and Coal Products, Vol. 3, Clarence Karr, Jr., ed., p. 247-275 (Academic Press, New York, NY, 1979).

[5c] Fruchter, J. S.; Laul, J. C.; Petersen, M. R.; Ryan, P. W.; Turner, M. E. High precision trace element and organic constituent analysis of oil shale and solvent refined coal materials, in Analytical Chemistry of Liquid Fuel Sources Tar Sands, Oil Shale, Coal, and Petroleum, P. C. Uden, S. Siggia, and H. B. Jensen, eds., p. 255-281 (American Chemical Society Advance in Chemistry Series 170, Washington, DC, 1978).

[6] Yen, T. F. The Role of Trace Metals in Petroleum. (Ann Arbor Science Publishers, Inc., Ann Arbor, 1975).

[7] Yen, T. F. The role of metal-heteroatom complexes in fossil fuel production, in DOE/NBS Workshop on Environmental Speciation and Monitoring Needs for Trace Metal-Containing Substances from Energy-Related Processes, F. E. Brinckman and R. H. Fish, eds., p. 9-20, (NBS Spec. Publ. 618, 1981).

[8] Filby, R. H. The nature of metals in petroleum, in The Role of Trace Metals in Petroleum. Yen, T. F. ed., (Ann Arbor Science Publishers Inc., Ann Arbor, 1975).

[9] Gluskoter, H. J. Mineral matter and trace elements in coal, in Trace Elements in Fuel, Babu, S. P. ed., (American Chemical Society Advances in Chemistry Series 141, Washington, DC, 1975).

[10] Finkelman, R. B. Modes of occurrence of trace elements in coal, Dissertation, University of Maryland, 1980.

[11] Bonnett, R.; Czechowski, F. Metals and metal complexes in coal, Phil. Trans. R. Soc. Lond. A, 300: 51-63; 1981.

[12] Bottino, N. R.; Cox, E. R.; Irgolic, K. J.; Maeda, S.; McShane, W. J.; Stockton, R. A.; Zingaro, R. A. Arsenic uptake and metabolism by the alga Tetraselmis Chui, in Organometals and Organometalloids Occurrence and Fate in the Environment, F. E. Brinckman and J. M. Bellama eds., p. 116-129 (American Chemical Society Symposium Series 82,Washington, DC, 1978).

[13] Weiss, C. S.; Jewett, K. L.; Brinckman, F. E.; Fish, R. H. Application of molecular substituent parameters for the speciation of trace organometals in energy-related process fluids by element-selective HPLC. in DOE/NBS Workshop on Environmental Speciation and Monitoring Needs for Trace Metal-Containing Substances from Energy-Related Processes, F. E. Brinckman and R. H. Fish, eds., p. 197-210, (NBS Spec. Publ. 618, 1981).

[14] Filby, R. H.; Sandstrom, D. R.; Lytle, F. W.; Greegor, R. B.; Khalil, S. R.; Ekambaram, U.; Weiss, C. W.; Grimm, C. A. Chemistry of trace element species in coal liquefaction processes, in DOE/NBS Workshop on Environmental Speciation and Monitoring Needs for Trace Metal-Containing Substances from Energy-Related Processes, F. E. Brinckman and R. H. Fish, eds., p. 21-38, (NBS Spec. Publ. 618, 1981).

[15] Khalil, S. R. The fate of trace elements in the Solvent Refined Coal processes SRC-I and SRC-II, Dissertation, Washington State University, 1979.

[16] Kland, M. J.; Fox, J. P. The partitioning of major, minor, and trace elements during in situ oil shale retorting, Energy and Environment Division, Lawrence Berkeley Laboratory Report LBID-034.

[17] Fish, R. H.; Brinckman, F. E.; Jewett, K. L. Fingerprinting inorganic arsenic and organoarsenic compounds in in situ oil shale process waters using a liquid chromatograph coupled with an atomic absorption spectrometer as a detector, Environ. Sci. & Tech., in press.

[18] Meyer, G. Ueber einiige anomale reaktionen, Ber. 16: 1439-1883.

[19] Curtin, D. J.; Dearth, J. D.; Everett, G. L.; Grosboll, M. P.; Myers, G. A. Arsenic an nitrogen removal during shale oil upgrading, Prepr. Pap. Am. Chem. Soc. Div. Fuel Chem. 23(4): 18-29, 1978.

[20] Jackson, J. A.; Blair, W. R.; Brinckman, F. E.; Iverson, W. P. Gas chromatographic speciations of methylstannanes in the Chesapeake Bay using purge and trap sampling with a tin-selective detector, Environ. Sci. Technol.; submitted.

[21] Olson, G. J.; Iverson, W. P.; Brinckman, F. E. Volatilization of mercury by Thiobacillus ferrooxidans, Cur. Microbiol. 5: 115-118; 1981.

[22] Kahn, N.; Van Loon, J. G. Atomic absorption spectrophotometry as a chromatography detector for copper-amino acid complexes in human serum, J. Liq. Chromatogr. 2: 23-36; 1979.

[23] Parks, E. J.; Brinckman, F. E.; Blair, W. R. Application of a graphite furnace atomic absorption detector automatically coupled to a high performance liquid chromatograph for speciation of metal-containing macromolecules, J. Chromatogr. 185: 563-72; 1979.

[24] Brinckman, F. E.; Blair, W. R.; Jewett, K. L.; Iverson, W. P. Application of a liquid chromatograph coupled with a flameless atomic absorption detector for speciation of trace organometallic compounds, J. Chrom. Sci. 15: 493-503, 1977.

[25] Jewett, K. L.; Weiss, C. S.; Brinckman, F. E. Signal enhancement of trace organometal compounds speciated by HPLC-GFAA using automated post-column treatment with transition metal salts, 183rd National American Chemical Society Meeting, Las Vegas, NV.

[26] Weiss, C. S.; Brinckman, F. E.; Parks, E. J. Incorporation of arsenic in metal-containing macromolecules in shale oils as determined by multielement-specific size exclusion chromatography, 183rd National American Chemical Society Meeting, Las Vegas, NV.

[27] Ediger, R. D. Atomic absorption analysis with the graphite furnace using matrix modification, Atomic Absorption Newsletter, 14: 127-130; 1975.

[28] Weiss, C. S. The detection of trace element species in solvent refined coal, Dissertation, Washington State University, 1980.

[29] Taylor, L. T.; Hausler, D. W. Organically bound metals in a solvent-refined coal: metallograms for a Wyomin subbituminous coal, Science 213: 644-646, 1981.

[30] Hagihara, N.; Sonogashira, K.; Takahashi, S. Linear polymers containing transition metals in the main chain, in Specialty Polymers, Cantow, H-J, et al. eds., p. 159 (Springer-Verlag, Berlin 1981).

24

BIOCHEMISTRY OF ARSENIC: RECENT DEVELOPMENTS

Ralph A. Zingaro
Department of Chemistry
Nestor R. Bottino
Department of Biochemistry and Biophysics
Texas A&M University
College Station, Texas 77843

This review covers the principal developments in this area reported during the past four years. The aquatic food web represents an important source of arsenic-containing biomolecules found in land animals. As they pass through the aquatic trophic levels, the arsenic concentrations undergo biological magnification and appear in aquatic organisms at much higher concentrations than in their natural environment. Algae, because they are located at the base of the aquatic food web, have been the subject of a great deal of study by arsenic biochemists. A number of studies underline the competition between phosphate and arsenate for the binding of transporters that make possible the passage of these ions through the algal cell membrane. The process appears to be dependent on the availability of photosynthetic energy. It has been reported that once inside the cell, measurable amounts of arsenic are incorporated into lipids. Evidence for the formation of arsoniumphospholipids is strong.

Among other arsenic containing molecules which have been identified are a dimethylarsenofuranoside from sea kelp and arsenobetaine from the western rock lobster. The methylation of inorganic oxyarsenic anions has been demonstrated among a wide range of living systems ranging from microbial to mammalian. Methylation appears to be one of, and perhaps the most fundamental of biochemical pathways. Recent work shows that the trivalent arsenicals, monomethylarsine oxide and monomethylarsine sulfide are converted to methylarsine by the yeast *Candida humicola*.

Both trivalent and pentavalent arsenic react with organic thiols to produce stable S-arsenic derivatives. This reaction may have important implications concerning the effect of arsenic compounds on thiol containing enzymes.

1. Introduction

During the last few years there has been a surge in interest in the metabolism of arsenic, particularly in the aquatic environment. One of the most recent reviews is that of Brinckman and Bellama [1].[1] The purpose of the present review is to cover salient developments that have taken place during the last four years or so. Initial work on the biochemical processes which arsenic undergoes in living organisms indicated that two major processes were utilized, namely, reduction of arsenic, from the pentavalent to the trivalent state and secondly, and often accompanying or following reduction, methylation of the element has been observed. These two processes, *viz.*, reduction and reduction-methylation are obvious detoxification mechanisms widely used by the living cells as a way of neutralizing the otherwise deleterious effects of arsenic and to facilitate its excretion into the environment.

This review is organized in the following manner. The first section will discuss some recent developments dealing with the metabolic fate of this element in algae. Based on this body of evidence, a model is proposed which could be used to describe the metabolism of the element in living cells. The next section will describe recent biochemical evidence which has been reported in other species and which supports the proposed model. Finally, a discussion of the reactions between pentavalent arsenicals and thiol groups, and the implications of these reactions is presented.

2. Arsenic Metabolism in Algae

Algae have been a preferred subject for arsenic metabolic studies during the last few years, probably due to two favorable factors. One is that algae can be cultured without great difficulty and experiments of many different types can be performed with them. The second factor concerns the ecological and nutritional importance of these organisms. Algae are located at the bottom of the aquatic food web. Thus, when arsenic-rich rocks or industrial wastes are washed into rivers and seas, arsenic, mainly in the form of arsenate [2,3] is picked up by algae and probably by other members of the lower trophic levels of the aquatic ecosystems. Algae exhibit concentrations of arsenic which are higher than those of the surrounding water [4,5], a process called bioaccumulation. Higher trophic levels do not contribute markedly to the accumulation so that one can say that whereas the aquatic ecosystem exhibits bioaccumulation, there is in fact no biomagnification. At any rate, members of higher trophic levels that feed on primary producers show a level of arsenic which is higher than that of their environment [4]. The knowledge of arsenic metabolism in algae is of paramount importance to understand the ecological significance of the element. Algae seem to pick up the arsenate from the water simply by

---

[1] Figures in brackets indicate the literature references at the end of this paper.

using phosphate transport systems located in the algal cell membranes [6,7,8]. See fig. 1. This process seems to be energy (light) dependant and to occur mostly when other energy requiring processes such as growth [9] and phosphate transport are not too active. Thus, arsenic uptake by algal cells is dependant on the phosphate concentration of the surroundings [6,7] and on the life stage of the cell [9]. In any case there seems to be an active regulatory mechanism which maintains the cellular arsenic concentration at non-toxic levels [10]. Once inside the algal cell, arsenate can remain as such or it can be reduced [6,11]. It is not clear what controls this difference in metabolic fate. Intact arsenate has been claimed to react with sugars enzymatically [12] or non-enzymatically [13] to eventually form hexose-6-arsenates, most probably glucose-6-arsenate. This compound seems to be able to participate in glycolysis or in the hexose-monophosphate shunt [13]. However, when glycolysis proceeds to the point of trioses, the proximity of arsenate (instead of phosphate) to the reacting group in the substrate seems to be recognized by the enzymes for triose degradation, and glycolysis stops while toxic symptoms develop [14]. Most of these facts have been known for a long time. The alternate pathway for arsenate, once it is inside the algal cell is probably dictated by the availability of reducing systems. Virtually nothing is known about the reducing agent involved or its mechanism. We do know that arsenate can be reduced to arsenite by certain algae [6,15]. Contrary to what has been observed in bacteria, arsenite seems to be less toxic than arsenate towards algae [9]. Furthermore, the reducing process does not seem to stop at arsenite. Instead, arsenic appears in the form of methylated arsonium derivatives replacing nitrogen in compounds such as betaine [16] phosphatidyl choline [17,18] and deoxyribosugars [19]. The mechanism of arsenic methylation is, at best, unclear. Methylating agents such as *S*-adenosyl methionine (SAM) [20] and cobalamin [21] have been proposed, SAM being the most favored candidate. The methylation of arsenic probably decides its fate and that of the cell. Thus, from what is presently known, the following hypothesis on arsenic effects on algal cells can be formulated (figure 1). At relatively low concentrations of arsenate in the aqueous surroundings, the algal cell can reduce most of it and send it through the reducing-methylating pathway to replace nitrogen and be excreted either as arsenobetaine or be stored temporarily as a phospholipid derivative in the membranes. It can then be gradually eliminated as arsenobetaine, methyl arsonate, dimethylarsinate, and possibly, other methylated species [22]. When arsenate reaches high levels in the surroundings, the reducing power of the algal cell becomes insufficient and some of the arsenate, instead of replacing nitrogen in amino acids or lipids, replaces phosphate in sugar metabolites, stops glycolysis and is toxic to the cell.

The pathways described above arise from investigations on algal cells. Whether they are also applicable to cells of higher organisms has yet to be determined. The information available suggests that this will be the case (discussion). In the following paragraphs we discuss experimental evidence which suggests that the alternate pathway hypothesis described above may apply to many other types of

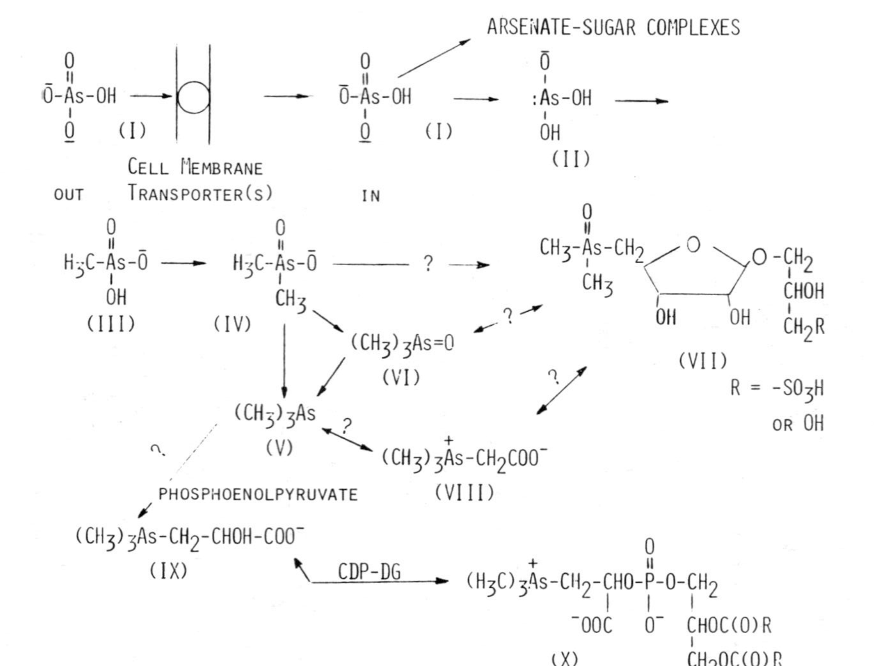

FIGURE 1. BIOCHEMICAL REACTIONS OF ARSENIC IN ALGAL CELLS. (I) = ARSENATE; (II) = ARSENITE; (III) = METHYL ARSONATE; (IV) = DIMETHYL ARSINATE; (V) = TRIMETHYLARSINE; (VI) TRIMETHYLARSINE OXIDE; (VII) = SULFOPROPYL(DIMETHYLARSENOSO)DEOXYRIBOSE OR DIHYDROXYPROPYL(DIMETHYLARSENOSO)-DEOXYRIBOSE; (VIII)= ARSENOBETAINE; (IX) - TRIMETHYLARSONIUM LACTATE; (X) = O-PHOSPHATIDYL-TRIMETHYLARSONIUM LACTATE; CDP-DG = CYTIDINE DIPHOSPHATE DIGLYCERIDE.

cells.

2.1. Arseno-lipids in Algae

Lunde [23,24,25] was probably the first to report the presence of arsenic in the lipids of algae and other aquatic organisms. Thus, arsenic can be found in both the lipid and non-lipid components of aquatic organisms as organoarsenic compounds, but has not been detected in the lipids of plants or land animals [23]. In 1977, Irgolic, et al. [1] found that the marine alga, *Tetraselmis chui*, was able to incorporate arsenic from a medium containing added arsenate and to form a lipid-soluble organoarsenic compound. Partial characterization of this compound and of that isolated from *Daphnia magna*, allowed the prediction that arsenocholine would be a part of the isolated lipids. Further experiments by the same group lend further support to that hypothesis [26]. This was confirmed in 1978 by Cooney, et. al. [18] who reported the isolation from the marine diatom, *Chaetoceros concavicornis*, grown in a medium containing ^{75}As-arsenate, of a phospholipid whose proposed structure corresponds to *O*-phosphatidyltrimethylarsonium lactate:

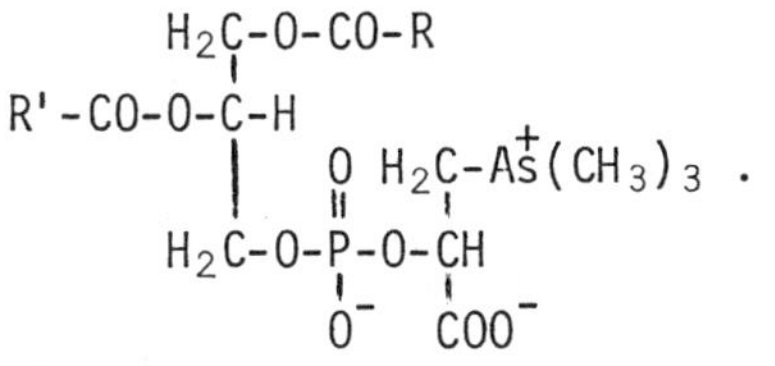

```
           H2C-O-CO-R
             |
 R'-CO-O-C-H
             |     O  H2C-As+(CH3)3 .     (I)
             |     ||    |
           H2C-O-P-O-CH
                   |     |
                   O-   COO-
```

The structure of this compound was derived from its behavior in two-dimensional paper chromatography and base-catalyzed deacylation followed by characterization of the products by acid or enzymic hydrolysis. The authors suggested that trimethylarsonium lactate might have been synthesized by the following pathway:

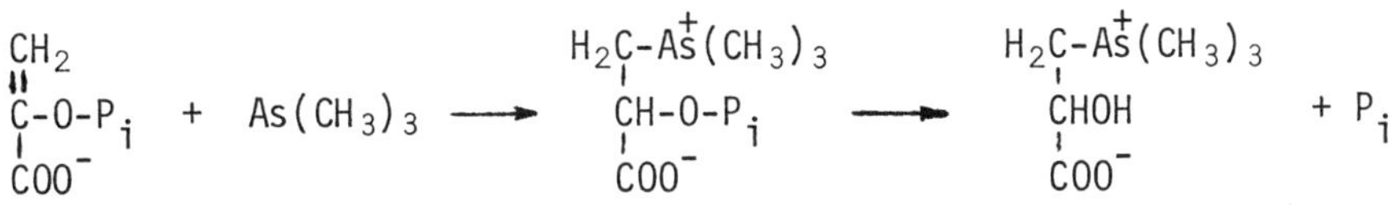

```
CH2                              H2C-As+(CH3)3          H2C-As+(CH3)3
||                                 |                      |
C-O-Pi  +  As(CH3)3  ---->       CH-O-Pi    ---->       CHOH          + Pi
|                                  |                      |
COO-                             COO-                   COO-
```

Thus, according to the suggestion of Cooney, et. al., phosphoenol pyruvate would react with trimethylarsine to yield trimethylarsonium-2-phospholactate which, in turn would lose phosphate to yield trimethylarsoniumlactate. No direct experimental evidence was given to support this view.

Since the oxidative chemical degradation of the novel arseno-lipid (I) yielded trimethylarsoniumbetaine, a compound that has been isolated from rock lobsters, *Panulirus cygnus* [16], Cooney, et al., suggested that a possible biological pathway for the degradation of the arsenolipid (I) might include arsenobetaine as well as methylarsonate and dimethylarsinate. They also proposed that this oxidation process may be mediated by surface bacterial action on the lipids of the algal membrane [22]. However, Andreae and Klumpp [11] cultured several marine phytoplankton species in bacteria-free media and found that the algae were able to produce and excrete into the medium no less than 12 organoarsenic compounds, including substantial amounts of methylarsonate, dimethylarsinate and arsenite.

3. Arsenic in Aquatic Invertebrates and Fish

Edmonds and Francesconi [16] were the first to isolate and characterize arsenobetaine from a natural source, the liver of the western rock lobster, *Panulirus cygnus* George. They are probably the only ones to achieve the unequivocal identification of such a compound. The techniques used are described in another section of this paper. More recently, Benson and Summons [27] analyzed various aquatic invertebrates and they detected in their tissues *O*-phosphatidyltrimethylarsonium lactate together with various other arsenic derivatives such as trimethylarsonium lactate. They also found relatively large amounts of other unidentified organoarsenic compounds. Since the total content of arsenic varied quite drastically among the invertebrates which were examined (from 5 to 1,000 ppm) the authors proposed that the arsenic content of the marine invertebrate cell may depend essentially on the temperature of the water, since animals from tropical waters, which exhibit low phosphate content, contained more arsenic than animals living in lower temperature waters. However, Ünlü and Fowler [28] had previously demonstrated that in laboratory experiments using the Mediterranean mussel, *Mytilus galloprovincialis* an increase in water temperature, without variation in its phosphorus content, affected favorably the uptake of arsenic. Salinity was also a factor in this process.

Similar to the invertebrates mentioned above, the green sunfish, *Lepomis cyanellus* exhibited temperature dependence on its uptake of arsenate [29,30,31]. The incorporation of arsenic into organic compounds of fish body tissues seems to occur more efficiently *via* the digestive tract than by injection or addition of arsenic in the water. This suggests that intestinal microorganisms may carry out some of the initial stages of the conversion [5,32 and references therein].

After absorption, arsenic was found in the gall bladder and, in decreasing proportions, in the liver, spleen, kidneys, ovaries, gills and testes of the green sunfish [32]. Similar studies were performed by other groups on rainbow trout, *Salmo gairdneri* using ^{74}As-arsenic acid given orally [33]. The distribution of the label among inorganic and organic compounds of erythrocytes, plasma, liver, kidneys, bile and muscle was determined 6, 12, 24 and 96 hours after the administration. In all cases, the major portion of the label in the organic fractions appeared in a substance that had properties reminiscent of those of arsenobetaine and related compounds. The ^{74}As content of this compound increased with time while the inorganic ^{74}As decreased. Most of the excretion of labeled arsenic seemed to be done by a route other than urine and bile, probably the gills.

The major organoarsenic excretion product from fish was later confirmed to be arsenobetaine by Cannon, et al. [34] in the dusky shark, *Carcarhinus obscurus* and by Edmonds and Francesconi [35] in the school whiting, *Sillago basensis*.

These studies on invertebrates and fish underline the similarity in the metabolic handling of arsenic by the various trophic levels of the aquatic ecosystem and expand the applicability of the hypothesis proposed above for algal cells to animal cells as well.

4. Methylation of Arsenic in Mammals

The transformation of arsenic compounds has been studied in a number of mammals including mice, dogs, humans and monkeys.

It is well known that rats are not suitable as experimental animals for arsenic metabolism studies because their arsenic metabolism differs considerably from that of other mammals. Vahter and Norin [36] selected mice as experimental animals because, unlike rats, they do not accumulate arsenic in their red blood cells and they exhibit a metabolism more like that of humans. Arsenate and arsenite were administered both orally and subcutaneously. These workers conclude that there is a difference in metabolism between penta- and trivalent inorganic arsenic and that it is dose-dependant. This difference is most apparent in liver and bile. In these organs, exposure to trivalent arsenic results in a much greater accumulation of the element than exposure to an equivalent amount of pentavalent arsenic.

The absorption of arsenic from the gastrointestinal tract is very high for both valance states. It was estimated that the initial absorption from the gastrointestinal tract is close to 100%. The whole-body retention of arsenic is about the same from the element in both valence states at low doses (0.4 mg/kg), but there is an obvious difference at high doses (4 mg/kg) as is evidenced by the greater retention of arsenite. The major metabolite is the same for both valence forms, presumably dimethylarsinate. The difference in retention at higher dose levels is attributed by these investigators to the fact that the organism's capacity for methylation has been exceeded. The [red cell-(As)/plasma-(As)] ratio was found to increase with increasing arsenic levels. Arsenic(III) was found to be retained to a higher degree than arsenic(V) in the skin.

When inorganic arsenic is administered to either the dog or cow, methylated forms of the element are found in their urine [37]. Dimethylarsinic acid has been identified as the major metabolite in the urine and plasma of the Beagle dog following the intravenous injection of inorganic arsenic [38]. These same investigators [39] were unable to detect any *in vitro* methylation of arsenic acid in the plasma, red blood cells or urine of dogs or humans.

Carbonneau and co-workers [40] studied the excretion of arsenic by monkeys fed "fish arsenic" or sodium arsenite. The "fish arsenic" was homogenized Atlantic grey sole containing 77 ppm arsenic. As has been observed in the case of human subjects, all of the excreted inorganic arsenic was eliminated in the urine. Individual metabolic response was manifested by the fact that while three animals excreted over 73 per cent of the arsenite ingested, a fourth excreted only 30 per cent. In contrast to the inorganic arsenic, although the major portion of the "fish arsenic" administered was voided in the urine, meaningful amounts (15.0, 15.9, 8.5 and 1 per cent, respectively) were eliminated in the feces. Material balance studies with the "fish arsenic" suggested that the excretion of these arsenicals is incomplete and that some of the arsenic may be retained in the body.

Crecelius [41] reported on the chemical forms of arsenic present in human urine following the ingestion of arsenite-rich wine, arsenate-rich drinking water and crab meat containing organoarsenic compounds whose identity was uncertain. About 10 per cent of the arsenite was excreted unchanged and some as arsenate, but the major portion was voided as dimethylarsinate and methylarsonate. The arsenite and arsenate levels increased rapidly following ingestion, but reached normal background levels after about 20 hours. The methylarsonate and dimethylarsinate levels remained elevated for 40 hours and approached background levels after about 85 hours. The ingestion of arsenate did not give rise to arsenite levels above the normal range, but a significant increase in arsenate levels occurred within the first 10 hours after ingestion. The concentration of dimethylarsinate reached a maximum 40 hours following ingestion and was still above background after 70 hours.

The organoarsenic compounds ingested by consumption of crab meat are excreted in two days and they are not converted to the inorganic forms by the body. The compounds appear to be excreted without any changes in chemical form.

Arsenite is subject to two removal processes. The first is probably a rapid assimilation into the blood followed by excretion through the kidneys. The second is slower and involves obvious methylation reactions which occur within the body. The biological half-life of arsenite is about 30 hours in humans.

The observations of Crecelius were largely corroborated by Tam, _et al_. [42], who utilized six adult human volunteers as subjects. These subjects received inorganic arsenic dispensed in gelatin capsules. Following oral dosing, 58 per cent of the ingested dose was recovered in the urine during the first five days. Of this, 51 per cent was in the form of dimethylarsinate, 21 per cent as methylarsonate and 27 per cent as inorganic arsenic. These authors suggest that methylarsonate may be a unique human metabolite. This is based on the observation that methylarsonate could be detected in dogs only when they are fed methylarsonate.

Further corroboration of the fate of arsenicals in humans is to be found in the results of Buchet and co-workers [43]. They administered sodium arsenite, methylarsonate and dimethylarsinate to groups of human subjects. The excretion rate increased in the order arsenite<dimethylarsinate<methylarsonate. After four days, the amount of arsenic present in the urine represented 46, 75 and 78 per cent of the ingested doses of these arsenicals, respectively. It was concluded that dimethylarsinate is recovered unchanged: 13 per cent of the methylarsonate is converted into dimethylarsinate while 75 per cent of the inorganic is methylated yielding about one-third as methylarsonate and two-thirds as dimethylarsinate.

In an attempt to ascertain whether the administration of methylating agents, *viz.*, methionine, choline, vitamin B_{12}, would stimulate arsenic methylation, a group of subjects was administered these reagents, together with inositol, before and after receiving sodium arsenite. However, no statistical difference was encountered in the chemical make-up of the arsenic metabolites. The failure to observe any change was attributed to the fact that the normal methylating capacity of the organism had not been saturated.

5. Arsenic Metabolism in Model Aquatic Ecosystems

During the period covered by this review, several attempts were made to study arsenic metabolism in somewhat simplified models of aquatic ecosystems. These models were composed of two or more trophic levels in which arsenic compounds, many times radiolabeled, were followed while passing from one level to another [5,55,45,46,47].

Unfortunately, for these kind of studies the complications of arsenic metabolism add to those of the aquatic ecosystems and the results are quite often difficult to interpret. After trying something of this kind in our laboratory [17], we are inclined to follow the more laborious task of searching one trophic level at a time and once the chemical nature and the dynamics of the organoarsenic compounds are more or less understood, move to the next upper trophic level. Once the situation at both levels is relatively clear, the movement of arsenocompounds through the system can be attacked favorably.

6. Replacement of Phosphate by Arsenate in Phosphorylitic Reactions

The critical role played by phosphate esters in carbohydrate metabolism cannot be over-emphasized. Because of the obvious chemical similarities which exit between phosphate and arsenate, phosphorylitic reactions in which arsenate replaces phosphate have long been a favorite subject for investigation [48]. The formation of sugar-arsenate esters has, and continues to be postulated in the interpretation of the experimental data. Nevertheless, in spite of repeated efforts on the part of many investigators, including ourselves, a sugar-arsenate ester has yet to be isolated and chemically characterized. This section of the present review discusses some of the recent developments in this area.

Several investigators have reported that sugar arsenate esters are formed when some monosaccharides are allowed to equilibrate with arsenate in aqueous solutions. However, the evidence is once again indirect. The arsenate ester is not isolated, but its presence in solution is inferred from enzyme assay.

Lagunas and Sols [12] note that a number of enzymes whose normal substrates are phosphorylated sugars show little or no activity on their corresponding non-phosphorylated substrates. They reported, however, that in the presence of arsenate, but not phosphate, the enzyme activity on non-phosphorylated substrates is greatly enhanced. The enzymes they studied included fructose-1-aldolase, glucose-6-phosphate dehydrogenase, glucosephosphate isomerase, α-glycerophosphate dehydrogenase and 1-phosphofructokinase. According to these investigators, "These observations suggest that the inducing effects of arsenate may involve an interaction of the arsenate with the sugar giving an ester-like intermediate which, like the corresponding phosphoric ester, can bind specifically to the enzyme in a form susceptible to catalytic attack."

The case of fructosediphosphate aldolase is especially interesting since its normal substrate is fructose-1,6-diphosphate. Arsenate was found not to increase the activity of this aldolase on fructose-1-phosphate, but it did elicit a marked activity of fructose-6-phosphate.

Long and Ray [49] prepared solutions containing arsenate and glucose and claimed the formation of an equilibrium mixture containing small amounts of glucose-6-arsenate and much smaller amounts of glucose-1-arsenate. The assay for glucose arsenate was based on earlier observations that phosphate from ^{32}P-labeled phosphoglucomutase is transferred to glucose in the presence of arsenate. In essence, the glucose arsenates were measured by means of "their rapid stoichiometric reaction with ^{32}P-labeled phosphoglucomutase." The products of the reaction are glucose-^{32}P-phosphates.

The observations, based largely on assays of solutions containing glucose and arsenate, were consistent with the following reactions:

$$\text{glucose} + A_i \rightleftharpoons \text{glucose-A}$$

$$\text{glucose-6-A} + E_p \longrightarrow \begin{cases} \text{glucose-1-P} + E_A \\ \text{glucose-6-A-1-P} + E_D \end{cases}$$

A_i - inorganic arsenate P = phosphate

E = enzyme E_D = dephosphoenzyme

Jaffé and Apitz-Castro [50] have investigated the reduction of dihydroxyacetone (DHA) mediated by glycerol-3-phosphate dehydrogenase (GPDH) in the presence of arsenate. The system, DHA-GPDH, is inactive unless arsenate is present in the reaction mixture. The fact that no inhibition of the reaction between the phosphorylated substrate, DHAP, was observed in the presence of DHA was taken to mean that DHA does not bind to the enzyme.

In the presence of arsenate, the value of 64 nM for DHA in the presence of arsenate is about 30 times higher than previously measured values for the phosphorylated substrates. Taken together with other observations it was concluded that arsenate is either acting on the substrate, DHA, directly, or through a ternary complex (As-E-DHA), but not on the enzyme directly. Based on stopped-flow experiments, *viz.*, the absence of a lag phase when arsenate and DHA were premixed before reaction with the enzyme and NADH, the formation of a DHA-arsenate adduct was postulated. It is this adduct which functions as the enzyme substrate. This adduct was considered to be "an ester-like product." A ^1H NMR measurement revealed that the α-protons of DHA are shifted up-field by 0.1 ppm upon the addition of arsenate, but the addition of phosphate brings about no detectable change.

Lagunas [13] has recently studied the oxidation of glucose and gluconate by glucose-6-phosphate dehydrogenase and 6-phosphogluconate dehydrogenase, resp., in the presence of arsenate. They postulate the existence of the following equilibria:

$$\text{glucose(G)} + \text{arsenate(A)} \rightleftharpoons \text{G6As} + H_2O$$
$$\text{gluconate(Gl)} + \text{arsenate(A)} \rightleftharpoons \text{6AsGl} + H_2O \ .$$

The arsenate ester concentrations were determined colorimetrically by the appearance of NADPH and the following reactions were suggested to explain the reduction of $NADP^+$:

$$\text{G6As} + NADP^+ \xrightarrow{E} \text{6-As-glucono-}\delta\text{-lactone} + \text{NADPH} + H^+$$
$$\text{6AsGl} + NADP^+ \xrightarrow{E'} \text{ribulose-5-As} + CO_2 + \text{NADPH} + H^+$$

E = glucose-6-phosphate dehydrogenase
E' = 6-phosphogluconate dehydrogenase

Hence, the existence of these arsenate esters is once again indirect. There is no question that, in the presence of arsenate, glucose and gluconate are oxidized by the dehydrogenases, but this does not constitute direct evidence for the formation of an arsenate ester. If, in fact, an arsenate ester were formed in solution, the equilibrium constant for the reaction, based on the data of Lagunas, is:

$$K = \frac{[\text{G6As}]}{[\text{G}][\text{A}]} = 3.6 \times 10^{-3}.$$

It is clear that direct evidence for the existence of sugar arsenate esters is still lacking. There is not doubt that arsenate can greatly affect the activity of a variety of phosphorylating enzymes. All of the evidence for the formation of sugar arsenates is indirect and is based on the fact that the formation of such intermediates is consistent with equilibrium and/or kinetic measurements. It has been our experience that the C-O-As bond is hypersensitive to moisture and the formation of a sugar arsenate ester in an aqueous medium is unlikely. The possibility that a hydrogen bonded species involving a sugar hydroxyl group and the arsenyl group merits some consideration. Such species may be recognized as substrates by the enzymes.

The esterification of arsonic and arsinic acids to form esters of the type shown below has been reported [51]. Their preparation requires the careful and continuous removal of water. They do not survive in

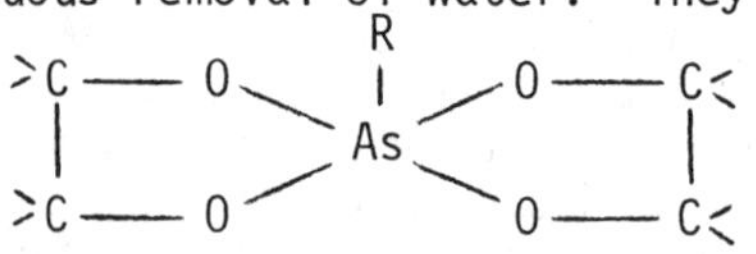

```
               R
>C —— O        |        O —— C<
|        \     |     /        |
|          >  As  <           |
|        /           \        |
>C —— O                 O —— C<
```

aqueous solution. Chan, et al. [52] isolated a substance from the incubation of rat liver mitochondria with ^{74}As-arsenate which they claimed possesses properties of a low molecular weight arsenate ester. However, its chemical characterization was greatly limited. Also, the substance, designated as As-X, did not dissociate into inorganic arsenate when treated for 10 minutes at 60° at pH 1-10. This stability, even at low pH's, makes it highly unlikely that As-X is a low molecular weight arsenate ester.

Conclusive evidence for the existence of sugar-arsenate esters and the nature of their reactions with the various phosphorylating enzymes must await the synthesis and characterization of such compounds. To date, this goal has eluded the chemist.

7. Some Recently Discovered Arsenic-containing Biomolecules

The conversion of arsenite and arsenate to the monoethyl and dimethylated derivatives has been demonstrated in a great variety of living species, from simple unicellular organisms through mammals. Recently, the marine environment, where a number of plant and animal species are known to concentrate arsenic, has furnished examples of new and interesting arsenic-containing biomolecules.

Edmonds and co-workers [16] were the first to isolate and identify arsenobetaine from the tail muscle of the western rock lobster, *Panuliris cygnus*. The compound was extracted by methanol extraction, evaporation, redissolution in water and extracton of some impurities with ether. The aqueous solution was acidified and shaken with phenol. Dilution of the phenol with ether was followed by water extraction. Ion exchange and chromatography on Zeokarb 225 (H^+) yielded, by recrystallization from acetone, crystals of arsenobetaine. The crystals were unequivocally identified by single crystal x-ray analysis. The structure of arsenobetaine is shown below.

$$(CH_3)_3\overset{+}{As}\text{-}CH_2COO^-$$

Arsenobetaine has also been shown to exist in the dusky shark, *Carcharhinus obscurus* [34] and the school whiting, *Sillago basensis* [35].

One of the most recent and most interesting reports on arsenic-containing biomolecules is that which describes the isolation of arsenosugars from the brown kelp, *Ecklonia radiata* [19]. These two water-soluble sugars account for 81 percent of the arsenic in this organism. Identification of these sugars was based primarily on ^1H and ^{13}C NMR spectral data. They have been assigned a ribo structure as follows:

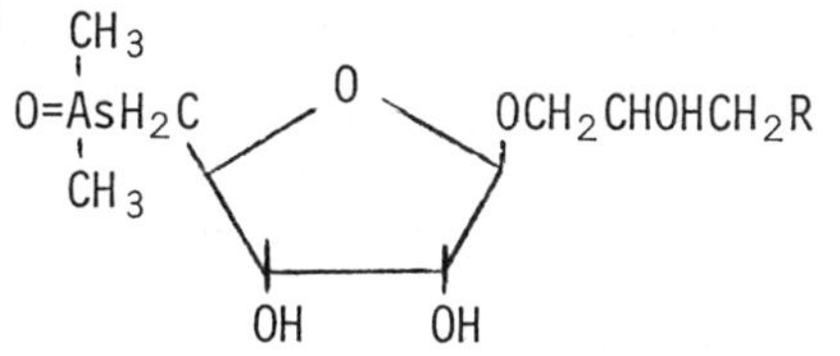

CH_3
$O{=}AsH_2C$ O $OCH_2CHOHCH_2R$
CH_3
OH OH

A, R = $-SO_3H$; B, R = $-OH$.

It is interesting to note that this derivative is not arsenate ester, i.e., there is no As-O-C bond, but has a direct carbon linkage to the ribose unit. This is not unexpected in view of the extreme hydrolytic instability of the C-O-As linkage. This molecule is of great interest because it bears two methyl groups and a methylene bridge. Hence, as suggested by Edmonds and Francesconi, such sugars could be metabolized further to arsenobetaine.

These reports on the presence of arsenobetaine and dimethylarsenosofuranosides in several marine species emphasize the most widely observed biochemical fate of arsenic in the environment, *viz.*, methylation.

8. Transformation of Arsenicals by Microorganisms

The methylation of arsenate by molds was observed as early as 1901 [53], but the identification of the product as trimethylarsine was reported by Challenger in 1945 [54]. A number of interesting results in this area have been reported during recent years by Cullen and co-workers. Virtually all of these studies have utilized the yeast, *Candida humicola* as the experimental organism. Actually, the first observation on the methylation of arsenicals by this organism was made by Cox and Alexander [55]. Cullen, et al. [56] were able to identify five distinct arsenic compounds following incubations of ^{74}As or ^{14}C labelled arsenic compounds with broken cell homogenates. When ^{74}As-arsenate was incubated with the cell homogenate, arsenite, methylarsonate and dimethylarsinate were identified; with ^{14}C-methylarsonate as substrate, dimethylarsinate and trimethylarsine oxide were produced. Replacement of the cell preparation by buffer failed to bring about any transformations. Hence, the various compounds identified represent probably intermediates in biosynthesis of trimethylarsine. This same group [20] grew cultures of this organism, as well as two others, aerobically in the presence of L-methionine-methyl-d_3. It was found that CD_3 was incorporated into the evolved methylated arsenic compounds. This indicates that *S*-adenosylmethionine or a related sulfonium compound is involved in the biological methylation. When *C. humicola* were preconditioned in $(CH_3)_2AsO_2^-$, the rate of Me_3As production from either arsenate or dimethylarsinate was more rapid than from non-preconditioned organisms [57]. However, the ability of the cell to assimilate arsenate was not affected by dimethylarsinate preconditioning.

It was found that the concentration of labeled arsenic increases in the cell during growth.

In some very recent work [58], this group has observed that trimethylarsine oxide is reduced by this yeast to trimethylarsine. The rate of the reduction was proportional to the cell concentration and no reduced product was formed by heated or broken cells. The reaction was inhibited by several electron transport inhibitors and uncouplers of oxidative phosphorylation. Pre-incubation of the cells in trimethylarsine oxide increased the rate of reduction dramatically. In this study, it was not possible to establish whether the reduction is enzymatic or whether it is the result of a chemical reaction between Me_3AsO with an enzymatically produced reducing agent. However, it was shown that a biologically intact organism was required to furnish the necessary reducing conditions. The stimulatory effect of methylated arsenicals on this particular reduction suggests the existence of a common pathway for the metabolism of methylarsonate, dimethylarsinate and trimethylarsine oxide.

Methylarsine sulfide $(MeAsS)_n$ was found to be considerably more toxic than $(MeAsO)_n$ towards several microorganisms [59]. Incubation of *C. humicola* with CH_3AsS showed that $(CH_3)_3As$ and CH_3AsH_2 were formed, but not $(CH_3)_2AsH$. Trimethylarsine was also produced from CH_3AsO. The formation of Me_3As from MeAsO and MeAsS suggests that these two methylated As(III) species may become involved in the Challenger pathway [54]. The failure to detect MeAsO or $MeAsH_2$ when *C. humicola* was incubated with arsenate and methylarsonate suggests that methylarsine oxide is not a "free" intermediate in the Challenger mechanism.

Cullen has expressed the opinion that methylcobalamin is not involved in natural systems in metal methylation reactions.

9. The Reaction of Pentavalent Arsenic Compounds with the Thiol Group

That an irreversible chemical reaction occurs between dimethylarsinic acid and thiol groups has been unequivocally established. The reaction is stoichiometric and is described by the following equation:

$$3\ HSCH_2\underset{NH_2}{C}H\overset{O}{\overset{\|}{C}}OH + (CH_3)_2\overset{O}{\overset{\|}{As}}OH \longrightarrow [SCH_2CH(NH_2)C(O)OH]_2$$

$$+\ (CH_3)_2AsSCH_2CH(NH_2)\overset{O}{\overset{\|}{C}}OH + 2\ H_2O\ .$$

In this reaction the thiol is oxidized to the disulfide and pentavalent arsenic is converted to the S-dimethylarsino derivative. In addition to cysteine, dimethylarsinic acid has been shown to react in an identical manner with glutathione and 1-thio-β-glucopyranose [60]. This fact has several important implications. It is essential that the use of cacodylate buffers in any enzyme studies be completely eliminated. This was previously suggested by Jacobson, et al. [61]. Furthermore, the use of thiol-containing reagents, such as xanthates, to distinguish between tri- and pentavalent arsenic for analytical purposes should not be used [62]. Such analytical procedures are especially unreliable when mammalian blood and urine samples are being analyzed since the conversion of both arsenite and arsenate to methylarsonate and dimethylarsinate has been incontrovertibly demonstrated in a number of mammalian species and the cacodylate is especially reactive towards the thiol group. Finally, the widely-held interpretation that the deactivation of enzymes by arsenicals *in vivo* involves, as a first step, their reduction to the trivalent oxide, is a concept that no longer needs to be invoked [63]. It is entirely possible for dimethylarsinate to deactivate an enzyme(E) in the manner shown below as suggested by Daniels and Zingaro [64].

$$Me_2AsO_2H + 3\ ESH \longrightarrow Me_2AsSE + (ES)_2 + 2\ H_2O$$

$$Me_2AsO_2H + E(SH)_3 \longrightarrow \begin{matrix} S \\ | \\ S \end{matrix}\!\!>\!E\text{-}SAsR_2 + 2\ H_2O\ .$$

The authors acknowledge the assistance of the following: the Robert A. Welch Foundation of Houston, Texas, which has supported the research of Ralph A. Zingaro for a number of years, the Texas Agricultural Experiment Station and the Chemical Manufacturers' Association. All have assisted in various ways in the preparation of this manuscript.

References

[1] Brinckman, F. E. and Bellama, J. M., Book, Organometals and Organometalloids, Amer. Chem. Soc. Symp. Ser. No. 82, 1978.

[2] Andreae, M. O., Distribution and speciation of arsenic in natural waters and some marine algae, Deep-Sea Res., 25, 391-402 (1978).

[3] Sanders, J. G., Effect of arsenic speciation and phosphate concentration on arsenic inhibition of *Skeletonema costatum* Bacillariophyceae), J. Phycol., 15, 424-428 (1978).

[4] Woolson, E. A., Bioaccumulation of arsenicals, in Arsenical Pesticides, E. A. Woolson, ed., pp. 97-107 (Amer. Chem. Soc. Symp. Ser. No. 7, Washington, D.C., 1975).

[5] Penrose, W. R., Conacker, H. B. S., Black, R., Méranger, J. C., Miles, W., Cunningham, H. M., and Squires, W. R., Implications of inorganic/organic interconversions on fluxes of arsenic in marine food webs, Environ. Health Persp., 19, 53-59 (1977).

[6] Blasco, F., Réduction partielle de l'arséniate en arsénite et exsorption de l'arsenic par *Chlorella pyrenoidosa*, Physiol. Vég, 13, 185-201 (1975).

[7] Planar, D. and Healey, F. P., Effects of arsenic on growth and phosphorus metabolism of phytoplankton, J. Phycol., 14, 337-341 (1978).

[8] Brunskill, G. J., Graham, B. W., and Rudd, J. W. M., Experimental Studies on the effect of arsenic on microbial degradation of organic matter and algal growth, Can. J. Fish., Aquat. Sci., 37, 415-423 (1980).

[9] Bottino, N. R., Newman, R. D., Cox, E. R., Stockton, R., Hoban, M., Zingaro, R. A., and Irgolic, K. J., The effects of arsenate and arsenite on the growth and morphology of the marine unicellular algae, *Tetraselmis chui* (chlorophyta) and *Hymenomas carterae* (chrysophyta), J. Exp. Mar. Biol. Ecol., 33, 153-168 (1978).

[10] Conway, H. L., Sorption of arsenic and cadmium and their effects on growth, micronutrient utilization, and photosynthetic pigment composition of *Asterionella formosa*, J. Fish. Res. Bd. Can., 35, 286-294 (1978).

[11] Andreae, M. O. and Klumpp, D., Biosynthesis and release of organoarsenic compounds by marine algae, Environ. Sci. Technol., 13, 738-741 (1979).

[12] Lagunas, R. and Sols, A., Arsenate induced activity of certain enzymes on their dephosphorylated substrates, FEBS Lett., 1, [1], 32-34 (1968).

[13] Lagunas, R., Sugar-arsenate esters: thermodynamics and biochemical behavior, Arch. Biochem. Biophys, 205, [1], 67-75 (1980).

[14] Lehninger, A. L., Biochemistry, 2nd. ed., Worth Publ., New York, 1975, p. 429.

[15] McBride, B. C., Merilees, H., Cullen, W. R., and Pickett, W., in Organometals and Organometalloids, F. E. Brinckman and J. M. Bellama, eds., Amer. Chem. Soc. Symp. Ser. No. 82, pp. 94-115 (1978).

[16] Edmonds, J. S., Francesconi, K. A., Cannon, J. R., Raston, C. L., Skelton, B. W., and White, A. H., Isolation, crystallization, crystal structure and synthesis of arsenobetaine, the arsenical constituent of the western rock lobster *Panulirus longipes cygnus*

George, Tetrahedron Lett., [18], 1543-1546 (1977).

[17] Irgolic, K. J., Woolson, E. A., Stockton, R. A., Newman, R. D., Bottino, N. R., Zingaro, R. A., Kearney, P. C., Pyles, R. A., Maeda, S., McShane, W. J., and Cox, E. R., Characterization of arsenic compounds formed by *Daphnia magna* and *Tetraselmis chui* from inorganic arsenate, Environ. Health. Persp., 19, 61-66 (1977).

[18] Cooney, R. V., Mumma, R. O., and Benson, A. A., Arsonium phospholipid in algae, Proc. Natl. Acad. Sci. USA, 75, 4262-4264 (1978).

[19] Edmonds, J. S. and Francesconi, K. A., Arseno-sugars from brown kelp (*Ecklonia radiata*) as intermediates in cycling of arsenic in a marine ecosystem, Nature, 289, [5798], 602-604 (1981).

[20] Cullen, W. R., Froese, C. L., Lui, A., McBride, B. C., Patmore, D. J., and Reimer, M., The aerobic methylation of arsenic by microorganisms in the presence of L-methionine-methyl-d_3, J. Organometal. Chem., 139, [1], 61-69 (1977).

[21] Wood, J. M., Chek, A., Dizikes, L. J., Ridley, W. P., Rakow, S., and Lakowicz, Mechanisms for the biomethylation of metals and metalloids, Fed. Proc., 37, 16-21 (1978).

[22] Benson, A. A., Cooney, R. V., and Herrera-Lasso, J. M., Arsenic metabolism in algae and higher plants, J. Plant. Nutr., 3, [1-4], 285-292 (1981).

[23] Lunde, G., The analysis of arsenic in the lipid phase of marine limnetic algae, Acta Chem. Scand., 26, 2642-2644 (1972).

[24] Lunde, G., The synthesis of fat and water-soluble arseno-organic compounds in marine and limnetic algae, Acta Chem. Scand., 27, 1586-1594 (1973).

[25] Lunde, G., A comparison of organoarsenic compounds from different marine organisms, J. Sci. Food Agr., 26, 1257-1262 (1975).

[26] Bottino, N. R., Cox, E. R., Irgolic, K. J., Maeda, S., McShane, W. J., Stockton, R. A., and Zingaro, R. A., Arsenic uptake and metabolism by the alga *Tetraselmis chui* (Organometals and Organometalloids, F. E. Brinckman and J. M. Bellama, eds., Amer. Chem. Soc. Symp. Ser. No. 82, Wash., D.C., 1978), pp. 116-129.

[27] Benson, A. A. and Summons, R. E., Arsenic accumulation in Great Barrier Reef invertebrates, Science, 211, 482-483 (1981).

[28] Ünlü, M. Y. and Fowler, S. W., Factors affecting the flux of arsenic through the mussel *Mytilus galloprovincialis*, Mar. Biol., 51, 209-219 (1979).

[29] Sorensen, E. M. B., Thermal effects on the accumulation of arsenic in green sunfish, *Lepomis cyanellus*, Arch. Environ. Contam. Toxicol., 4, 8-17 (1976).

[30] Sorenson, E. M. B., Toxicity and accumulation of arsenic in green sunfish, *Lepomis cyanellus* exposed to arsenate in water, Bull. Environ. Contam. Toxicol., 13, 756-761 (1976).

[31] Sorensen, E. M. B., Ultrastructural Changes in the hepatocytes of green sunfish, *Lepomis cyanellus Rafinesque* exposed to solutions of sodium arsenate, J. Fish. Biol., 8, 229-240 (1976).

[32] Sorensen, E. M. B., Henry, R. E., and Ramirez-Mitchell, R., Arsenic accumulation, tissue distribution and cytotoxicity in teleosts following indirect aqueous exposures, Bull. Environ. Contam. Toxicol., 21, 162-169 (1979).

[33] Oladimeh, A. A., Qadri, S. U., Tam, G. K. H., and de Freitas, S. W., Metabolism of inorganic arsenic to organoarsenicals in rainbow trout (*Salmo gairdneri*), Ecotoxicol. Environ. Safety, 3, 394-400 (1979).

[34] Cannon, J. R., Edmonds, J. S., Francesconi, K. A., and Langsford, J. B., Management and Control of Heavy Metals in the Environment, *Int. Conf. London*, 283-286 (CEP Consultants, Edinburgh, 1979).

[35] Edmonds, J. S. and Francesconi, K. A., Mar. Pollut. Bull., in the press.

[36] Vahter, M. and Norin, H., Metabolism of ^{74}As-labeled trivalent and pentavalent inorganic arsenic in mice, Environ. Res., 21, [2], 446-457 (1980).

[37] Lakso, J. U. and Peoples, S. A., Methylation of inorganic arsenic by mammals, J. Agr. Food Chem., 23, [4], 674- (1975).

[38] Tam, G. K. H., Charbonneau, S. M., Bryce, F., and Lacroix, G., cited in [39] of this review, but not located in journal, Bull. Envoron. Contam. Toxicol. 1978b.

[39] Tam, G. K. H., Charbonneau, S. M., Lacroix, G., and Bryce, F., *In vitro* methylation of ^{74}As in urine, plasma and red blood cells of human and dog, Bull. Environ. Contam. Toxicol., 22, 69-71 (1979).

[40] Charbonneau, S. M., Spencer, K., Bryce, F., and Sandi, E., Arsenic excretion by monkeys dosed with arsenic-containing fish or with inorganic arsenic, Bull. Environ. Contam. Toxicol., 20, [4], 470-477 (1978).

[41] Crecelius, E. A., Changes in the chemical speciation of arsenic following ingestion by man, Environ. Health. Persp., 19, 147-150 (1977).

[42] Tam, G. K. H., Charbonneau, S. M., Bryce, F., Pomroy, C., and Sandi, E., Metabolism of inorganic arsenic (^{74}As) in humans following oral ingestion, Toxicol. Appl. Pharm., 50, 319-322 (1979).

[43] Buchet, J. P., Lauwerys, and Roels, H., Comparison of the urinary excretion of arsenic metabolites after a single oral dose of sodium arsenite, monomethylarsonate, or dimethylarsinate in man, Int. Arch. Occup. Environ. Health, 48, 71-79 (1981).

[44] Woolson, E. A., Isensee, A. R., and Kearney, P. C., Distribution and isolation of radioactivity from ^{74}As-arsenate and ^{14}C-methanearsonic acid in aquatic model ecosystem, Pest. Biochem. Physiol., 6, 261-269 (1976).

[45] Bohn, A., Trace metals in fucoid algae and purple sea urchins near a high arctic lead/zinc ore deposit, Mar. Pollut. Bull., 10, 325-327 (1979).

[46] Spehar, R. L., Fiandt, J. T., Anderson, R. L., and DeFoe, D. L., Comparative toxicity of arsenic compounds and their accumulation in invertebrates and fish, Arch. Environ. Contam. Toxicol., 9, 53-63 (1980).

[47] Sanders, J. G., Arsenic cycling in marine systems, Mar. Environ. Res., 3, 257-266 (1980).

[48] Cohn, M., in The Enzymes, Vol. 5, second ed., P. D. Boyer, H. Lardy, and K. Myrbäck, eds., pp. 179-206 (Academic Press, New York, N.Y., 1961).

[49] Long, J. W. and Ray, W. J., Jr., Kinetics and thermodynamics of the formation of glucose arsenate. Reaction of glucose arsenate with phosphoglucomutase, Biochem., 12, [20], 3932-3937 (1973).

[50] Jaffé, K. and Apitz-Castro, R., Studies on the mechanism by which inorganic arsenate facilitates the enzymatic reduction of dihydroxyacetone by α-glycerophosphate dehydrogenase, FEBS Lett., 80, [1], 115-118 (1977).

[51] Levskaya, G. S., and Kolomiets, A. F., Esterification of arsonic and arsinic acids, Zh. Obshch. Khim., 37, [4] 905-907 (1967).

[52] Chan, T-L., Thomas, B. R., and Wadkins, C. L., The formation and isolation of an arsenylated component of rat liver mitochondria, J. Biol. Chem., 244, [11], 2883-2890 (1969).

[53] Gasio, B., Arch. Ital. Biol., 35, 201 (1901).

[54] Challenger, F., Biological methylation, Chem. Rev., 36, [3], 315-361 (1945).

[55] Cox, D. P. and Alexander, M., Production of trimethylarsine gas from various arsenic compounds by three sewage fungi, Bull. Environ. Contamin. Toxicol., 9, [2], 84-88 (1973).

[56] Cullen, W. R., McBride, B. C., and Pickett, A. W., The transformation of arsenicals by *Candida humicola*, Can. J. Microbiol., 25, [10], 1201-1205 (1979).

[57] Cullen, W. R., McBride, B. C., and Reimer, M., Induction of aerobic methylation of arsenic by *Candida humicola*, Bull. Environ. Contam. Toxicol., 21, [1,2], 157-161 (1979).

[58] Pickett, A. W., McBride, B. C., Cullen, W. R., and Manji, H., The reduction of trimethylarsine oxide by *Candida humicola*, Can. J. Microbiol, in the press (1981).

[59] Cullen, W. R., Biotransformation of organic compounds, paper presented to the Bioorganometallic/Environmental section, Tenth International Conf. on Organometallic Chemistry, Toronto, Can., August 10, 1981.

[60] Banks, C. H., Daniel, J. R., and Zingaro, R. A., Biomolecules bearing the S- or $SeAsMe_2$ function: amino acid and steroid derivatives, J. Med. Chem., 22, [5], 572-575 (1979).

[61] Jacobson, K. B., Murphy, J. B., and Das Sarma, B., Reaction of cacodylic acid with organic thiols, FEBS Lett., 22, [1], 80-82 (1972).

[62] Crawford, T. B. B. and Storey, I. D. E., A quantitative micromethod for the separation of inorganic arsenite from arsenate in blood and urine, Biochem. J., 38, [2], 195-198 (1944).

[63] Arsenic, report of the sub-committee on arsenic, committee on medical and biologic effects of environmental pollutants, National Academy of Sciences, Washington, D.C., pp. 133-136 (1977).

[64] Daniel, J. R. and Zingaro, R. A., Dimethylarsinous acid esters of 1-thio- and -selenogalactose. A new class of potential carcinostatic agents, Phosphorus and Sulfur, 4, [2], 179-185 (1978).

DISCUSSION

E. A. Woolson: Do you think that the plants that become tolerant to high levels of inorganic arsenate in soil use the arsenate as a substrate instead of phosphate since if they're put into soil with normal phosphate levels, the plants die? Do you think this is an enzyme adaptation?

R. A. Zingaro: It surely sounds like it to me. I didn't know about this form of adaptation, but I think it fits nicely into this pattern.

A. Schwerdtle: In synopsis, would you say that the sugars (possibly as phosphate and nucleic acid derivatives) and the sulfhydryl (-SH) group are the two main arsenic biological transformation pathways, or do you have some other hypotheses?

R. A. Zingaro: One thing gives us a clue. Dimethylarsinate reacts easily and rapidly with thiol groups. Yet, dimethylarsinates appear to be non-toxic, which means that they must go right out of the kidneys without ever getting into the system. Of course, the biochemist or the physiologist can offer an alternate explanation, namely, it does inactivate thiol-containing proteins, because of the configuration of the protein, and the ability to actually chemically interact with the protein. On the other hand, I feel that the toxic effects due to large amounts of inorganic arsenic appear to be interfering with the phosphate cycle. The ability of arsenate in the absence of phosphate to affect the functions of such enzymes without the actual formation of an ester is something that merits a great deal of study. One of the things I'd like to look at is arsenate polyhydroxy compound interaction. My feeling is that you don't form an ester. I can't conceive of a carbon oxygen arsenic compound surviving in ordinary water. But we know that the O-As bond can form strong hydrogen bonds and hydroxyl from sugar to arsenyl oxygen could certainly represent an activated complex which seems to me thermodynamically a lot more reasonable.

25

UPTAKE, TRANSLOCATION AND PHYTOTOXICITY OF ARSENIC IN PLANTS

R. Don Wauchope
U.S. Department of Agriculture - Agricultural Research Service
Southern Weed Science Laboratory
P.O. Box 225
Stoneville, MS 38776

Arsenic is a natural trace constituent of soils and plants. At normal levels (1-20 ppm As) in soils which support plant growth, it occurs mainly in the inorganic AS(V) (arsenate) form, which is tightly bound to soil minerals, particularly the colloidal metal hydrous oxides, by ionic bonds. Under these conditions it is only slightly available for root uptake and both native and cultivated plants seldom show arsenic concentrations greater than 1 part per million in tissue. Once arsenic enters a plant it is, like phosphorus, freely transported both actively and passively into all active tissues, and tissue levels are simply proportional to arsenic availability. Toxicity in plants occurs when (a) abnormally high arsenic levels are produced in soil, either deliberately or accidently by man's activities, (b) a change in soil conditions increases availability, (c) plant foliage is exposed to arsenical compounds. The latter is typically the result of a deliberate agrichemical application for insecticidal, defoliation or weed control purposes, the major compounds being arsenic acid and its salts, and salts of methanearsonate and dimethylarsinic acid. The toxicity of arsenic compounds in plants is expressed in a wide variety of symptoms including effects on photosynthesis, respiration, growth regulation, and reproduction. Apparently a broad spectrum of biochemical processes is affected and definitive research on which process or processes are most important, even in herbicidal uses, has not been done.

1. Introduction

This paper presents the results of a literature review on (a) arsenic levels found in soils, either as a natural constituent or as a result of contamination by man, (b) the factors in soils which affect the availability of soil arsenic to plants, (c) arsenic levels found in plants whether from soil sources or from arsenical pesticide applications, (d) arsenic movement (translocation) within plants, (e) arsenic toxicity in plants. A detailed presentation of these topics would be very long; the author will summarize as briefly as possible what is and what is not known, and provide a bibliography which will be a useful introduction to the literature.

2. Arsenic Levels in Soils

2.1. Uncontaminated soils

Total-arsenic levels in ppm (μg/g dry soil) as reported in surveys are given in table 1.

Table 1. Arsenic levels reported in survey papers, 1940-1979.

Soils description[a] (crop or site)	As levels range, ppm	Reference
Field soils	0.5-14	[161]
Corn belt soils	0.1-23	[31]
Uncontaminated soils	0.3-38	[153]
Potato soils, Wisconsin	2-26	[130]
Noncropland soils	0.1-54	[151]
Cotton/vegetable soils	1.5-54	[132]
Grain/root crop soils	1.4-27	[132]
Ontario soils (including orchards)	1-121	[101]
Urban soils	0.1-112	[28,30]
Uncontaminated semiarid soils	7-655	[123]
Contaminated soils	11-550	[153]
Cropland soils	.09-180	[29,150,151]
Smelter-contaminated soils	114-315	[41,161]
Orchard soils	4-2600	[161]
"	11-300	[69]
"	10-124	[19]
"	40-233	[99]
"	13-219	[132]

[a]Soils are from US sites unless otherwise described.

The crustal abundance of arsenic is 5 ppm [102][1] but it is concentrated in base-metal ores; igneous rocks, sandstones and limestones contain 1.5-2 ppm. Shales and deep-sea sediments contain 10 ppm [107], similar to weathered soils. This suggests that the widespread, higher arsenic concentrations in soils (relative to parent materials) is due to volcanic activity [57]. Table 1 can be summarized to say that typical uncontaminated agricultural soils will contain 1-20 ppm arsenic (this range is also based on many single values reported in the literature), that most soils will fall between 1 and 50 ppm, and that natural levels between 0.1 and several hundred can occur, though rarely [102].

[1]Figures in brackets indicate the literature references at the end of this paper.

The volcanic origin of native soil arsenic explains a major feature of its characteristics in soils; it is mainly associated with metal hydrous oxides, principally of iron and aluminum, that are on the surface of mineral particles rather than with the soil parent mineral matrix itself. In nonflooded, aerobic soils the main arsenic species is the arsenic(V) oxoanion or arsenate, which is thermodynamically stable with respect to lower oxidation states and with respect to various organoarsenic species which can occur due to soil microbial activity [9,23,57,140,157]. Arsenate soil chemistry is generally similar to phosphate (see below): it has similar pKa values for its three deprotonation steps in water, and both oxoanions will exist at normal soil-water pH principally as the mono- or dianions (for the second deprotonation step, the values are 7.21 for phosphate and 7.09 for arsenate [143]). Both form insoluble salts and bond in similar ways to soil surfaces.

Unlike phosphate, arsenate does not appear to form complex polymeric organospecies such as are associated with soil organic matter [82], and, arsenate is reducible to lower oxidation states and arsenic undergoes simple biological methylation-demethylation reactions in soil. This is discussed further below. For uncontaminated soils, then, we are mainly concerned with arsenate and the availability of arsenate for plant uptake. Unlike the geochemically similar element selenium, which is much more available and is concentrated in plants, arsenic as native arsenate, though a trace component of all soils and plants, has never been reported to reach levels in plants which are toxic to animals.

2.2. Contaminated soils

Soils unintentionally contaminated with enough arsenic to cause toxicity to non-target species have been found in many areas of North America (table 1) and are nearly always associated with massive, long-term use of arsenical insecticides.[2] In addition there are a few cases of significantly elevated soil residues from the use of arsenite salts as potato defoliants [126,130,135]. Inorganic arsenic salts were determined to be effective insecticides before the turn of the century [97]. Prior to the introduction of the organochlorine insecticides in the 1940's, lead and calcium arsenates were major "economic poisons," and were used at rates up to 100 pounds per acre per year in orchards and cotton fields. Studies in the 1930's showed that soils of these two crops could become practically sterile near the surface, depending on intensity and length of arsenic use, soil texture, and soil iron and aluminum content [4,35,56,99,103,139]. Assuming the usual estimate that an acre plow-layer (seven inches depth) of soil weighs about 2 million pounds, then each pound of arsenic mixed uniformly in an acre plow-layer of soil contributes about 1/2 ppm arsenic, or from 3 to 50% of the 1-20 ppm already there. Lead and calcium arsenates contain 20 to 40% arsenic as

[2]Industrial contamination of soils or water (smelting, mining, coal-powered generator plant wastes, etc.) is a problem in a few localities but will not be dealt with here.

formulated; thus it is not difficult to account for the several hundred ppm arsenic found in orchard soils, some of which received arsenic treatments for 20-40 years or more [5,19,69,85].

There is continuing concern over the potential for arsenic buildup in soils with current uses of arsenical pesticides. According to one estimate [96], 82% of domestic arsenic use is in agriculture. It should be recognized, however, that the majority of this use is now in the form of methanearsonate herbicides for selective weed control in noncrop areas and cotton [11,71]. Although large acreages are involved, the annual application rates used are much smaller than were used for inorganic arsenical insecticides: 2-4 pounds per acre per year of the methanearsonates MSMA[3] or DSMA are typical, commercial formulations of which contain about 20% arsenic. Thus, only a fraction of a pound of elemental arsenic is being added to these soils each year. Using the "plow-layer" assumptions above, this, in the absence of dissipation, will increase natural arsenic levels by only a few tenths of a ppm per year.

3. Factors Affecting Soil Arsenic Availability to Plants

3.1. Arsenate

It was recognized quite early that the total arsenic found in contaminated soils is not a reliable indicator of potential plant toxicity [3,38,39,40,62,114,139]. As with phosphate, arsenate is tightly bound to soil surfaces. This is evidenced by the very slight downward mobility of arsenic from contaminated surface zones to deeper soil layers even when heavily contaminated soils are exposed to years of percolating rainfall [5,69,135]. The amounts of soil arsenic extracted by water alone are always a very small fraction of the total found by exhaustive digestion [2,19,45,102, 162]. This small fraction does leach downward: water-soluble arsenic fractions are higher in soils from low-rainfall areas [161].

Arsenate availability has been studied by two complementary techniques. Extraction studies show that arsenate may be extracted by reagents known to provide measures of phosphate availability and location with specific soil components. Adsorption studies show that soil properties such as clay content and amorphous oxide content are correlated with the degree of soil adsorption of arsenate and with the phytotoxicity of added arsenate.

Extraction studies generally confirm the close similarity of phosphate and arsenate as to location in soil "fractions." Water-soluble arsenic, as mentioned above is only a few percent of the total and tends to be correlated with total-arsenic only in soils recently treated with arsenic [32,45,78,162]. In soils with old arsenic residues, or native arsenic, there is little correlation

---

[3]Abbreviations used in the text are the common names of these herbicides accepted by the Weed Science Society of America [148].

[19,102]. "Available" arsenic, as determined by extraction with .025 N HCl/.03 N NH_4F (Bray P-1) solution or 1 N ammonium acetate [78], 0.5 N sodium bicarbonate or .05 N HCl/.025 N H_2SO_4 [160], does correlate with plant response and with added arsenic amounts [32,160]. Extractants which remove phosphate associated with iron or aluminum hydrous oxides [76,109] work similarly with arsenate [79,82,84,160, 161], and these extractants will extract the bulk of soil arsenate which is not so tightly bound as to require fusion or concentrated acid digestion. The iron fraction tends to have the greater amount of arsenic unless the amount of active iron oxides is very low [57, 160,162].

Under mildly reducing conditions, ferric oxide can be reduced to ferrous, iron(II) oxides. Apparently the ferrous species form a less tightly-bound complex with arsenate and the ferrous ion itself is less acidic and more soluble. Under these conditions arsenate is released [45,68]. For a soil-immersed platinum versus standard calomel electrode pair this process becomes significant in the neighborhood of 100 mv electochemical potential (Eh).

Adsorption and bioassay studies in the 1930's and earlier confirmed field observations that arsenic was "fixed" like phosphate, by high-clay soils [38,62,98] and soils that were high in iron [40, 49,114], and that arsenite was much more toxic than arsenate [26,56, 139]. These and more recent studies also make it clear that the initial rapid adsorption of added arsenate by soils is only the first step in a slow fixation process which requires months [79,105, 140,162], so that successive crops grown on arsenic treated soils exhibit decreasing arsenic response and residues [66,69,158]. Like phosphate, arsenate will take part in soil anion exchange almost instantaneously [22,42,152] but is less strongly bound than phosphate by this mechanism [42,43,134]. Both anions then slowly diffuse into and complex with the outer layers of hydrous oxide soil particle coatings. If the most active outer layer of these coats is removed by extraction, adsorption decreases dramatically [79]. There are some indications that aluminum hydrous oxides are the most active adsorbents initially but that iron oxides are the favored matrix in time [57,161,162].

It should be remembered that the surfaces of the hydrous oxides of the higher oxidation states of aluminum, iron, manganese, etc., are themselves not at equilibrium [81], but undergo extremely complex, polymeric, slow solution and precipitation reactions which are effected by soil wetting/drying cycles and pH [125]. The true solution species in these reactions are unknown even in such highly-studied cases as iron [137]. This explains the tendency of pure-salt solubility equilibrium constants, such as for iron and aluminum arsenate and phosphate compounds to underpredict arsenate/phosphate concentrations in soil water [92], even when amorphous-phase solubility values are used [67]. In summary, adsorption studies are most useful for examining the rapid, but not the ultimate (if there is such) activity-determining adsorption-desorption processes of these anions.

3.2. Arsenite

As stated above, under mildly reducing conditions (Eh ≈ 100) arsenate is released by ferric-ferrous oxide reduction. If the Eh is reduced to slightly more reductive values (near zero or negative values -- not uncommon for microbiologically active, flooded soils), arsenate itself is reduced to arsenite [57]. Arsenite is 4 to 100 times more toxic than arsenate to plants [25,26,35,94], and its mammalian toxicity has restricted its agricultural use to a few specialized situations. If arsenite is added to well-drained aerobic soils it can be oxidized to arsenate by microbial action within a few days [110]. Purely electrochemical oxidation, however, is slow [57]. Like arsenate, arsenite availability is effected by soil texture [40,59,77] and the presence of iron [40] and is slowly fixed in time [38,77]. It is somewhat more easily leached than arsenate [131].

Arsenite is implicated in a sterility disorder of rice originally called "rice blight" [70]. The disorder is commonly observed only at maturity: otherwise normal-appearing plants are found to have panicles with empty kernels and the panicles stand up rather than being bowed with the weight of the kernels as with normal plants. For this reason the disorder is commonly called "straighthead" [27]. Straighthead generally occurs only on light, coarse soils, or soils which are high in organic matter. It occurs only with flooded rice, and can be alleviated by draining paddies prior to panicle initiation, or by breeding for resistance [138,149]. The tendency for straighthead to occur in anaerobic soil environments suggests that arsenite release may be the cause, but straighthead symptoms can be induced by foliar or root contact with other arsenical compounds [56,112, 149], and, indeed, other nonarsenic chemicals [27]. Straighthead is also observed when rice is grown in rotation with cotton which has received arsenical applications [56]. Greenhouse studies of the sterility problem are complicated by the fact that rice growth in general is relatively sensitive to arsenicals [120,121]. The disorder is difficult to induce experimentally [27], and is complicated, in very anaerobic conditions, by iron and sulfide toxicity [14,50].

3.3. Methylarsenic species

In addition to arsenate and arsenite, arsenic undergoes microbiological methylation-demethylation reactions in soils. Species found include methanearsonate, $CH_3AsO_3^{2-}$; dimethylarsinate, $(CH_3)_2AsO_2^{2-}$; trimethylarsineoxide $(CH_3)_3AsO$; dimethylarsine $(CH_3)_2AsH$; and trimethylarsine, $(CH_3)_3As$ [23,48,157]. Generally the amounts present in soil are assumed to be small relative to arsenate [157], but "speciation" techniques for detection of specific arsenical compounds are relatively recent and studies are few [24,111,159,164].

However, methanearsonate and dimethylarsinate are applied to crops and soils, the former principally as MSMA and DSMA in cotton, and the latter, known as cacodylic acid, or the cacodylate ion, as a non-selective, postemergence herbicide. In soils, adsorption of these anions is correlated with the same soil properties that determine the degree of arsenate or phosphate adsorption: active iron/aluminum content, and clay content [1,47,82,121,142]. Unlike most herbicides, which are slightly-soluble, nonionic, organic compounds, MSMA and cacodylic acid adsorption are not correlated with soil organic matter. Indeed, the reverse is true; soil organic matter appears to deactivate the clay/oxide surfaces [142]. These anions are somewhat less adsorbed and more easily extracted than arsenate and phosphate [82,121,142], and have less preference for iron over aluminum oxide surfaces [82]. Like arsenate, they become less available in time [53,74,120], they are slightly more mobile by leaching than arsenate [47,71,72]. They are generally considered inactive as herbicides in the soil, but if soil concentrations are high (many times normal application rate), plant uptake and toxicity can occur [14,54,65,74,120,121].

4. Amounts and Toxic Levels of Arsenic in Plants

4.1. Plant uptake and levels observed in tissue

Arsenic toxicity in plants is a four-stage process: 1 -- adsorption of arsenic onto plant surfaces; 2 -- movement of arsenic from the exterior to the interior of roots or tops; 3 -- translocation to the site of action; 4 -- a toxic biochemical reaction. In roots, all anions are strongly adsorbed to the membrane surface, followed by a metabolically-driven, selective transfer to the symplasm and transport [58]. With arsenate or arsenite the adsorption to the exterior root surface is quite rapid and intense leading to very high arsenic concentrations in (actually on) roots for solution culture experiments [89,94,106]. The adsorption-absorption process is most rapid for arsenate, followed by arsenite, methylarsenate, and dimethylarsinate, in that order, in bean plants [117].

Although trace element uptake in general depends on both plant species and element availability [152] in both solution and soil culture it appears that, for arsenate and arsenite, there is little difference in total uptake between species of higher plants. The amounts of arsenic found in plant tissue (unless limited by toxicity effects, which are species-specific) are generally proportional to solution or available soil concentration, and are similar for different species grown in the same solution [94] or the same soil [60,61,66,85,104,123,128,133,158,160]. This generalization applies to plant total shoot concentrations -- roots and seeds may have higher or lower arsenic concentrations, respectively, because of contact with soil or low translocation [27,40,94,99,140].

The ranges of arsenic levels reported in plants are summarized in table 2. In uncontaminated soils there is a fair degree of

Table 2. Arsenic levels reported in higher plants.

Crop or plant group	Specific plants[a] analyzed	As levels found[b]								Reference
		Tops or foliage		Roots		Edible parts[c]		Toxic levels		
		natural[d]	contam.[e]	natural	contam.	natural	contam.	foliage	edible parts	
Citrus	grapefruit, lemon, orange, mandarin	nd[f]-3	0.8-116	0.2-113	600-1200	0.01-1				[91]
Cucurbits	cucumber, pumpkin, squash					0.02-2.5				[8,15,91]
Legumes	alfalfa, clover vetch, trefoil	0.02-5	1-14	0.8-16	2-83					[4,8,14, 15,34,60, 64,66,78, 82,83,91, 94,95, 144,153, 158]
	beans, peas, soybeans, snapbeans, lima beans, cowpeas	0.01-3	0.3-14	0.1-10	0.6-40	0.01-0.9	0.06-7	1-111	0.3-0.9	
Grasses and sedges	Sudan grass, "grasses", "sedges", brome, nut-sedge, orchardgrass, timothy, canary grass, smooth cordgrass, sourgrass	.02-4	2-340	0.3-263	7-1100			70		[51,52,63, 66,73,91, 94,153]
Grains	barley, corn, millett, oats, wheat, sorghum, rice	tr[f]-3	3-156	tr-80	1-1200	0.03-3	0.07-250	5-250		[4,34,48, 50,55,60, 77,82,83, 88,90,91, 112,153, 162]

Table 2 - cont.

Crop or plant group	Specific plants[a] analyzed	As levels found[b]								Reference
		Tops or foliage		Roots		Edible parts[c]		Toxic levels		
		natural[d]	contam.[e]	natural	contam.	natural	contam.	foliage	edible parts	
Leafy vegetables	broccoli,cabbage, cauliflower,celery, endive,kale, lettuce,parsley, spinach,chard, asparagus	0.01-4	0.4	0.4-0.5	11-18			3.4-10		[15,23,60, 83,91, 158]
Nuts	almond,chestnut, filbert,macadamia	4-6	18	1-9	7-22	0.1-0.3				[91]
Root crops	beet,carrot,onion, parsnip,potato, radish	0.01-3.1	1-20	0.01-1.8	0.1-54			44	0.1-1.2	[14,34,55, 60,64,78, 91,130, 131,158]
Small fruits	blueberry,blackberry, grapes,strawberry, currant	.02-2.3	0.2-1.2			0.02-0.8	nd-25			[5,6,15, 65,91]
Tree fruits	apple,apricot,banana, cherry,peach,pear, plum	tr-2.7	5-13			0.07-0.8		0.3-8.6		[15,83,91]
Fruiting vegetables	eggplant,pepper, tomato	tr-7	6-330	0.3	13-1700	0.01-3	0.2-1.4	5-100		[8,60,64, 93,158]

Table 2 - cont.

Crop or plant group	Specific plants[a] analyzed	As levels found[b]								Reference
		Tops or foliage		Roots		Edible parts[c]		Toxic levels		
		natural[d]	contam.[e]	natural	contam.	natural	contam.	foliage	edible parts	
Aquatic plants	watercress, water-hyacinth, "marine macrophytes," duckweed, alligatorweed	0.5-2.1	15-1200							[7,91]
Native small vascular plants	many species such as ragweed, thistle, etc.									[153]
Ferns	unspecified	.7-3.5								[63]
Mosses	unspecified	2-5	2-23							[63]
Trees	unspecified, hemlock	0.2-0.4	0.7-5	1.4-2	7.5-138					[63,126]
Tobacco		0.8-2	1-78							
Cotton						0.02-2	0.1-40	4.4	0.3	[13,14,82,54]

[a]Not all plants listed are represented by all ranges given.
[b]Most data are ppm dry weight, or μg/As/g dry material; some references do not specify.
[c]Edible parts which are not foliage or roots.
[d]Levels found in plants growing in soil not known to be contaminated.
[e]Levels found in plants growing in soil known to be contaminated, or exposed to arsenical sprays.
[f]tr - "trace" or nd = "none detected" are terms used in older literature, where detection limits were ≈ 0.1 ppm [12].

agreement between most higher plants: foliar levels of 0.01-5 ppm dry weight, root levels up to an order of magnitude higher. Given the range of natural levels of arsenic in soils, differences between levels appear to be mainly regulated by availability. Indeed, it appears that foliar fresh-weight arsenic concentrations (≈ 1/10 of dry weight levels [15,153]) and soil solution arsenic concentrations may be nearly equal for a given plant/soil combination.

Very high foliar arsenic concentrations (several thousands of ppm) have been reported for Douglas fir needles [44] and for plants growing on mine wastes [16]. These may be nothing more than extreme examples of the several-hundred ppm values often observed with contaminated soils.

4.2. Maximum levels and implications for human and animal diets

That some plants may accumulate thousands of ppm arsenic in their tissue -- levels at which most of the plants in table 2 would be dead -- dramatically illustrates that plant species do differ in their tolerance, if not their uptake tendency, for arsenic. This has been confirmed by many soil tests, in which the response of different plants to increasing levels of arsenic is observed. Reports of relative sensitivity (in order of increasing sensitivity) include:

<u>Arsenate</u>

asparagus, tomato, potato, carrot, tobacco, dewberry, grape, raspberry, rye sudan grass	<	strawberry sweet corn beet squash	<	snap bean, lima bean, onion, pea cucumber, alfalfa, [91] legumes

cabbage < tomato ≈ radish < lima beans ≈ spinach < green beans [158]

barley << peas [25]

wheat < potatoes < beans < peas < radishes [133]

<u>Arsenite</u>

cotton < soybean [46]

potatoes < peas < corn < snap beans [78]

<u>Methanearsonate</u>

peanut < sorghum < tomato < watermelon [74]

wheat < cotton ≈ corn < oats < soybeans < rice [121]

cotton < potatoes < snapbeans ≈ soybeans < rice [14]

There are disparities between rankings for the different arsenic species above, but most plants retain their relative positions. Of course, studies in soils are confounded by availability factors.

There is a real need for definitive toxicity studies comparing arsenic species and plant species, and defining threshold toxic levels in solution. The study by Sachs and Michaels [117] is an excellent example. Solution studies show that arsenate concentrations from 0.2 to 100 ppm are toxic and that arsenite is toxic to the same plants at 1/4 to 1/100 the levels of arsenate [36,89,94,106,149].

Relating solution-culture toxic threshold levels to soil concentrations is not straightforward. "Water soluble" levels in soils that appear to cause toxicity are lower than in solution culture, typically 0.1 to 30 ppm [39,56,149], reflecting that "available" fractions are not counted. Toxicity symptoms and even crop yields may be correlated with "available" amounts (see above), with toxic levels in the range 2 to 50 ppm for many crops [78,139, 141,158]. Again, careful toxicity studies may allow one to relate "available" toxic values to solution toxic values.

Arsenic levels in vegetables, grain, and other crops at the consumer level are low, and have decreased in recent years [80,83]. Amounts found are far below tolerances set for the use of arsenical pesticides [80]. The usual statement (e.g., [85]) that toxicity limits plant arsenic uptake to safe levels is not, however, confirmed in table 2. Under conditions of exposure to threshold levels in the soil, the statement appears to be true. If, however, crops are exposed to a large pulse of arsenic, as with arsenical sprays, they may accumulate residues which are unacceptable -- this can occur if label restrictions on timing of application during the growing season are violated [13,54,112,145].

5. Translocation of Arsenic Species in Plants

Root uptake and translocation of arsenic into leaves was demonstrated as early as 1845 [33]. Translocation from foliar contact was observed in experiments with arsenite compounds in the 1930's [37], and is still an active research area in weed control. Regardless of the mode of entry, the result is the same: arsenic, particularly as arsenic(V) compounds, is flushed throughout the plant within a few hours, moving freely both symplastically (active cell cytoplasm-to-cytoplasm transport, including phloem movement), and apoplastically (extracellular transport, including xylem movement). The mode of entry of course determines the initial transport mechanism, but cross-over soon occurs, the paths being root → xylem → leaves → phloem and leaves → phloem → roots, tops, and xylem [39,58].

5.1. Translocation after root contact

All the arsenicals are strongly adsorbed to root surfaces from solution. This adsorption is apparently limited only by availability. For this reason observed arsenic concentrations in roots are very high in hydroponic experiments [27,71] and are higher than in other plant parts in most soil-grown plants [94,117,140]. Arsenite transport from roots is limited by its high toxicity to root membranes

[94,117]; arsenate is less soluble in soil but more readily adsorbed and translocated, because its toxicity to roots is less. Indeed, where non-lethal amounts are available translocation may result in comparable concentrations in foliage and in roots [91,99]. Sachs and Michaels [117] demonstrated that root adsorption by beans increased in the order cacodylate < methanearsenate < arsenite < arsenate. However, a much higher proportion of adsorbed cacodylate was translocated, and that which was translocated was more toxic than similar methanearsonate levels in the plant tops. Other studies [53,74,163] have also shown that cacodylate, usually thought of as a contact herbicide [11,97] is quite active *via* root entry and xylem transport. The overall toxicity of solutions of arsenicals to roots is still arsenite > arsenate > cacodylate > MSMA [117].

5.2. Translocation after foliar contact

Translocation from foliage is a fundamental research problem in herbicide mode-of-action studies and the use of ^{74}As as a tracer provided some of the first insights into the process [11,39]. Again, arsenite toxicity is immediate and kills foliar tissue before much translocation can occur, although some is observed [11,37,116]. Arsenate behaves similarly, but symplastic translocation can lead to elevated arsenic levels throughout plant tops [58]. This suggests that it may be fortunate that the early heavy insecticidal uses of arsenates involved mostly very insoluble salts.

Cacodylate absorption by leaves is greater than that of arsenate or arsenite [71,116,117], but it is still considered mainly a contact herbicide. In contrast to root exposure, it is less absorbed and translocated in leaves than methanearsonate. Methanearsonate gives the least contact toxicity and shows rapid symplastic translocation throughout plants [71,73,117,122], even exuding some of the applied material from the roots [48]. Generally, a significant fraction of the MSMA or DSMA contacting plant leaves is absorbed into the leaves, but only a few percent of that absorbed moves out of the treated area into the other areas of the treated leaf, or to other parts of the plant [51,71,116,117]. Absorption and translocation of methanearsonate can both be increased by surfactants [100,154], and increasing temperature generally increases translocation and toxicity [86,87,116]. Yield effects and residues of arsenic found in harvested seed in cotton, soybeans and rice are a function of the nearness to harvest of exposure of the plants to methanearsonate spray [21,54, 112,145,146]. There is some evidence that the counterions and/or pH of a methanearsonate spray solution can effect leaf penetration [10,124,145,154], and these effects are compounded by differences in temperature effects between arsenicals [86,87].

Methanearsonate translocation is apparently quite similar to phosphate translocation. Leaves may import it or export it depending on their stage of growth [39], transport is toward active growing areas [51,71] or storage areas [51,73], making it particularly effective against rhizomatous plants. There is some evidence also for redistribution, as with phosphate, as the plant grows: in rice, residues can occur in seed which develop weeks after the initial contact and translocation [146].

6. Mechanisms of Toxicity

The mode of action of arsenical herbicides has been the subject of several recent reviews [11,97,156]. McEwan and Stephenson [97] classified modes of action of herbicides as follows: (a) contact poisoning, (b) mitotic inhibition, (c) photosynthetic inhibition, (d) respiratory inhibition, (e) nucleic acid and protein synthesis interference. The toxicity of arsenic compounds to plants is so general that each of these modes of action has at one time or other been suggested. The similarity of uptake, translocation, and distribution of these compounds, especially methanearsonate, to phosphate, and the absence of any specific localization in plants (if indeed this could be used as evidence for a specific mechanism) is probably the main reason for a lack of progress in this area.

The picture is clearest perhaps for arsenite and cacodylate: arsenite is so toxic to all tissues that it simply destroys the first membranes encountered, probably by reaction with protein sulfhydryl groups [129,147] causing disruption of root functions on contact [75,94,103,108] and rapid necrosis on foliar contact [11,119]. Arsenite can be detoxified in soils because of this affinity for sulfur [56]. Cacodylate also reacts with sulfhydryl groups, although the effect is not as rapid and cacodylate translocation complicates the picture [97,117]. It may be that the dimethyl substitution for two OH groups in cacodylate provides a subtle steric or hydrophobic control over which proteins are effected, allowing for some translocation.

Arsenate does not react with SH groups and does not exhibit the rapid membrane disruption effects of arsenite [11]. It is a well-known decoupler of phosphorylation in mitochondria [136], can inhibit leaf uptake of other chemicals [119], and can inhibit seed germination, though high levels are required [26,129].

Methanearsonate, though a major herbicide, is also not well understood. Effects on amino acid content [122], respiration [18, 113], chlorophyll synthesis [11,65,88], and, in rice, a growth regulator effect [27], have all been observed or suggested. Methanearsonate is a selective herbicide and the source of this selectivity for grassy weeds is (not surprisingly in view of the lack of understanding of its toxic mechanism) also a mystery, though one suggestion has been made [18]. Some "metabolite" formation in specific plants has been found [51,117,122]. All studies agree that little or no carbon-arsenic bond breakage occurs in plants [48, 51].

7. Some Research Needs

It is remarkable that so little is known about the biochemical behavior of arsenate and methanearsonate in plants. The problem is complicated because their behavior and general distribution in plants are so much like phosphate, and analytical methods for natural levels of arsenic have not, until rather recently, been very good. Methanearsonate is a nonproprietary herbicide and studies which might normally be expected with a major herbicide have not been done, such as determining precise localization of arsenic residues, if any, within plant cells and inherent toxicity (bioassay) measurements to better define selectivity.

Low levels of arsenate [49,140,155,160], methanearsonate [20] and even arsenite [94] have been reported to stimulate plants, but these observations have always been secondary results of toxicity studies. This problem certainly deserves attention on its own, especially since reliable trace analytical methods are now available, and since low levels of arsenic are found in nearly all soils.

The interaction of phosphate with arsenate toxicity is an interesting and unresolved problem. Since phosphate and the organoarsenicals have the same (or nearly so) fixation sites in soil (in fact, phosphate is slightly preferred [142]) it has often been suggested that soil arsenical toxicity might be increased by phosphate fertilization. In fact, arsenic and MSMA are made more <u>available</u> by phosphate treatment [1,155] and in some cases toxicity and plant arsenic residues increase [77,115,128,155]. At the same time, however, phosphate very effectively competes with arsenic for root adsorption [35,115] and Sanders [118] has shown that arsenate is toxic to algae at very low solution concentrations when the phosphate concentration is also very low. At higher soil pH, where phosphate availability is low, phosphate addition can decrease arsenic toxicity [17]. Some bioassay studies using the various arsenic species with phosphate in solution culture, and using modern speciation and trace-level monitoring would be a significant contribution in this area.

References

[1] Akins, M. B. and Lewis, R. J., The sorption and leaching of 74-arsenic labeled disodium methanearsonate, Agron. Abstr. 1974, 119 (1974).

[2] Akins, M. B. and Lewis, R. J., Chemical distribution and gaseous evolution of arsenic-74 added to soil as DSMA - ^{74}As, Soil Sci. Soc. Am. J. 40, 655-658 (1976).

[3] Albert, W. B. and Arndt, C. H., The concentration of soluble arsenic as an index of arsenic toxicity to plants, S.C. Agric. Exp. Stn., 44th Ann. Rep. (1931).

[4] Albert, W. B. and Paden, W. R., Calcium arsenate and unproductiveness in certain soils, Science 73, 622 (1931).

[5] Anastasia, F. B. and Kender, W. J., The influence of soil arsenic on the growth of lowbush blueberry, J. Environ. Qual. 2, 335-337 (1973).

[6] Anderson, A. C., Abdelghani, A. A., Hughes, J., and Mason, J. W., Accumulation of MSMA in the fruit of the blackberry (*Rubus* sp.), Bull. Environ. Contam. Toxicol. 74, 124-126 (1980).

[7] Anderson, A. C., Abdelghani, A. A., and McDonell, D., Screening of four vascular aquatic plants for uptake of monosodium methanearsonate (MSMA), Sci. Total. Environ. 16, 95-98 (1980).

[8] Anderson, L. W. J., Pringle, J. C., and Raines, R. W., Arsenic levels in crops irrigated with water containing MSMA, Weed Sci. 26, 370-373 (1978).

[9] Andreae, M. O., Distribution and speciation of arsenic in natural waters and some marine algae, Deep-Sea Res. 25, 391-402 (1978).

[10] Arle, H. F. and Hamilton, K. C., Topical applications of DSMA and MSMA in irrigated cotton, Weed Sci. 19, 545-547 (1971).

[11] Ashton, F. M. and Crafts, A. S., in *Mode of Action of Herbicides*, pp. 147-162 (Wiley-Interscience, New York, N.Y., 1973).

[12] Association of Official Analytical Chemists, *Official Methods of Analysis of the ADAC*, 11th ed., pp. 399-403 (AOAC, Washington, D.C., 1970).

[13] Baker, R. S., Arle, H. F., Miller, J. H., and Holston, J. T. Jr., Effects of organic arsenical herbicides on cotton response and chemical residues, Weed Sci. 17, 37-40 (1969).

[14] Baker, R. S., Barrentine, W. L., Bowman, D. H., Hawthorne, W. L., and Pettiet, J. V., Crop response and arsenic uptake following soil incorporation of MSMA, Weed Sci. 24, 322-326 (1976).

[15] Barudi, W. and Bielig, H. J., Gehalt an Schwermetallen (Arsen, Blei, Cadmium, Quecksilber) in oberirdisch wachsenden Gemuse- und Obstarten, Z. Lebensm. Unters. Forsch. 170, 254-257 (1980).

[16] Beckert, W. F., Mercury, lead, arsenic and cadmium in biological tissue: the need for adequate standard reference materials, U.S. Env. Prot. Agency EPA-600/4-78-051 (Aug. 1978).

[17] Benson, N. R., Effect of season, phosphate, and acidity on plant growth in arsenic-toxic soils, Soil Sci. 76, 215-224 (1953).

[18] Benson, A. A. and Knowles, F. C., Arsenic metabolism and photosynthetic productivity, 5th Int. Congr. Photosyn. (1980) (in press).

[19] Bishop, R. F. and Chisholm, D., Arsenic accumulation in Annapolis Valley orchard soils, Can. J. Soil Sci. 42, 77-80 (1962).

[20] Blythe, T. O., Grooms, S. O., and Frans, R. E., Determination and characterization of the effects of fluometuron and MSMA on Chlorella, Weed Sci. 27, 294-295 (1979).

[21] Bode, L. E. and McWhorter, C. G., Toxicity of MSMA, fluometuron and propanil to soybeans, Weed Sci. 25, 101-105 (1977).

[22] Bohn, H., McNeal, B., and O'Connor, G., in Soil Chemistry, pp. 171-192 (Wiley-Interscience, New York, N.Y., 1979).

[23] Braman, R. S., Molecular forms of arsenic in the environment, in Proc. Workshop Toxic. Biota Met. Forms Nat. Water, R. W. Andrew, P. V. Hudson, and D. E. Komasewich, eds., pp. 249-261 (Great Lakes Reg. Off., Int. Util. Comm., Windsor, Ont., 1978).

[24] Braman, R. S. and Forebeck, C. C., Methylated forms of arsenic in the environment, Science 182, 1247-1249 (1973).

[25] Brenchley, W. E., On the action of certain compounds of zinc, arsenic, and boron on the growth of plants, Ann. Bot. 28, 283-305 (1914).

[26] Brenchley, W. E., in Inorganic Plant Poisons and Stimulants, pp. 51-64 (Cambridge Univ. Press, Cambridge, U.K., 1927).

[27] Burdette, D. L., MSMA toxicity to rice: the straighthead disorder, Thesis, Univ. Arkansas, Fayetteville, Ark. (1979).

[28] Carey, A. E., Douglas, P., Tai, H., Mitchell, W. G., and Wiersma, G. B., Pesticide residue concentrations in soils of five United States cities, 1971 - urban soils monitoring program, Pestic. Monit. J. 13, 17-22 (1979).

[29] Carey, A. E., Gowen, J. A., Tai, H., Mitchell, W. G., and Wiersma, G. B., Pesticide residue levels in soils and crops, 1971 - national soils monitoring program (III), Pestic. Monit. J. 12, 117-136 (1978).

[30] Carey, A. E., Wiersma, G. B., and Tai, H., Pesticide residues in urban soils from 14 United States cities, 1970, Pestic. Monit. J. 10, 54-60 (1976).

[31] Carey, A. E., Wiersma, G. B., Tai, H., and Mitchell, W. G., Organochlorine pesticide residues in soils and crops of the cornbelt region of the United States - 1971, Pestic. Monit. J. 6, 369-376 (1973).

[32] Carrow, R. N., Rieke, P. E., and Ellis, B. G., Growth of turfgrasses as effected soil phosphorus and arsenic, Soil Sci. Soc. Am. Proc. 39, 1121-1124 (1975).

[33] Chatin, A., Etudes de physiologie vegetale faites au moyen de l'acide arsenieux, Compt. Rend. 20, 21-28 (1845).

[34] Chisholm, D., Lead, arsenic, and copper content of crops grown on lead arsenate-treated and untreated soils, Can. J. Plant Sci. 52, 583-588 (1972).

[35] Clements, H. F. and Munson, J., Arsenic toxicity studies in soil and culture solution, Pac. Sci. 1, 151-171 (1947).

[36] Collier, G. F. and Greenwood, D. J., The influence of solution concentration of aluminum, arsenic, boron and copper on root growth in relation to the phytotoxicity of pulverized fuel ash, J. Sci. Food Agric. 28, 145-151 (1977).

[37] Crafts, A. S., The use of arsenical compounds in the control of deep-rooted perennial weeds, Hilgardia 7, 361-372 (1933).

[38] Crafts, A. S., The toxicity of sodium arsenite and sodium chlorate in four California soils, Hilgardia 9, 461-498 (1935).

[39] Crafts, A. S. and Crisp, C. E., in Phloem Transport in Plants (W. H. Freeman and Co., San Francisco, Calif., 1971).

[40] Crafts, A. S. and Rosenfels, R. S., Toxicity studies with arsenic in eighty California soils, Hilgardia 12, 177-199 (1939).

[41] Crecelius, E. A., Johnson, G. C., and Hofen, G. C., Contamination of soil near a copper smelter by arsenic, antimony, and lead, Water Air Soil Pollut. 3, 337-342 (1974).

[42] Dean, L. A. and Rubins, E. J., Anion exchange in soils: I. Exchangeable phosphorus and the anion exchange capacity, Soil Sci. 63, 377-387 (1947).

[43] Deb, D. L. and Datta, N. P., Effect of associating anions on phosphorus retention in soil. I. Under variable phosphorus concentration, Plant Soil 26, 303-316 (1967).

[44] Delavault, R. E. and Manson, R. J., Spectroscopic determination of arsenic in geochemical samples, in Geochemical Exploration (Proc. 3rd Int. Symp. Geochem. Explor., C.I.M. Special vol. #11, 1971).

[45] Deuel, L. E. and Swoboda, A. R., Arsenic solubility in a reduced environment, Soil Sci. Soc. Am. Proc. 36, 276-278 (1972).

[46] Deuel, L. E. and Swoboda, A. R., Arsenic toxicity to cotton and soybeans, J. Environ. Qual. 1, 317-320 (1972).

[47] Dickens, R. and A. E. Hiltbold, Movement and persistence of methanearsonates in soil, Weeds 15, 299-304 (1967).

[48] Domir, S. C., Woolson, E. A., Kearney, P. C., and Isensee, A. R., Translocation and metabolic fate of monosodium methanearsenic acid in wheat (Triticum aestirum L.), J. Agric. Food Chem. 24, 1214-1217 (1976).

[49] Dorman, C. and Coleman, R., The effect of calcium arsenate upon the yield of cotton in different soil types, J. Am. Soc. Agron. 31, 966-971 (1939).

[50] Duah-Yentumi, S., Tsutsumi, M., and Kurihara, K., Intensification of arsenic toxicity to paddy rice by hydrogen sulfide and ferrous ion. II. Effects of ferric sulfate and ferric hydroxide application on arsenic toxicity to rice plants, Soil Sci. Plant Nutr. 26, 571-580 (1980).

[51] Duble, R. L., Holt, E. C., and McBee, G. G., The translocation of two organic arsenicals in purple nutsedge, Weed Sci. 16, 421-424 (1968).

[52] Edwards, A. C. and Davis, D. E., Effects of an organic arsenical herbicide on a salt marsh ecosystem, J. Environ. Qual. 4, 215-219 (1975).

[53] Ehman, P. J., Effect of arsenical build-up in the soil on subsequent growth and residue content of crops, Proc. South. Weed Sci. Soc. 18, 685-687 (1965).

[54] Ehman, P. J., Residues in cottonseed from weed control with methanearsonates, Proc. South. Weed Sci. Soc. 19, 540-541 (1966).

[55] Elfving, D. C., Haschek, W. M., Stehn, R. A., Bache, C. A., and Lisk, D. J., Heavy metal residues in plants cultivated on and in sites indigenous to old orchard soils, Arch. Environ. Hlth. 33, 95-97 (1978).

[56] Epps, E. A. and Sturgis, M. B., Arsenic compounds toxic to rice, Soil Sci. Soc. Am. Proc. 4, 215-218 (1939).

[57] Ferguson, J. F. and Gavis, J., A review of the arsenic cycle in natural waters, Water Res. 6, 1259-1274 (1972).

[58] Foy, C. L. and Yamaguchi, S., Mechanisms of root absorption of organic molecules (Proc. 7th Ann. Symp., So. Sect. Am. Soc. Plant Physiol., Emory Univ., Atlanta, Ga., April 16, 1964), Book, Absorption and Translocation of Organic Substances in Plants, J. Hacskaylo, ed., pp. 5-28 (Wallace Printing Co., Bryan, Tex., 1965).

[59] Frans, R. E., Scogley, C. R., and Ahlgren, G. H., Influence of soil type on soil sterilization with sodium arsenite, Weeds 4, 11-14 (1956).

[60] Furr, A. K., Kelly, W. C., Bache, C. A., Gutenmann, W. H., and Lisk, D. J., Multielement uptake by vegetables and millet grown in pots on fly ash amended soil, J. Agric. Food Chem. 24, 885-888 (1976).

[61] Geisman, J. R., Carey, W. E., Gould, W. A., and Alban, E. K., Distribution of arsenic residues by activation analysis, J. Food Sci. 34, 295-298 (1969).

[62] Gile, P. L., The effect of different colloidal soil materials on the toxicity of calcium arsenate to millet, J. Agric. Res. 52, 477-491 (1936).

[63] Girling, C. A., Peterson, P. J., and Minski, M. J., Gold and arsenic concentrations in plants as an indication of gold mineralization, Sci. Total Environ. 10, 79-85 (1978).

[64] Grant, C. and Dobbs, A. J., The growth and metal content of plants grown in a soil contaminated by a copper/chrome/arsenic wood preservative, Environ. Pollut. 14, 213-226 (1977).

[65] Gur, A., Gil, Y., and Bravdo, B., The efficacy of several herbicides in the vineyard and their toxicity to grapevines, Weed Res. 19, 109-116 (1979).

[66] Gutenmann, W. H., Pakkala, I. S., Churey, D. J., Kelly, W. C., and Lisk, D. J., Arsenic, boron, molybdenum, and selenium in successive cuttings of forage crops grown on fly ash amended soil, J. Agric. Food Chem. 27, 1393-1395 (1979).

[67] Hess, R. E. and Blanchar, R. W., Arsenic stability in contaminated soils, Soil Sci. Soc. Am. J. 40, 847-852 (1976).

[68] Hess, R. E. and Blanchar, R. W., Dissolution of arsenic from a waterlogged and aerated soil, Soil Sci. Soc. Am. J. 41, 861-865 (1976).

[69] Hess, R. E. and Blanchar, R. W., Arsenic determination and arsenic, lead, and copper content of Missouri soils, Univ. Missouri Agric. Exp. Stn. Res. Bull. 1020, Nov., 1977. Columbia, MO (1977).

[70] Hewitt, J. L., Rice blight, Univ. Ark. Agric. Exp. Stn. Bull. #110 (1912).

[71] Hiltbold, A. E., Behavior of organoarsenicals in plants and soils, in Arsenical Pesticides, E. A. Woolson, ed., pp. 53-69 (Am. Chem. Soc. Symp. Ser., 7, ACS, Washington, D.C., 1975).

[72] Hiltbold, A. E., Hajek, B. F., and Buchanan, G. A., Distribution of arsenic in soil profiles after repeated applications of MSMA, Weed Sci. 22, 272-275 (1974).

[73] Holt, E. C., Faubion, J. L., Allen, W. W., and McBee, G. G., Arsenic translocation in nutsedge tuber systems and its effect on tuber vitality, Weeds 15, 13-15 (1967).

[74] Horowitz, M., Activity and degradation of arsonates in soil, Spec. Publ. Agric. Res. Org., Volcani Cent., Bel Dagan, Israel 82, 20-26 (1977).

[75] Isensee, A. R., Jones, G. E., and Turner, B. C., Root absorption and translocation of picloram by oats and soybeans, Weed Sci. 19, 727-731 (1971).

[76] Jackson, M. L., Phosphorus determinations for soils, in Soil Chemical Analysis, pp. 134-182 (Prentice Hall, Inc., Englewood Cliffs, N.J., 1958).

[77] Jacobs, L. W. and Keeney, D. R., Arsenic-phosphorus interactions on corn, Soil Sci. Plant Anal. 1, 85-93 (1970).

[78] Jacobs, L. W., Keeney, D. R., and Walsh, L. M., Arsenic residue toxicity to vegetable crops grown on Plainfield sand, Agron. J. 62, 588-591 (1970).

[79] Jacobs, L. W., Syers, J. K., and Keeney, D. R., Arsenic sorption by soils, Soil Sci. Soc. Am. Proc. 34, 750-754 (1970).

[80] Jelinek, C. F. and Corneliussen, P. E., Levels of arsenic in the United States food supply, Environ. Hlth. Perspect. 19, 83-87 (1977).

[81] Jenne, E. A., Trace element sorption by sediments and soils -- sites and processes, in Symposium on Molybdenum in the Environment, vol. 2, W. Chappell and K. Peterson, eds., pp. 425-533 (Marcel Dekker, Inc., New York, N.Y., 1977).

[82] Johnson, L. R. and Hiltbold, A. E., Arsenic content of soil and crops following use of methanearsonate herbicides, Soil Sci. Soc. Am. Proc. 33, 279-282 (1969).

[83] Johnson, R. D., Manske, D. D., and Podrebarac, D. S., Pesticide, metal and other chemical residues in adult total diet samples -- (XII) -- August, 1975-July, 1976, Pestic. Monit. J. 15, 54-69 (1981).

[84] Johnston, S. E. and Barnard, W. M., Comparative effectiveness of fourteen solutions for extracting arsenic from four Western New York soils, Soil Sci. Soc. Am. J. 43, 304-308 (1979).

[85] Jones, J. S. and Hatch, M. B., Spray residues and crop assimilation of arsenic and lead, Soil Sci. 60, 277-288 (1945).

[86] Keeley, P. E. and Thullen, R. J., Cotton response to temperature and organic arsenicals, Weed Sci. 19, 297-300 (1971).

[87] Keeley, P. E. and Thullen, R. J., Control of nutsedge with organic arsenical herbicides, Weed Sci. 19, 601-606 (1971).

[88] Kenyon, D. J., Elfving, D. C., Pakkala, I. S., Bache, C. A., and Lisk, D. J., Residues of lead and arsenic in crops cultured on old orchard soils, Bull. Environ. Contam. Toxicol. 22, 221-223 (1979).

[89] Knop, W., Uber die Aufname verschiedener Substanzen durch die Pflanze, welche nicht zu den Nahrstoffen gehoren, Jahreb. Agric. Chem. 8, 138-140 (1884).

[90] Liebhart, W. C., The arsenic content of corn grain grown on a coastal plain soil amended with poultry manure, Commun. Soil Sci. Plant Anal. 7, 169-174 (1976).

[91] Liebig, G. F., Arsenic, in Diagnostic Criteria for Plants and Soils, H. D. Chapman, ed., pp. 13-23 (Univ. Calif. Div. Agric. Sci., Davis, Calif., 1966).

[92] Lindsay, W. L. and Moreno, E. C., Phosphate phase equilibria in soils, Soil Sci. Soc. Am. Proc. 24, 177-182 (1960).

[93] Luh, M.-D., Baker, R. A., and Henley, D. E., Arsenic analysis and toxicity -- a review, Sci. Total Environ. 2, 1-12 (1973).

[94] Machlis, L., Accumulation of arsenic in the shoots of Sudan grass and bush beans, Plant Physiol. 16, 521-544 (1941).

[95] MacPhee, A. W., Chisholm, D., and MacEachern, C. R., The persistence of certain pesticides in the soil and their effect on crop yields, Can. J. Soil Sci. 40, 59-62 (1960).

[96] May, T. W. and McKinney, G. L., Cadmium, lead, mercury, arsenic, and selenium concentrations in fresh water fish, 1976-77 -- national pesticide monitoring program, Pestic. Monit. J. 15, 14-38 (1981).

[97] McEwen, F. L. and Stephenson, G. R., The Use and Significance of Pesticides in the Environment (Wiley-Interscience, New York, N.Y., 1979).

[98] McGeorge, W. T., Fate and effect of arsenic applied as a spray for weeds, J. Agric. Res. 5, 459-463 (1915).

[99] McLean, H. C., Weber, A. L., and Jaffe, J. S., Arsenic content of vegetables grown in soils treated with lead arsenate, J. Econ. Entomol. 37, 315-316 (1944).

[100] McWhorter, C. G., Toxicity of DSMA to johnsongrass, Weeds 14, 191-194 (1966).

[101] Miles, J. R. W., Arsenic residues in agricultural soils of southwest Ontario, J. Agric. Food Chem. 16, 620-622 (1968).

[102] Mitchell, R. L., Trace elements in soils, in Chemistry of the Soil, 2nd ed., F. Bear, ed., pp. 320-368 (ACS Monogr. Ser., Reinhold Publ. Crop., New York, N.Y., 1964).

[103] Morris, H. E., Injury to growing crops caused by the application of arsenical compounds to the soil, J. Agric. Res. 34, 59-78 (1927).

[104] Morrison, J. L., Distribution of arsenic from poultry litter in broiler chickens, soil, and crops, J. Agric. Food Chem. 17, 1288-1290 (1969).

[105] Neiswander, C. R., Duration of the effectiveness of lead arsenate applied to turf for white grub control, J. Econ. Entomol. 44, 221-224 (1951).

[106] Nobbe, F., Buessler, P., and Will, H., Untersuchungen uber die Giftwirkung des Arsens, Blei, und zinc in Pflanzlichen Organismus, Landw. Vers. Sta. 30, 380-423 (1884).

[107] Onishi, H. and Sandell, E. B., Geochemistry of arsenic, Geochim. Cosmochim. Acta 7, 1-33 (1955).

[108] Orwick, P. L., Schrieber, M. M., and Hodges, T. K., Adsorption and efflux of chloro-s-triazines by Setaria roots, Weed Res. 16, 139-144 (1976).

[109] Peterson, G. W. and Corey, R. B., A modified Chang and Jackson procedure for routine fractionation of inorganic soil phosphorus, Soil Sci. Soc. Am. Proc. 30, 563-565 (1966).

[110] Quastel, J. H. and Scholefield, P. G., Arsenite oxidation in soil, Soil Sci. 75, 279-285 (1953).

[111] Ricci, G. R., Shepherd, L. S., Colobus, G., and Hester, N. E., Ion chromatography with atomic absorption spectrometric detection for determination of organic and inorganic arsenic species, Anal. Chem. 53, 610-613 (1981).

[112] Richard, E. P., Hurst, H. R., and Wauchope, R. D., Effects of simulated MSMA spray drift on rice (Oryza sativa) growth and yield, Weed Sci. 29, 303-308 (1981).

[113] Richardson, J. T., Frans, R. E., and Talbert, R. E., Reactions of Euglena glacilis to fluometuron, MSMA, metribuzin, and glyphosate, Weed Sci. 27, 619-624 (1979).

[114] Rosenfels, R. S. and Crafts, A. S., Arsenic fixation in relation to the sterilization of soils with sodium arsenite, Hilgardia 12, 203-229 (1939).

[115] Rumburg, C. B., Engel, R. E., and Meggitt, W. F., Effect of phosphorus concentration on the absorption of arsenate by oats from nutrient solution, Agron. J. 52, 452-453 (1960).

[116] Rumburg, C. B., Engel, R. E., and Meggitt, W. F., Effect of temperature on the herbicidal activity and translocation of arsenicals, Weeds 8, 582-588 (1960).

[117] Sachs, R. M. and Michaels, J. L., Comparative phytotoxicity among four arsenical herbicides, Weed Sci. 19, 558-564 (1971).

[118] Sanders, J. G., Effects of arsenic speciation and phosphate concentration on arsenic inhibition of Skeletenoma costatum (Bacillariophyceae), J. Phycology 15, 424-427 (1979).

[119] Sargent, J. A. and Blackman, G. E., Studies on foliar penetration. IV. Mechanisms controlling the rate of penetration of 2,4-dichlorophenoxyacetic acid (2,4-D) into leaves of Phaseolus vulgaris, J. Exp. Bot. 20, 542-555 (1969).

[120] Schweizer, E. E., Toxicity of DSMA residues to cotton and rotational crops, Weeds 15, 72-76 (1967).

[121] Schweizer, E. E., Effects of residue of MSMA in soils on cotton, soybeans, and cereal crops, Miss. State Univ. Agric. Exp. Stn. Bull., No. 736, Miss. State Univ., State College, Miss. (1967).

[122] Sckerl, M. M. and Frans, R. E., Translocation and metabolism of MAA-14C in johnsongrass and cotton, Weed Sci. 17, 421-427 (1969).

[123] Sharma, R. P. and Shupe, J. L., Lead, cadmium, and arsenic residues in animal tissues in relation to those in their surrounding habitat, Sci. Total Environ. 7, 53-62 (1977).

[124] Shepherd, C. G., Standifer, L. C., Sloane, L. W., Melville, D. R., and Wright, M. E., A preliminary report on the effect of pH on herbicidal activity of sodium salts of methane arsenic acid, Proc. South. Weed Sci. Soc. 19, 542 (1966).

[125] Sillen, L. G., Quantitative studies of hydrolytic equilibria, Quart. Rev. 13, 146-168 (1959).

[126] Sinclair, W. A., Stone, E. L., and Scheer, C. F., Toxicity to hemlocks grown in arsenic-contaminated soil previously used for potato production, HortScience. 10, 35-36 (1975).

[127] Singh, R. K. N. and Campbell, R. W., Herbicides on bluegrass, Weeds 13, 170-171 (1965).

[128] Small, H. G. and McCants, C. B., Influence of arsenic applied to the growth media on the arsenic content of flue-cured tobacco, Agron. J. 54, 129-133 (1962).

[129] Speer, H. L., The effect of arsenate and other inhibitors on early events during the germination of lettuce seeds (Lactuca sativa L.), Plant Physiol. 52, 142-146 (1973).

[130] Steevens, D. R., Walsh, L. M., and Keeney, D. R., Arsenic residues in soil and potatoes from Wisconsin potato fields, Pest. Monit. J. 6, 89-90 (1972).

[131] Steevens, D. R., Walsh, L. M., and Keeney, D. R., Arsenic phytotoxicity on a Plainfield sand as effected by ferric sulfate or aluminum sulfate, J. Environ. Qual. 1, 301-303 (1972).

[132] Stevens, L. L., Collier, C. W., and Woodham, D. W., Monitoring pesticides in soils of regular, limited, and no pesticide use, Pestic. Monit. J. 4, 145-166 (1970).

[133] Stewart, J. and Smith, E. S., Some relations of arsenic to plant growth. II., Soil Sci. 14, 119-126 (1927).

[134] Swenson, R. M., Cole, C. V., and Sieling, D. H., Fixation of phosphate by iron and aluminum and replacement by organic and inorganic ions, Soil Sci. 32, 3-22 (1949).

[135] Tammes, P. M. and deLint, M. M., Leaching of arsenic from soil, Neth. J. Agric. Sci. 17, 128-132 (1969).

[136] TerWelle, H. F. and Slater, E. C., Uncoupling of respiratory-chain phosphorylation by arsenate, Biochim. Biophys. Acta 143, 1-17 (1967).

[137] Tyree, S. Y., Metal hydrolysis and oxide-hydroxide precipitation, Croat. Chem. Acta 45, 195-198 (1973).

[138] U.S. Dept. of Agriculture, Rice in the United States: varieties and production, Agric. Hndbk No. 289 (rev.), USDA, Washington, D.C. (1973).

[139] Vandecaveye, S. C., Horner, G. M., and Keaton, C. M., Unproductiveness of certain orchard soils as related to lead arsenate spray accumulations, Soil Sci. 42, 203-215 (1936).

[140] Walsh, L. M. and Keeney, D. R., Behavior and phytotoxicity of inorganic arsenicals in soils, in Arsenical Pesticides, E. A. Woolson, ed., pp. 35-52 (Am. Chem. Soc. Symp. Ser., 7, ACS, Washington, D.C., 1975).

[141] Walsh, L. M., Sumner, M. E., and Keeney, D. R., Occurrence and distribution of arsenic in soils and plants, Environ. Hlth. Perspect. 19, 67-71 (1977).

[142] Wauchope, R. D., Fixation of arsenical herbicides, phosphate, and arsenate in alluvial soils, J. Environ. Qual. 4, 355-358 (1975).

[143] Wauchope, R. D., Acid dissociation constants of arsenic acid, methylarsonic acid (MAA), dimethylarsinic acid (cacodylic acid) and N-(phosphonomethyl)glycine (glyphosate), J. Agric. Food Chem. 24, 717-721 (1976).

[144] Wauchope, R. D., Selenium and arsenic levels in soybeans from different production region of the United States, J. Agric. Food Chem. 26, 226-228 (1978).

[145] Wauchope, R. D. and McWhorter, C. G., Arsenic residues in soybean seen from simulated MSMA spray drift, Bull. Environ. Contam. Toxicol. 17, 165-167 (1977).

[146] Wauchope, R. D., Richard, E. P., and Hurst, H. R., Effects of simulated MSMA drift on rice (Oryza sativa). II. Arsenic residues in foliage and grain and relationships between arsenic residues, rice toxicity symptoms, and yields, Weed Sci. 30, (1982) (in press).

[147] Webb, J. L., in Enzyme and Metabolic Inhibitors, vol. 3 (Academic Press, New York, N.Y., 1966).

[148] Weed Science Society of America, Herbicide Handbook, 4th ed., (Weed Sci. Soc. Am., Champaign, Ill., 1979).

[149] Wells, B. R. and J. T. Gilmour, Sterility in rice cultivars as influenced by MSMA rate and water management, Agron. J. 69, 451-454 (1977).

[150] Wiersma, G. B., Sand, P. F., and Schultzmann, National soils monitoring program - six states, 1967, Pestic. Monit. J. 5, 223-227 (1971).

[151] Wiersma, G. B., Tai, H., and Sand, P. F., Pesticide levels in soils, FY 1969 - national soils monitoring program, Pestic. Monit. J. 6, 194-228 (1972).

[152] Wiklander, L., Cation and anion exchange phenomena, in Chemistry of the Soil, 2nd ed., F. Bear, ed., pp. 163-205 (ACS Monogr. Ser., Reinhold Publ. Corp., New York, N.Y., 1964).

[153] Williams, K. T. and Whetstone, R. R., Arsenic distribution in soils and its presence in certain plants, U.S. Dept. Agric. Tech. Bull., No. 732 (1940).

[154] Wills, G. D., Effects of inorganic salts on the toxicity of dalapon and MSMA to purple nutsedge, Weed Sci. Soc. Am. Proc. 1971, 84-85 (1971).

[155] Woolson, E. A., Effects of fertilizer materials and combinations on the phytotoxicity, availability, and content of arsenic in corn (maize), J. Sci. Food Agric. 23, 1477-1481.

[156] Woolson, E. A., Organoarsenical herbicides, in Herbicides-Chemistry, Degradation, and Mode of Action, pp. 741-776 (Marcel Dekker, Inc., New York, N.Y., 1976).

[157] Woolson, E. A., Fate of arsenicals in different environmental substrates, Environ. Hlth. Perspect. 19, 73-81 (1977).

[158] Woolson, E. A., Arsenic phytotoxicity and uptake in vegetable crops, Weed Sci. 21, 524-527 (1973).

[159] Woolson, E. A. and Aharonson, N., Separation and detection of arsenical pesticide residues and some of their metabolites by high pressure liquid chromatography-graphite furnace atomic absorption spectrometry, J. Assoc. Off. Anal. Chem. 63, 523-528 (1980).

[160] Woolson, E. A., Axley, J. H., and Kearney, P. C., Correlation between available soil arsenic, estimated by six methods, and response of corn (*Zea mays* L.), Soil Sci. Soc. Am. Proc. 35, 101-105 (1971).

[161] Woolson, E. A., Axley, J. H., and Kearney, P. C., The chemistry and phytotoxicity of arsenic in soils: I. contaminated field soils, Soil Sci. Soc. Am. Proc. 35, 938-943 (1971).

[162] Woolson, E. A., Axley, J. H., and Kearney, P. C., The chemistry and phytotoxicity of arsenic in soils: II. Effects of time and phosphorus, Soil Sci. Soc. Am. Proc. 37, 254-259 (1973).

[163] Woolson, E. A. and Kearney, P. C., Persistence and reactions of ^{14}C-cacodylic acid in soils, Environ. Sci. Technol. 7, 47-50 (1973).

[164] Yamamoto, M., Determination of arsenate, methanearsonate, and dimethylarsinate in water and sediment extracts, Soil Sci. Soc. Am. Proc. 39, 859-861 (1975).

DISCUSSION

C. R. Coggins: You mentioned that you thought if there were arsenic levels in plant tissues which were toxic to humans, then they would also be toxic to the plants themselves. Do you have a maximum level for human consumption of arsenic in plant tissue?

R. D. Wauchope: In every case where an organo-arsenical or an inorganic arsenical is used as a pesticide, there is a tolerance for residues. The value, for instance, for MSMA residues, in cottonseed is about 0.7 ppm as arsenic trioxide or about 0.6 ppm arsenic. We get higher levels than that in some of our experiments--we're getting several parts per million when we simulate label violations. Of course, it's not at all unusual to get values that are higher than that even in natural situations. I think that the accepted value for typical vegetation levels of arsenic is around 0.2 parts per million arsenic dry weight. If you exceed 0.6 parts per million arsenic in any of these MSMA use situations, it's grounds for seizure.

E. A. Woolson: The straighthead problem can be alleviated in patty rice cultures if the patty is drained so that the soil becomes aerobic soon enough in the growth cycle and is, therefore, aerobic before initiation begins. Do you feel that the problem is one of stress or of a reduced arsenic or iron or something else related to the condition in the patty?

R. D. Wauchope: Do you mean is the straighthead due to the anaerobic conditions themselves?

E. A. Woolson: Yes.

R. D. Wauchope: I don't think anyone knows the answer to that question. There is interesting work coming out of Japan where they're talking about sulfide and ferrous ion toxicities as being complicating factors. The farmers knew that you could drain a patty if you did it prior to panicle initiation and they've known that for fifty years. They used to think it was some kind of disease. They bred rice for resistance to straighthead, and they really didn't have any idea what it was. However, they did know that you could drain it and stop the problem.

E. A. Woolson: Why don't they normally drain the field then to prevent any possible problem?

R. D. Wauchope: It costs a lot of money to pump all that water. You're talking about a tremendous quantity of water.

W. Peters: EPA has listed arsenic as a hazardous air pollutant as of June 5, 1981, but we have only listed the inorganic arsenic as the hazardous air pollutant and not total arsenic. We are now studying various source categories of inorganic arsenic. We have looked at and identified the significant sources of inorganic arsenic and have some sense of what controls are on existing plants, of emissions estimates, and some preliminary public exposure estimates. We're about ready to look at (1) the quality of controls required by existing standards within the states and the regulatory requirements on existing controls; (2) the ease of standards development; and, (3) the feasibility of controls. At this point we are looking at the primary copper, lead and zinc smelters; the secondary lead smelters; cotton gins that handle arsenic acid dessicated cotton; glass manufacturing; and arsenic chemical plants. The exceptions to this discussion are the primary copper smelters that are processing high arsenic ores. EPA had completed all studies on the two plants in this source category and were about to propose a regulation, however, one plant has shut down and the other is negotiating with EPA to put on best available technology. The development of the proposed standard is on hold.

If EPA were to go ahead and begin developing regulations for the source categories mentioned earlier, it would take about one and a half to two years or more for a proposal to be in the Federal Register.

Our Environmental Criteria Assessment Office is developing a health document for arsenic. It has been drafted several times and is now undergoing another draft and will be ready for the Science Advisory Board perhaps by early spring, 1982. At that time, there will be a chance for review and public comment.

R. D. Wauchope: Is there information in the literature or somewhere in your government files on arsenic air levels in cotton gins?

W. Peters: We know only of one study and this was done by the Texas Air Quality Control Board back in 1973 by Cyril Durenberger. As a matter of fact, we have found out that the Texas Air Quality Control Board has all those high-volume filters still in storage. They have reanalyzed the filters and are going to have them forwarded to EPA for another analysis. Indications are, by the way, that the arsenic levels, according to x-ray fluorescence, have remained the same over a period of almost eight years. That's about the only study we know of. I believe we will try to initiate something next fall.

26

BIOTRANSFORMATION OF ARSENIC IN THE MARINE ENVIRONMENT

Meinrat O. Andreae
Department of Oceanography
Florida State University
Tallahassee, FL 32306

Three major modes of arsenic species biotransformation have been found to occur in the environment: (1) redoxtransformation between arsenic(III) and arsenic(V), (2) the reductive methylation of arsenic to volatile and highly toxic methylarsines, and (3) the biosynthesis of structurally complex organoarsenic compounds by marine algae and subsequent oxidation and breakdown reactions of these compounds. Arsenite can be produced directly from arsenate by many marine organisms; it may also occur as an intermediate in the transformation of organoarsenic compounds back to arsenate. No evidence for reductive biomethylation has been observed in the marine environment, in contrast to some earlier hypotheses. No volatile organoarsenic compounds are formed in seawater; biomethylation of arsenic therefore does not contribute to the atmospheric arsenic cycle. On the other hand, the biosynthesis of organoarsenic compounds by marine planktonic algae and subsequent reactions involving these compounds are dominant features in the marine biogeochemistry of arsenic. The concentration of arsenate in surface seawater approaches that of the required nutrient, phosphate, in many regions of the oceans, and these two ions are chemically very similar. Probably as a result of this problem, marine algae possess mechanisms to transform arsenate into less toxic forms, among them dimethylarsinic acid and large organoarsenic compounds (e.g. trimethylarsonium-lactic acid-phospholipid). These compounds are handed along the food chain and are the probable precursors of organoarsenic compounds found in marine crustaceans and fish (e.g. arsenobetaine). These processes lead to the presence of considerable amounts of methylarsenic acids in the biologically active euphotic zone in the marine surface waters.

1. Introduction

The toxicity of arsenic, like that of all elements, is highly dependent on its chemical form. This connection between chemical speciation and toxicological properties is considered self-evident for most elements, including arsenic's closest neighbor in the periodic system, phosphorus: compare for example the toxicity of phosphate and phosphine (PH_3). In much the same way, the toxic action of arsenic differs both qualitatively and quantitatively between different species, e.g. arsenate, arsenite, trimethyl arsine, dimethylarsinic acid, etc. [1, 2][1]. Furthermore, different target organisms, e.g. phytoplankton, fish, man, respond differently to the different chemical species, so that the detoxifying reaction of one organism may produce a compound which is more toxic to another organism than the original compound; an example for this may be seen in the transformation of arsenate by some molds into trimethyl arsine, a deadly poison to humans [3]. It is therefore important that we understand the biological and chemical processes which are able to transform the chemical species of arsenic into each other.

Three main types of transformations of arsenic species have been identified in the environment: (1) the simple redox-reaction which changes arsenite into arsenate and which can either occur abiotically or be biologically catalyzed, (2) the reductive biomethylation of arsenic to volatile methylarsines, and (3) the biosynthesis of structurally complex organoarsenic compounds and their subsequent degradation. We will discuss these processes and their role in the marine environment in the following paragraphs.

2. Redoxtransformations Between Arsenic(III) and Arsenic(V)

In seawater containing free dissolved oxygen, arsenic(V) is the thermodynamically stable form of the element [4]. Thus in most of the ocean (with the exception of a few anoxic basins) arsenate should be the only detectable species. And indeed, we do find that arsenate dominates the chemical speciation of arsenic in the ocean, and that arsenite is present in amounts exceeding those of arsenate only in the reduced, oxygen-free pore waters of sediments [4] and in anoxic basins like the Baltic Sea (Andreae, unpublished data). However, significant amounts of arsenite (up to 10% of total As) are found in the surface and deep waters of the oceans and conversely, some arsenate is still present in anoxic waters. The redox pair arsenate/arsenite is therefore not in thermodynamic equilibrium anywhere in the marine environment.

Several mechanisms may be responsible for the presence of arsenite in oxic seawater: abiotic reduction of arsenate by organic matter, the reduction of arsenate by bacteria or phytoplankton, or the release of arsenite as a result of the breakdown of organoarsenic compounds. Uthe and Reinke [5] have shown that arsenate is reduced rather effectively by a large variety of non-living biological materials, including tissue samples of marine fish and invertebrates. Maybe more pertinent to the speciation of arsenic in seawater, Johnson [6] demonstrated the reduction of arsenate to arsenite in seawater medium by a mixed population of bacteria from Narragansett Bay during log-phase growth.
The reduction was essentially complete after a few hours, and no detectable removal of arsenic from solution either as volatile forms or by bacterial uptake was observed. No formation of organoarsenic compounds was detected in these experiments. A coral species (*Pocillopora verrucosa*) is also able to reduce arsenate to arsenite [7]. In

---

[1]Figures in brackets indicate the literature references at the end of this paper.

this case it is not clear if the reduction is due to the coral animal or to the symbiotic zooxanthellae. Blake and Johnson [8] demonstrated the reduction of arsenate to arsenite by clumps of the pelagic alga Sargassum sp., containing the total associated community. The role of algae in the reduction of arsenate was first suggested by Blasco et al. [9], who found that it occurred in cultures of Chlorella, a freshwater green alga. Andreae and Klumpp [10] observed the release of arsenite in a number of cultures of marine phytoplankton species (table 1). As most of these cultures were bacteria-free, the reduction has to be due to the activity of the algae themselves rather than to bacterial contaminants or symbionts. We conclude from these observations that the ability to reduce arsenate to arsenite is a rather general property of marine organisms.

Table 1. Arsenic Species in Phytoplankton Growth Media After Log-Phase Growth (concentrations in nanomoles/liter).

Algal species	As(III)	As(V)	MMAA	DMAA	Total As
Skeletonema costatum	<0.1	0.4	0.2	1.3	1.9
Cricosphaera carterae	8.6	0.1	0.5	1.9	11.1
Gonyaulax polyedra	0.4	10.4	<0.1	0.1	10.8
Platymonas suecica	<0.1	1.9	trace	1.9	3.8
Sterile control	<0.1	13.9	0.1	0.02	14.0

The reverse reaction, the oxidation of arsenite to arsenate, occurs in seawater at a rather slow rate: Johnson and Pilson [11] estimate a turnover of the arsenite pool about once every six months in surface ocean waters. In deep waters, the oxidation rate is difficult to predict due to the large extrapolation required from the experimental conditions to those in the deep ocean. Bacterial catalysis of the oxidation of arsenite has been shown to exist for many species of bacteria (e.g. Alcaligenes sp. [12,13]), but Johnson and Pilson's [11] experiments did not show a significant difference in initial oxidation rates between sterile and non-sterile seawater.

In summary, the redox speciation of arsenic in seawater appears to result from a dynamic equilibrium between biogenic reduction of arsenate (by a variety of micro- and macroorganisms) which occurs largely in the surface ocean, and the slow abiotic oxidation of arsenite throughout the ocean.

3. The Reductive Biomethylation of Arsenic

This process, which leads to the formation of the volatile and extremely toxic methylarsines, was investigated by Challenger's group, who became interested in this phenomenon as a consequence of poisoning incidents in England and Germany. Their work (reviewed in [3]) laid the foundation to the study of the biological methylation of metals and metalloids. They identified the mold Scopulariopsis brevicaulis as an organism able to produce trimethyl arsine from arsenite via a biochemical pathway in which successive reactions with carbonium ions yield methylarsonic acid, dimethylarsinic acid, and trimethylarsine oxide. A final reduction gives trimethyl arsine. This process was also observed to occur in many other species of fungi [14]. It takes place under aerobic conditions, and while the toxic trimethyl arsine is subject to oxidation by air, it is stable enough to persist in indoor atmospheres long enough to create a hazard.

The biomethylation of arsenic under anoxic conditions by methanogenic bacteria was investigated by McBride and Wolfe [15]. This process proceeds by two methylating steps to dimethylarsinic acid like in the aerobic system, but then reduction to dimethyl arsine occurs rather than further methylation. Dimethyl arsine is a highly reactive substance; it oxidizes rapidly in air and is added easily to disulfide bonds. This explains the tendency for dimethyl arsine to become strongly adsorbed to materials like red rubber and to be rapidly transformed into nonvolatile compounds in sewage sludge [15].

In 1974, Wood [16] had suggested a biogeochemical cycle of arsenic, including the generation of volatile arsines in sediment pore waters and anoxic basins of oceans and lakes, similar to the cycle of mercury described in the same paper. The existence of such a biogeochemical cycle would have two important consequences: (1) toxic methylarsines could be generated from inorganic arsenic in biologically active environments and lead to an enhancement of the potential danger of environmental arsenic contamination, and (2) volatile methylarsines could be released to the atmosphere and lead to the enrichment of arsenic in atmospheric aerosol particles which has been observed throughout the World Atmosphere [17]. Some experimental support for this proposed cycle was provided by the work of Wong et al. [18], who observed the production of methylarsonic acid, dimethylarsinic acid, and trimethylarsonium oxide when lake sediments and water from Ontario lakes (some of which contained high levels of industrial arsenic contamination) were incubated with nutrient broth and additions of arsenic compounds. In bacterial cultures, they observed the production of di- and trimethylarsines in arsenic-enriched media in addition to the compounds mentioned above.

In view of the potential importance of these processes, I developed methods to determine the presence of the methylarsenic species suggested by Wood [16] at the part-per-trillion level in environmental samples [19] and investigated the speciation of arsenic in seawater from oxic and anoxic environments, in sedimentary pore waters of coastal and open ocean sediments and in rain water, as well as in marine planktonic and benthic algae and algal culture media [4,10,20,21]. However, in none of the samples investigated was there any evidence for the presence of the free methylarsines. Therefore, these methylarsines are either not formed in the marine environment, or they exist in concentrations too small to be detected with our very sensitive methods, in which case they cannot be expected to play a significant role toxicologically or geochemically.

Our observations were supported by the work of McBride et al. [22], who studied the biosynthesis of methylarsines by methanogenic bacteria in sewage sludge and marine mud. Their data show evidence for the production of methylarsines in sewage sludge; however, these compounds were never present at detectable concentrations in solution but were absorbed immediately and irreversibly to unknown phases. In marine muds, on the other hand, they were not able to detect even the formation of methylarsines or related compounds. The reductive biomethylation of arsenic in the form suggested by Wood [16] can thus be ruled out as a significant process in the marine environment.

4. The Biosynthesis of Structurally Complex Organoarsenic Compounds by Marine Algae

4.1. The speciation of arsenic in seawater

While I had observed that the biomethylation of arsenic did not occur in the anoxic marine environment, I did find methylated arsenic species, methylarsonic acid and dimethylarsinic (cacodylic) acid, to be present consistently in surface seawaters. These compounds had also been observed in coastal surface waters by Braman and Foreback [23].

In vertical profiles through the marine water column, I observed a characteristic distribution of arsenic species, which is shown here at the example of the station AS3-2 in the eastern North Pacific off Southern California (figure 1). The same general type of vertical distribution was found at all the stations investigated so far, in the Pacific Ocean [4,20], the Atlantic [24], and the Gulf of Mexico (Andreae, unpublished results). Arsenate is the predominant species, both in the deep waters and in the surface layer. At station AS3-2, which is in the Santa Catalina basin of the continental borderland off Southern California, oxygen is almost completely depleted below a depth of about 500 m. This does not, however, influence the arsenate/arsenite ratio (thermodynamic calculations predict that arsenate is predominant whenever detectable amounts of oxygen are present). Arsenate shows a nutrient-like profile, with depletion in the synthetically active surface waters, and a sharp concentration increase between 70 and 250 m depth due to the regeneration of arsenate during biomass decomposition and due to upward mixing of more arsenate-rich deep waters. The increase in arsenate in the deepest sample may be due to release of arsenic from the sediment surface. Arsenate behaves thus much like phosphate, the most significant difference being that in the case of phosphate the surface water depletion reaches almost 100%, while in the case of arsenate 20-30% depletion is typical [20].

Arsenite, methylarsonate, and dimethylarsinate have maxima in the surface layer. This behavior is most pronounced in the case of dimethylarsinate, which follows in its distribution closely that of phytoplankton productivity, in this example estimated by the concentration of particulate chlorophyll a. The methylarsenicals drop to very low concentrations below the level of 1% light penetration (the euphotic depth), which at this station was at about 65 m depth. The monomethylated form, methylarsonate, is usually about a factor of ten less abundant than the dimethyl species. No evidence was ever found for the presence of trimethyl arsenic species in seawater.

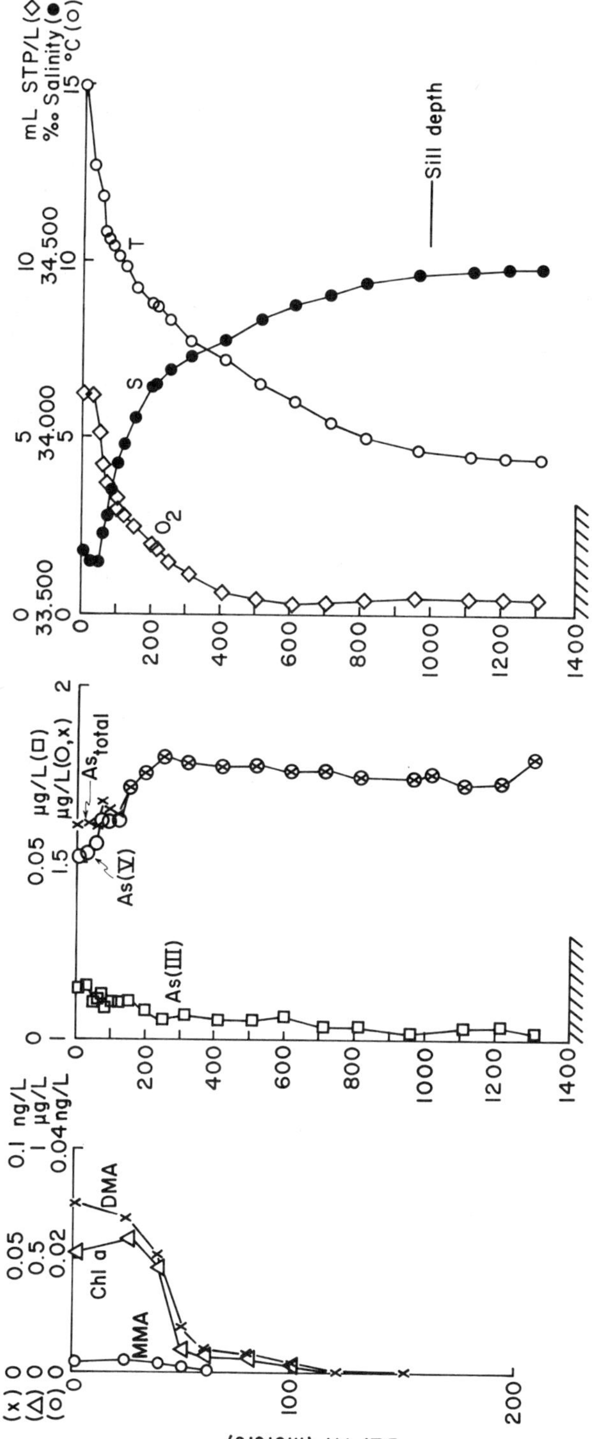

Figure 1. Arsenic species distribution at Station AS3-2 in the Northeastern Pacific.

These observations suggest that arsenate is taken up by phytoplankton in the euphotic surface waters together with the chemically closely related phosphate, and subsequently converted to arsenite and methylarsenicals and released back into the water column. The vertical distribution of the methyl species, especially their sharp decrease at the base of the euphotic zone, supports a direct release of these compounds by live algal cells; if they were the result of bacterial decomposition of algal matter they should be present down to considerably greater depths. Arsenite, on the other hand, has a broader distribution extending below the euphotic depth; in some cases it shows a sharp maximum at the level of maximum regeneration just below the euphotic zone [20]. Therefore, its production may be connected with bacterial and zooplankton activity. This is in agreement with the fact that the ability to reduce arsenate to arsenite is quite widespread among marine organisms. Thus, arsenite could be formed both directly by reduction of arsenate, and as a product of the demethylation of organoarsenic compounds.

The marine food chain shows a tendency for discrimination against arsenic relative to phosphate. While some of the most primitive marine organisms, e.g. marine yeasts, are not able to differentiate between phosphate and arsenate in their uptake mechanism [25], our investigations on algae show that they are able to take up phosphate preferentially [10]. Together with the release of organoarsenic species, this results in an arsenic/phosphorus ratio in algae which is about ten times lower than that in seawater (biodiminution). Table 2 shows the arsenic and phosphorus concentrations and their ratios at different levels of the marine food chain. This ratio decreases sharply with increasing trophic level; thus the arsenic contained in the primary biomass is not completely and readily absorbed into cellular constituents by the consumers, whereas phosphorus is retained to a large extent in the heterotrophic biomass. This is further substantiated by the observation that the regeneration of arsenic in the marine water column always occurs at considerably shallower depths than the regeneration of phosphate [20].

Table 2. Arsenic and Phosphorus in Seawater and Organisms at Increasing Trophic Levels (concentrations in ppm for seawater and mg/kg dry weight for biological materials).

	As	P	As/P	Ref.
Surface seawater	0.0015	0.016	9.4×10^{-2}	[20]
Marine algae	30	4,250	7.1×10^{-3}	[26]
Crustacea	1.5	9,000	1.7×10^{-4}	[1,26]
Pisces	1.7	18,000	9.3×10^{-5}	[26,27]
Mammalia	0.2	43,000	4.6×10^{-6}	[26]

4.2. Biosynthesis of organoarsenic species by marine phytoplankton

To further substantiate the role of marine planktonic algae in the biotransformation of arsenic species in the ocean, we investigated the biochemistry of arsenic in pure, axenic cultures of marine phytoplankton [10]. Nine different cultures were selected from the culture collection of the Scripps Institution of Oceanography, including species of diatoms, coccolithophorids, dinoflagellates, and green algae. These algae were grown in standard culture media without addition of arsenic above that contained in the seawater which was used as the base of the medium. After the end of the log-phase growth the cells were filtered off and the medium analyzed for arsenic speciation. In all instances (with the exception of two poorly growing species), substantial changes had occurred. The concentration of total dissolved arsenic had decreased to 14-80% of the original value, and significant levels of arsenite, methylarsonic acid, and dimethylarsinic acid were found. Table 1 shows some of these results. It is of particular interest that the proportions of the arsenic species produced are quite variable for the different organisms. While some algae, e.g. _S. costatum_, took up most of the arsenic in solution, and released some back in the form of organoarsenicals, others (especially the coccolithophorids) took up very little arsenic and converted most of the arsenate to arsenite. Similar variability in arsenic retention by phytoplankton species was observed by Bottino et al. [28].

In order to investigate further the biochemistry of these processes we added radioactive ^{74}As as tracer to some of these cultures. Uptake began immediately, and after an initial rise a nearly constant arsenic-to-biomass ratio was maintained during growth. Again, the amount taken up by the cells varied considerably from one algal species to another (from 5 to 78% of the arsenic contained in the medium). The cells were harvested and extracted with a water/methanol/chloroform system. A considerable amount of arsenic remained in the insoluble residue (19-96%). Again, the coccolithophorid _Cricosphaera carterae_ was unique in the respect of having incorporated very little arsenic, most of it attached to the insoluble residue, and having produced almost no organoarsenicals. A considerable amount of the arsenic in the insoluble fraction could be made soluble by digestion in warm, dilute hydrochloric acid. Most of it was then present in the form of arsenite. The production of arsenite via a similar mechanism (algal reduction and subsequent hydrolysis) may be responsible for some of the arsenite in surface seawater. In this context it is of interest that _C. carterae_, which produced most arsenite in culture, was isolated from the North Central Pacific Gyre, where I also found the highest concentration of dissolved arsenite.

The other species had produced large amounts of organoarsenicals both in the water/methanol fraction and in the chloroform fraction. These extracts were subjected to paper chromatography (water/methanol fraction) and thin-layer chromatography (chloroform fraction). These procedures resulted in the separation of up to 12 organoarsenical metabolites. As arsenobetaine, $(CH_3)_3As^+CH_2COO^-$, had been isolated from _Panulirus longipes_, the Australian rock lobster [29], we suspected that this compound or the closely related arsenocholine, $(CH_3)_3As^+CH_2CH_2OH$, were important compounds in these extracts. We synthesized these compounds in the laboratory and attempted to match them chromatographically with the algal products. However, none of the algal compounds cochromatographed with arsenobetaine; the substance which cochromatographed with arsenocholine in the paper chromatograms was separated from it by low-voltage paper electrophoresis. By analysis with the borohydride method [19], some arsenate, monomethylarsonate, and dimethylarsinate was found in the water/methanol extracts. The amounts of the methylarsenicals can be significantly increased by base hydrolysis of the extracts. This procedure also results in the production of some trimethylarsine oxide. This suggests that configurations with up to three methylgroups attached to arsenic like those in arsenobetaine and -choline are indeed present in some of the natural products.

4.3. The structure of the algal organoarsenic compounds

The presence of these and similar structures in marine organisms had also been suggested by other workers: Penrose [30] found evidence for tri- or tetramethylarsonium structures in tissues of witch flounder (*Glycocephalus cynoglossus*), and Bottino et al. [31] had found in the phospholipid fraction of a marine green alga arsenic compounds which they suspected to be related to arsenocholine. Lunde ([32] and references therein) had observed organoarsenic compounds in freshwater and marine algae and in fish; these compounds have chemical properties similar to those found in our work.

The most significant progress in our understanding of the marine biochemistry of arsenic came with the studies of Edmonds et al. [29], who demonstrated the presence of arsenobetaine in rock lobster, and the work of Cooney et al. [33], who identified O-phosphatidyltrimethylarsoniumlactic acid as the major organoarsenic compound in the lipid extract of *Chaetoceros concavicornis*, a marine diatom:

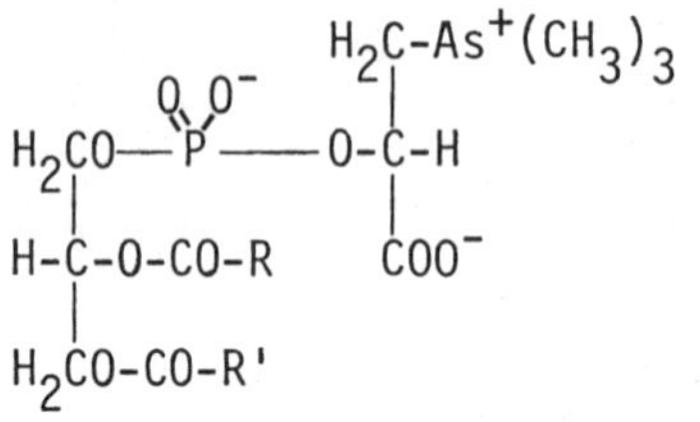

```
                     H2C-As+(CH3)3
           O  O-       |
H2CO——P————O-C-H
   |                   |
H-C-O-CO-R          COO-
   |
H2CO-CO-R'
```

4.4. Fate of the algal organoarsenic compounds in seawater and in the food chain

Trimethylarsonium lactate gives arsenobetaine upon oxidative chemical degradation involving selective decarboxylation and oxidation of the α-hydroxyacid [33]. Hydrolysis of arsenobetaine is rapid in alkaline solutions and leads to mono- and dimethylarsinate and trimethylarsine oxide. As the simple mono- and dimethylarsinates are only minor constituents in biological materials, but were the only organoarsenicals found in natural waters by the borohydride method, we investigated if the larger organoarsenicals constituted a significant fraction of the arsenic content of natural waters by subjecting previously analyzed samples to rigorous base hydrolysis. The samples were then reanalyzed, and the results compared with the first analysis. In no case was a significant increase in the total amount of arsenic detected, which precludes the presence of organoarsenicals of the type found in the algal extracts in river or surface ocean waters at levels higher than one or two percent of the total arsenic content. This suggests that either the organoarsenicals are very short-lived outside the cell and are rapidly hydrolyzed to the simple methylarsenic acids once they are released, or that the complex organoarsenicals are first enzymatically hydrolyzed within the cell and the resulting methylarsenicals then rapidly excreted.

Benson et al. [34] have suggested that extracellular oxidases and other hydrolytic enzymes of marine bacteria associated with the phytoplankton attack hydrophilic trimethylarsoniumlactate groups at the surface of the algal cell membrane, liberating trimethylarsoniumlactic acid and/or dimethylarsinate into the environment. This mechanism may explain the absence of detectable intermediates in seawater, but at least in culture other mechanisms must prevail, as most of our phytoplankton cultures were bacteria-free. We suggest that the precursors of the mono- and dimethylarsinates are enzymatically converted within the cell to the simple methylarsenicals which are then excreted.

The fate of the organoarsenic compounds synthesized by marine algae in the food chain is not clear. While Edmonds et al. [29] had detected arsenobetaine in lobster, Benson and Summons [35] did not see evidence of significant conversion of the trimethylarsonium lactate in the giant clam *Tridacna maxima*. On the basis of alkaline degradation experiments, H. Norin (1981, personal communication) suggests that arsenobetaine is the dominant compound in fishmeal, while in shrimp two thirds of the organoarsenic is present as arsenobetaine, the rest as other organoarsenicals, possibly arsenocholine. Crecelius [36] showed that organoarsenicals of marine origin are not significantly metabolized after ingestion by man.

4.5. The biological rationale of the metabolic cycle of arsenic

The biological rationale for the existence of this relatively complex metabolic cycle of arsenic in marine algae is not clear at this time. It may be a detoxification system designed to cope with the problem of the relatively high arsenate to phosphate ratios in surface seawater (0.1 to $\sim$1: Andreae [20]). As mentioned above, these two ions are very similar to each other and therefore difficult to separate by biological uptake systems. The primary producers may thus deal with this problem either by attempting to convert the arsenate after its initial uptake into a chemically different form so it can be immediately reexcreted (e.g. arsenite in the case of the coccolithophorids), or by converting them into relatively inert, non-toxic organoarsenicals, which are then either incorporated into structural parts of the cell or released as methylarsenicals after hydrolysis . The detoxification hypothesis is supported by the work of Sanders [37], who observed an increase of arsenate toxicity with decreasing phosphate concentrations. In contrast, the methylarsenicals (at concentrations up to 250 times higher than found in the ocean) were not toxic even at low levels of phosphate. As no significant conversion of organoarsenic compounds to inorganic arsenic occurs along the food chain, the higher trophic levels would thus profit from the algal detoxification mechanism. On the other hand, arsenic has been shown to be an essential element in a number of higher organisms, including goats and chickens [38]. No information is available on the possibility of arsenic essentiality in phytoplankton, as all experiments have been conducted with seawater-based media containing natural arsenic levels or with artificial media containing unknown levels of arsenic [28].

5. Biomethylation in the Ocean as a Source of Atmospheric Arsenic

While our investigations into the biogeochemistry of arsenic in the marine environment had led to the discovery of significant biotransformations of arsenic species, they had not shown evidence for the generation of volatile arsines which would make the large scale transfer of arsenic into the atmosphere possible. Sanders and Windom [39] verified the possibility of a volatilization of arsenic by algal biomethylation in culture experiments by analysis of the headspace gas over the culture; they did not find any evidence for such a process to occur. In view of the chemical stability of dimethylarsinic acid, this compound should be detectable in atmospheric particulates and precipitation. In fact, Johnson and Braman [40] found some evidence for the presence of methylarsenic species in the particulate and gas phase in urban and suburban environments. However, in a large number of analyses of marine rain, I did not find methylated species present above the detection limit of 0.2 ng As/L [21]. Together with the data of Walsh et al. [41] on the arsenic content of marine aerosols this precludes the existence of a significant ocean-to-atmosphere transfer by biomethylation reactions. The inorganic arsenic concentrations which I measured in marine rain were much lower than those previously published: the average value for Pacific rain in my samples was 19 ng As/L, as compared to the value of 1.6 μg/L given by Kanamori and Sugawara [42]. This difference is probably due to the fact that their samples were taken in the more polluted environment around Japan. Using these high rainwater values, mass balance models have been designed [43] which postulate very large fluxes of methylated arsenic from the oceans to the atmosphere. On the basis of our data, on the other hand, it can be shown that such a mechanism does not exist, and that a global mass balance can be constructed for the atmospheric arsenic cycle which is dominated by anthropogenic inputs and in which there is no place for a significant biogenic source of arsenic to the atmosphere [41,21].

[1] Schroeder, H.A. and Balassa, J.J., Abnormal trace metals in man: arsenic, J. chron. Dis. 19, 85-106 (1966).

[2] Lewis, R.J. and Tatken, R.L., Registry of toxic effects of chemical substances, U.S. Dept. of Health, Education and Welfare, Cincinnati, Ohio (1978).

[3] Challenger, F., Biological methylation, Chem. Reviews 36, 315-361, (1945).

[4] Andreae, M.O., Arsenic speciation in seawater and interstitial waters: the influence of biological-chemical interactions on the chemistry of a trace element, Limnol. Oceanogr. 24, 440-452 (1979).

[5] Uthe, J.F. and Reinke, J., Arsenate ion reduction in non-living biological materials, Environ. Lett. 10, 83-88 (1975).

[6] Johnson, D.L., Bacterial reduction of arsenate in sea water, Nature 240, 44-45 (1972).

[7] Pilson, M.E.Q., Arsenate uptake and reduction by Pocillopora verrucosa, Limnol. Oceanogr. 19, 339-341 (1974).

[8] Blake, N.J. and Johnson, D.L., Oxygen production-consumption of the pelagic Sargassum community in a flow-through system with arsenic additions, Deep Sea Res. 23, 773-778 (1976).

[9] Blasco, F., Gaudin, C., and Jeanjean, R., Absorption des ions arséniate par les chlorelles: réduction partielle de l'arséniate en arsenite, C. R. Hebd. Seances Acad. Sci. Ser. D Sci. Nat. 273, 812-815 (1971).

[10] Andreae, M.O. and Klumpp, D.K., Arsenic uptake and metabolism by marine phytoplankton, Env. Sci. Tech. 13, 738-741 (1979).

[11] Johnson, D.L. and Pilson, M.E.Q., The oxidation of arsenite in seawater, Environ. Lett. 8, 157-171 (1975).

[12] Osborne, F.H. and Ehrlich, H.L., Oxidation of arsenite by a soil isolate of Alcaligenes, J. Appl. Bacteriol. 41, 295-305 (1976).

[13] Phillips, S.E. and Taylor, M.L., Oxidation of arsenite to arsenate by Alcaligenes faecalis, Appl. and Environ. Microbiol. 32, 392-399 (1976).

[14] Cox, D.P., Microbiological methylation of arsenic, in Arsenical Pesticides, E.A. Woolson, ed., Am. Chem. Soc. Symposium Series No. 7, 81-76 (Washington, D.C., 1975).

[15] McBride, B.C. and Wolfe, R.S., Biosynthesis of dimethylarsine by methanobacterium, Biochemistry 10, 4312-4317 (1971).

[16] Wood, J.M., Biological cycles for toxic elements in the environment, Science 183, 1049-1052, 1974.

[17] Rahn, K.A., The chemical composition of the atmospheric aerosol, Technical Report, Graduate School of Oceanography, University of Rhode Island, Kingston, R.I., July 1976.

[18] Wong, P.T.S., Chau, Y.K., Luxon, L., and Bengert, G.A., Methylation of arsenic in the aquatic environment, Proc. 11th Annual Conf. on Trace Substances in Environmental Health, June 1977, Columbia, Missouri, D.D. Hemphill, ed., 100-106 (1977).

[19] Andreae, M.O., Determination of arsenic species in natural waters, Anal. Chem. 49, 820-823 (1977).

[20] Andreae, M.O., Distribution and speciation of arsenic in natural

waters and some marine algae, Deep Sea Res. 25, 391-402 (1978).

[21] Andreae, M.O., Arsenic in rain and the atmospheric mass balance of arsenic, J. Geophys. Res. 85, 4512-4518 (1980).

[22] McBride, B.C., Merilees, H., Cullen, W.R., and Pickett, W., Anaerobic and aerobic alkylation of arsenic, in Organometals and Organometalloids, Occurrence and Fate in the Environment, F.E. Brinckman and J.M. Bellama, eds., Am. Chem. Soc. Symposium Series No. 82, 94-115 (Washington, D.C., 1978).

[23] Braman, R.S. and Foreback, C.C., Methylated forms of arsenic in the environment, Science 182, 1247-1249 (1973).

[24] Waslenchuk, D.G., The budget and geochemistry of arsenic in a continental shelf environment, Marine Chem. 7, 39-52 (1978).

[25] Button, D.K., Dunker, S.S., and Morse, M.L., Continuous culture of Rhodotorula rubra: kinetics of phosphate-arsenate uptake, inhibition, and phosphate-limited growth, J. Bacteriol. 113, 599-611 (1973).

[26] Bowen, H.J.M., Trace Elements in Biochemistry (Academic Press, N.Y., 1966).

[27] LeBlanc, P.J. and Jackson, A.L., Arsenic in marine fish and invertebrates, Mar. Pollut. Bull. 4, 88-90 (1973).

[28] Bottino, N.R., Newman, R.D., Cox, E.R., Stockton, R., Hoban, M., Zingaro, R.A., and Irgolic, K.J., The effects of arsenate and arsenite on the growth and morphology of the marine unicellular algae Tetraselmis chui (Chlorophyta) and Hymenomonas carterae (Chrysophyta), J. exp. mar. Biol. Ecol. 33, 153-168 (1978).

[29] Edmonds, J.S., Francesconi, K.A., Cannon, J.R., Raston, C.L., Skelton, B.W., and White, A.H., Isolation, crystal structure, and synthesis of arsenobetaine, the arsenical constituent of the western rock lobster Panulirus longipes cygnus George, Tetrahedron Lett. 18, 1543-1546 (1977).

[30] Penrose, W.R., Conacher, H.B.S., Black, R., Méranger, J.-C., Miles, W., Cunningham, H.M., and Squires, W.R., Implications of inorganic/organic interconversion on fluxes of arsenic in marine food webs, Int. Conf. on Arsenic, Fort Lauderdale, Florida (October 1976).

[31] Bottino, N.R., Cox, E.R., Irgolic, K.J., Maeda, S., McShane, W.J., Stockton, R.A., and Zingaro, R.A., Arsenic uptake and metabolism by the alga Tetraselmis chui, in Organometals and Organometalloids, Occurrence and Fate in the Environment, F.E. Brinckman and J.M. Bellama, eds., Am. Chem. Soc. Symposium Series No. 82, 116-129 (Washington, D.C., 1978).

[32] Lunde, G., Occurrence and transformation of arsenic in the marine environment, Environ. Health Perspectives 19, 47-52 (1977).

[33] Cooney, R.V., Mumma, R.O., and Benson, A.A., Arsoniumphospholipid in algae, Proc. Natl. Acad. Sci. USA 75, 4262-4264 (1978).

[34] Benson, A.A., Cooney, R.V., and Summons, R.E., Arsenic metabolism: a way of life in the sea, Proc. 3rd Internat. Symp. on Arsenic and Nickel, Jena, M. Anke, H.-J. Schneider, and Chr. Brückner, eds., 139-145 (July 1980).

[35] Benson, A.A. and Summons, R.E., Arsenic accumulation in Great Barrier Reef invertebrates, Science 211, 482-483 (1981).

[36] Crecelius, E.A., Changes in the chemical speciation of arsenic following ingestion by man, Environ. Health Perspectives 19, 147-150 (1977).

[37] Sanders, J.G., Effects of arsenic speciation and phosphate concentration on arsenic inhibition of Skeletonema costatum (Bacillariophyceae), J. Phycol. 15, 424-428 (1979).

[38] Anke, M., Schneider, H.-J., and Brückner, C., eds., Proc. 3rd Internat. Symp. on Arsenic and Nickel, Jena, German Democratic Republic (July 1980).

[39] Sanders, J.G. and Windom, H.L., The uptake and reduction of arsenic species by marine algae, Estuarine and Coastal Marine Sci. 10, 555-567 (1980).

[40] Johnson, D.L. and Braman, R.S., Alkyl- and inorganic arsenic in air samples, Chemosphere 6, 333-338 (1975).

[41] Walsh, P.R., Duce, R.A., and Fasching, J.L., Considerations of the enrichment, sources, and flux of arsenic in the troposphere, J. Geophys. Res. 84, 1719-1726 (1979).

[42] Kanamori, S. and Sugawara, K., Geochemical study of arsenic in natural waters; 1: Arsenic in rain and snow, J. Earth Sci. 13, 23-35 (1965).

[43] Lantzy, R.J. and MacKenzie, F.T., Atmospheric trace metals: global cycles and assessment of man's impact, Geochim. Cosmochim. Acta 43, 511-525 (1979).

DISCUSSION

R. D. Wauchope: What do you consider to be the dominant organo-arsenical in marine waters?

M. O. Andreae: Definitely dimethylarsinic acid. We've worried about whether arsenobetaine or related compounds are present. They're reasonably stable compounds and have an important role. We've taken seawater that we had previously analyzed for arsenic speciation and applied some alkaline digestion procedures to it. This should show if there are any compounds like arsenobetaine, arsenocholine or any of the other organo-arsenicals present which are all sensitive to base hydrolysis. In fact, if they are present, they must be there at levels smaller than about two or three percent of total arsenic, which is the precision to which you can do this experiment.

R. D. Wauchope: I've got a second question. I'm used to thinking of plants doing things to increase their chances of survival and everyone here has been talking about how arsenite is toxic and in fact, is the most toxic species. So why are we seeing reduction by marine plants?

M. O. Andreae: This is one of the things that I skipped because I didn't have time. It turns out that the critical variable in this context, is the relationship between arsenate and phosphate. In surface seawater, the arsenate to phospate ratio tends to approach unity especially under conditions of phosphate depletion. As the plants are dependent on the uptake of phosphate from seawater, this high level of phosphate contamination with arsenate starts to become increasingly stressful as phosphate levels drop. Since arsenite and the methylarsenicals are significantly different in chemical properties and in their uptake properties from arsenate, this reductive step can help the plant cope with the arsenate problem. You can show that the toxicity of arsenate increases with decreasing phosphate levels, whereas this is not the case with the toxicity of arsenite and the organo-arsenicals.

Name not given: From your data, it appears that some specie of algae have a different uptake than others. Have you done more work on that and can one make a generalization?

M. O. Andreae: With the dinoflagellates, there's a little bit of a problem. They grow very slowly, so I wouldn't look too closely at those data. But if you look at the other organisms which have high and mutually comparable growth rates so that you end up with a comparable biomass, you can see that there's a tendency for diatoms to incorporate a very considerable amount of arsenic from their environment. Other groups, like the coccolithophorids, carbonate forming algae, tend not to take up much arsenic and rather convert most of the arsenate in their environment into arsenite.

27

MAN'S PERTURBATION OF THE ARSENIC CYCLE

E.A. Woolson
U.S. Department of Agriculture
Bldg. 050
BARC-West
Beltsville, Maryland 20705

The natural arsenic cycle is not greatly disturbed, on a global basis, by man's activities. Locally, however, it may be altered through the smelting of ores or the use of arsenical compounds as pesticides. When natural sulfidic minerals are processed or weathered, arsenic is changed to a chemical form that becomes available for movement and biotransformation within the environment. The highest environmental concentrations have occurred in air and soils around smelter operations, while coal combustion puts arsenic into air at much lower concentrations over a wider area. The quantity of arsenic used in pesticide production has remained relatively constant although the compounds produced have changed. Arsenic, whether applied in inorganic or organic forms, forms arsenate; this in turn reacts with hydrous iron and aluminum oxides in soil or sediment to form very insoluble compounds and complexes. Before they are metabolized, the organoarsenicals also form analogous compounds and complexes, which are more soluble than the inorganic forms. Organoarsenical herbicides, feed additives, or waste products are then metabolized by soil microorganisms to form arsenate. Under some conditions, part of the inorganic arsenate is converted to methylated arsenicals, which are then reduced biologically to volatile alkylarsines. These gases are released from soil, are reoxidized, adsorbed on particulate matter, and removed in rainfall. This cycle repeats itself until the arsenic ultimately resides in oceanic sediments . The arsenic, at any point of its movement toward oceanic sediments, may be intercepted by plant or animal uptake delaying the time required to complete the cycle, but not its eventual outcome.

1.1. Natural Sources of Arsenic

Arsenic is a mobile, ubiquitous element which continually changes form and location in the environment. The rates of change and movement vary depending on the substrate where the arsenical is found.

Natural sources of arsenic in air have been estimated [42] (table 1). The table includes global production, emission factors, and total emissions. The total atmospheric arsenic burden totals 7,900 tons As/year. However, arsenic does not accumulate in air because of dust settling and rainfall. Studies with isotopic radon revealed a residence time in air between 8 and 13 days depending on the hemisphere and geographical location [39]. For arsenic, residence time is estimated to be 9 days.

TABLE 1. Estimate of Arsenic Emissions From Natural Sources[a]

Source	Global Production or Consumption, 10^{12} g/yr	Arsenic Emission Factor, g/g source	Total Arsenic Emission ($d \leq 5$ μm), 10^9 g/yr
Ocean			
Bubble bursting	1000	5.7×10^{-7}	0.028
Gas exchange			0.084
Earth's crust			
Particle weathering	800	2.0×10^{-6}	0.24
Direct volatilization			0.0007
Volcanoes	25	2.8×10^{-4}	7.0
Forest wild fires	320	5.0×10^{-7}	0.16
Terrestrial biosphere			0.26
Natural source total			7.8

[a]Walsh et al, [39]

The major natural source comes from volcanoes followed by terrestrial biosphere, and particle weathering of the earth's crust. The volcanic emissions were estimated based on As content of ash and enrichment factors due to condensation of volatile arsenic on the volcanic dust. The terrestrial biosphere emissions are estimated from submicrometer particle and vapor release by plants from data available on zinc and lead. Emissions from particle weathering arise from dust emitted to the atmosphere.

Other natural emission sources include forest wild fires, oceanic gas exchange and bubble bursting, and direct volatilization from the earth's crust.

1.2. Anthropogenic Sources of Arsenic

Man, through need for various minerals, removes arsenic as a by-product or waste product from its natural source. Anthropogenic sources include such activities as producing various metals, burning fuels, unwanted vegetation and wastes, controlling pests and mining of ores and minerals. Estimates of the emissions thus generated are found in table 2.

TABLE 2. Estimate of Arsenic Emissions From Anthropogenic Sources[a]

Source	Global Production or Consumption, 10^{12} g/yr	Arsenic Emission Factor, g/g source	Total Arsenic Emission ($d \leq 5$ μm), 10^9 g/yr
Coal	3245	1.7×10^{-7}	0.55
Light fuels	585	1.2×10^{-10}	0.00007
Residual fuels	956	4.3×10^{-9}	0.0041
Wood fuel	1200	5.0×10^{-7}	0.60
Agricultural burning	1120	5.0×10^{-7}	0.56
Waste incineration	540	1.6×10^{-6}	0.43
Iron/steel production	1220	7.0×10^{-6}	4.2
Copper production	8.7	3.0×10^{-3}	13.
Lead/zinc production	9	5.0×10^{-4}	2.2
Mining mineral ore	2500	1.0×10^{-8}	0.013
Arsenic/chemicals	0.040	1.0×10^{-2}	0.20
Arsenic/agriculture	0.037	5.0×10^{-2}	1.9
Cotton ginning	14	3.3×10^{-6}	0.023
Anthropogenic source total			23.6

[a]Walsh et al, [39]

Production of various metals found as sulfidic minerals releases the greatest amount of arsenic. The total emissions for iron/steel/copper/lead/and zinc production are 82% of the anthropogenic source and 62% of all emissions. Agricultural pesticide use releases the next largest amount, followed by individual combustion sources (i.e. coal, fuels, agricultural burning, and waste incineration). Other minor emission sources include cotton ginning and arsenic for chemicals. Anthropogenic emission sources are estimated to be greater than natural sources by a factor of three.

The production of pesticide products containing arsenic is presented in table 3. The largest amounts of arsenicals are used for cotton desiccation and wood preservation. Other arsenicals are used as herbicides, growth regulators, or feed additives. Historically, production has been relatively constant except for use in wood treatment, which has increased in recent years.

Table 3. Estimated Production of Arsenical Agricultural Chemicals

Compound	Use	Production
		10^9 g As
Arsenic acid	Wood	5.1 - 7.5
	Cotton	3.2 - 5.1
Methanearsonate	Cotton, Turf	1.9 - 4.2
Cacodylate	Cotton	0.43
Arsanilate	Feed	0.43
4-Ureidophenyl arsonic acid	Feed	0.41
3-Nitro	Feed	0.23
Others	Grape, Grapefruit	0.06
Total Produced		11.76 - 18.36

2.1. Residues in Air

The emission of volatile arsenical gases from smelter and mining activities raises atmospheric levels in the vicinity of the source. The level measured varies with distance from the source, height of the stack, and wind speed. Coal combustion in large cities contributes significant amounts of arsenic to urban air when compared to air in rural settings. Although air measurements in individual cities may deviate, it has been suggested that order of magnitude differences exist for residues found in air of rural, urban areas, and cities containing smelters. The air averages 0.002, 0.02, and 0.2 microgram As/m^3 respectively in each setting. Individual samples may be much higher than the averages above, especially in cities containing smelters.

2.2. Residues in Soil

Arsenic is deposited on soil from air pollution, directly as a pesticide, or as waste materials from mining activities. Table 4 indicates the relative magnitude of the maximum values reported in the literature from several sources.

Table 4. Soil Residues at Various Sites

Site	Residue ppm As	Reference
Uncontaminated	1 - 40	[43]
Pesticide, inorganic	1 - 2,550	[49]
Pesticide, organic	1 - 19	[32,29]
Industry, mining	100 - 41,200	[30]
Industry, smelting	10 - 21,213	[18]

Normal background soil levels are a reflection on the parent material for any particular soil. Thus, a soil weathered from a sulfidic mineral will generally contain higher levels than other types of soil. Most soils contain 1 to 5 ppm As.

Large amounts of lead arsenate were applied for many years to control codling moths in apple orchards. Residues averaged 200 - 300 ppm As in most orchard soils which were tested. The organic herbicides add much lower quantities of arsenic on a yearly basis and do not accumulate as rapidly. Mine tailings contribute to some of the highest soil levels observed, while smelters contribute somewhat less. Soil residues, resulting from aerial deposition, drop quickly as the distance from the stack increases, declining to background levels within 8 to 16 km of the emission source [2,26]. Residues decrease with the square of the distance from the stack [18].

2.3. Residues in Plants

Arsenic is taken into plants through root uptake or foliar absorption. Generally, the larger the amount of arsenic available for plant uptake, the larger the plant residue will be. However, available As is not proportional to total As. A low (10 - 50 ppm As) As content in a sandy soil may be more phytotoxic (i.e. available) than much higher levels (200 - 500 ppm As) in a heavier clay soil [41] and the plant grown on such soil will contain greater residue levels.

Several generalizations can be made about As content of plants resulting from only root uptake:

1. Plants grown on sands and sandy loams have higher residues at equivalent soil levels than those grown on heavier textured soils (silts, clays).
2. Fruit has lower levels than leaves, stems, or roots.
3. Roots contain the highest levels.
4. The skin of root crops has higher residues than the inner flesh.
5. Residues are low unless plant growth is severely affected.
6. Crops are different in their uptake of As.

Residual arsenic levels in plants from several sources are presented in table 5. The most striking feature is the residue level of up to 6,640 ppm As for plants grown on waste piles from smelters and mining activities. Little bluestem, <u>Andropogon scoparius</u> Michx., possessed a tolerance to arsenic after growing on contaminated soil (levels from 100 - 41,200 ppm As) which plants growing on uncontaminated soil did not possess [30]. Arsenic tolerance appeared to be an evolved characteristic under genetic control.

Residue levels in plants generally follow levels found in soils. The maximum value found in the literature is reported as well as the approximate average residue. Since the arsenicals are no longer sprayed on food crops, residue values reflect only root uptake and not foliar deposition. The addition of sludge or fly ash to soil does not seem to pose an unacceptable risk of plant contamination by As.

Table 5. Plant Residues

Site	Residue Found ppm As (dry wt)	Reference
Pesticide (uptake)	up to 20; $\bar{x}$ = 1	[22]
Smelter, mine	up to 6,640; $\bar{x}$ = 20	[28]
Sludge/soil	up to 1.23; $\bar{x}$ = 0.2	[38]
Fly ash/soil	up to 5.8; $\bar{x}$ = 0.2	[36]

3.1. Reactions of Soil Residues

Arsenic reacts with soil components regardless of the source or form of the arsenical. Adsorption onto and reaction with hydrous iron and aluminum oxides, which coat soil particles predominate. Heavier soils with a higher clay content and a resulting higher surface area and hydrous oxide content adsorb more arsenic than do lighter sandier soils with low clay content. Arsenicals also react with ions in solution. The most common of these are Fe, Al, Ca, and Mg, but may also include Mn and Pb in some soils. Each ion removes a portion of the arsenical based on the solubility product of the compound and the amount of reactants present. Thus, a soil may be saturated relative to some compounds and not others [17]. The solubility profiles of these compounds are also affected by pH; therefore changing the soil pH may affect each arsenical's solubility.

Two forms of oxidation are active in transforming the arsenicals environmentally; one destroys the carbon/arsenic bond and is associated with microbial activity and the other a change in oxidation state, which may or may not be affected by microbial activity.

Organoarsenical herbicides or feed additives are oxidized in soil to arsenate [40,46,47]. Naturally occurring organoarsenical compounds are also probably oxidized, although no actual studies have been conducted on these compounds.

Some arsenate may be reduced to arsenite under the proper environmental conditions. Arsenic in sediments or in flooded anaerobic soil can be reduced as a function of redox potential [15]. Additionally, microbes may play some role.

After arsenic has been in the soil for some time, a hydroxy apatite type mineral may be formed. The basis for this hypothesis lies in the apparent non-availability of what should be toxic residues and the accumulation of residues in orchards where large amounts of arsenical insecticides were once applied. Volatility losses, to be discussed later, apparently are quite low from these areas. There is also a gradual decrease in extractable residue after a 4 to 12 week initial reaction period [50]. This would seem to indicate a gradual shifting of chemical form.

3.2. Methylation of Soil Arsenic

Methylation of inorganic arsenicals can occur in soil. When various arsenicals are added to soil, the actual species found varies based on the compound added and time. Table 6 indicates the identity of the arsenic species found after application of different arsenic herbicides. Cacodylic acid degraded oxidatively to methanearsonate and arsenate. Some methanearsonate was methylated to cacodylate as well as oxidized to arsenate. Small amounts of methanearsonate and cacodylate were found from arsenite addition after a 26 week summer growing period in a field experiment (Woolson, unpublished). This is indicative of the two competing mechanisms. Under summer conditions, oxidation is faster than methylation. However, in the fall, winter, and spring when soils are saturated, one would expect to find larger amounts of the organo forms because reduction and methylation reactions would be favored.

Table 6. Residues Speciated in Soil Found After 26 Weeks

Applied	Arsenate	Methanearsonate ppm As	Cacodylate
Arsenite	20.0	0.3	0.2
Methanearsonate	9.6	0.7	0.3
Cacodylate	31.9	1.2	0.3

3.3. Volatilization of Arsenic

Methylation occurs in soils and the methylated compounds can be reduced and volatilized. As one might expect, the addition of methylated arsenical herbicides results in a larger volatility loss than the addition of inorganic arsenate or arsenite [48] since only reduction is required and not methylation followed by reduction.

In recent years, several researchers have examined air above soil or soil/grass surfaces to detect volatile or particulate arsenicals. Braman [6] detected both dimethyl- and trimethylarsines above grass treated with arsenite, arsenate, MAA, or cacodylic acid as well as above grass treated only with lake water containing naturally occurring inorganic arsenic (0.76 ppb As), MAA (0.07 ppb As), and cacodylic acid (0.16 ppb As). Volatilized arsines were continuously trapped on silver coated glass beads. Woolson [48] also detected dimethyl- and trimethylarsines evolved from soil treated with arsenate, MAA, or cacodylic acid. Air above the soil was purged continuously, but the evolved arsines were trapped in a $KI\text{-}I_2$ solution. Neither author detected arsine or methylarsine.

On the other hand, others have examined air above culture-solutions in order to detect alkylarsines. Cheng and Focht [9] and Shariatpanahi et al [37] analyzed gases collected in the head space above soil cultures.

They found arsine, methylarsine, dimethylarsine, and/or trimethylarsine in the head gas depending on the starting material and microorganisms present. Cox and Alexander [11] however, also examined head gas above microbes isolated from sewage and found only trimethylarsine--regardless of the starting material.

Table 7 summarizes the organisms (where identified), the substrate, and the gas produced from several experiments.

The number of microbial species isolated and the substrates examined are too limited to make sweeping generalizations. Different microorganisms may metabolize each arsenical differently. In the only direct comparison between the same fungi, Shariatpanahi et al [37] found all three alkylarsines, while Cheng and Focht [9] only found arsine. Both analyzed head space gas, the former group by GC analysis and the latter by GC-MS. The cultures of Cheng and Focht were made anaerobic with helium, while Shariatpanahi et al just sealed their flasks without purging. This may account for the observed differences. Neither group measured the solution Eh. The possibility of disproportionation of alkylarsines in trapped head gas can not be ruled out.

4. Arsenic Cycle

4.1. Local

Several cycles describing local arsenic movements have been presented in the literature in recent years. These have included estuaries, agricultural fields, lakes, marine, and microbial cultures.

The basis for arsenic's chemical transformations lies in the microbial methylation/reduction cycle. Challenger [8] first proposed a series of reductions and methylations to explain the appearance of trimethylarsine oxide and trimethylarsine above cultures. McBride and Wolfe [25] later showed that a methyl-donor was required for the methylation of arsenate or arsenite to cacodylic acid and dimethylarsine. Braman [6] summarized the environmental chemistry of arsenic where fungi, bacteria, pH, and Eh are shown to influence the equilibriums involved at each step of the cycle. Depending on various environmental conditions, the reactions proceed toward either methylation or demethylation.

4.2. Agricultural Fields

Sandberg and Allen [31] proposed a cycle for arsenic flux within the context of agricultural arsenical useage. They examined fluxes from air, water, plant, soil, fertilizer, dust, and pesticide application. The main sink was soil within their ecosystem model. The net loss, after pesticide application of inorganic arsenic, was estimated to be 0.046 to 0.096 %/day; for methanearsonate--0.055 to 0.070 %/day; for cacodylate--0.052 to 0.101 %/day. At the maximum recommended application rates, total soil residues from cacodylate would increase at the rate of 2.6 to 3.3 ppm As/ha/year, while soil residues from methanearsonate would increase at 1.5 to 1.9 ppm As/ha/year. Losses to the air would be primarily as alkylarsines, although loss through dust is possible.

Table 7. Alkylarsines Isolated in Various Substrates

Organism	Substrate	Compound Added[a]	Found[b] A	MA	DMA	TMA	Ref
Penicillium	Isolate from sewage	+3	-[c]	-	-	-	[11]
		+5	-	-	-	-	
		MAA	-	-	-	x	
		CA	-	-	-	x	
Gliocladium roseum	do	+3	-	-	-	-	
		+5	-	-	-	-	
		MAA	-	-	-	x	
		CA	-	-	-	x	
Candida humicola	do	+3	-	-	-	x	
		+5	-	-	-	x	
		MAA	-	-	-	x	
		CA	-	-	-	x	
Unidentified	Soils	+3	x	-	-	-	[9]
	(Hanford sl,	+5	x	-	-	-	
	Altamont cly l,	MAA	x	x	-	-	
	Domino si l)	CA	x	-	x	-	
Pseudomonas	Soil isolate	+3	x	-	-	-	
		+5	x	-	-	-	
Alcaligenes	Soil isolate	+3	x	-	-	-	
		+5	x	-	-	-	
Unidentified	Soil	MAA	Unidentified				[1]
Unidentified	Soil	+5	-	-	x	x	[48]
		CA	-	-	x	x	
Proteus sp.	Soil isolates	+5	-	x	x	-	[37]
E. coli	do	+5	-	x	x	-	
Flavobacterium sp.	do	+5	-	-	x	-	
Corynebacterium sp.	do	+5	-	-	x	x	
Pseudomonas sp.	do	+5	-	x	x	x	
Unidentified	grass/soil	+3	-	-	x	x	[6]
		MAA	-	-	-	x	
		CA	-	-	x	x	
		PAA	-	-	x	x	

[a]+3 - arsenite; +5 - arsenate; MAA - methanearsonate; CA - dimethylarsinate; PAA - phenylarsonate

[b]A - arsine; MA - methylarsine; DMA - dimethylarsine; TMA - trimethylarsine.

[c]Not detected

4.3. Lakes and Streams

Ferguson and Gavis [15], Wood [44], and Woolson [47] have proposed various schematic diagrams to illustrate the movement and/or metabolism within the air/water/sediment phases of the aqueous ecosystem. The models indicate that most arsenic is adsorbed to sediments where two fates await it: 1.) desorption back into solution or 2.) methylation/reduction to form organoarsenical compounds. The methylarsonate can be further methylated to cacodylic acid which is then either reduced to dimethylarsine or methylated to trimethylarsine oxide (TMAO). The TMAO can be reduced to its arsine. The alkylarsines present in the aqueous phase then volatilize to the air or are oxidized back to their respective oxides, which are adsorbed back onto the sediment. The formation of arsenosulfides may also occur in the sediments when sufficient sulfur compounds are present.

Crecelius [12] reported on a geochemical cycle in Lake Washington which receives dust input from the ASARCO copper smelter in Tacoma, Washington. The input consists of partially soluble arsenic-rich dust from the smelter and removal of dissolved As from the lake water by bacteria or an inorganic reaction. Budget analysis shows that half the As comes from the atmosphere and half from rivers. Forty-five percent of the input is removed by outflow and 55% by accumulation in the sediments. The input from rain was arsenate while the interstitial water contained arsenite (55%), arsenate (45%), and trace amounts (1 ppb As) of cacodylic acid.

4.4. Estuaries

The cycling of arsenic in the Puget Sound region was investigated by Crecelius and Carpenter [13] and Carpenter et al [7]. Puget Sound received large amounts of arsenic from stack dust and from liquid and solid effluent from the ASARCO copper smelter. Levels were elevated in soil near the plant and on islands in Puget Sound downwind from the smelter. The smelter fallout contributed an amount equal to inputs from rivers feeding the Sound and the liquid effluent from the smelter (300 tons/year from each source). However, the major source is from inflowing sea water. Most dissolved arsenic flows out through the Strait of Juan de Fuca. Sedimentation processes remove less than 15% of the dissolved arsenic input. No large flux from sediments into solution was observed. Vigorous tidal action quickly dilutes the anthropogenic arsenic to levels equal to the natural background. Elevated residues are found only in close proximity to the smelter.

4.5. Oceanic

Sanders [34] examined the cycling of arsenic in the Georgia Bight area. The largest input is from the Gluf Stream's intrusion. This source accounts for greater than 99% of arsenic inputs. Other sources include river run-off, atmospheric deposition, and diffusion from sediments. The intrusion arsenic is present almost entirely as arsenate. Some arsenite is formed in anaerobic sediments or by cellular action [20] of bacteria.

Biota contains only 0.1% of the total As present, but greatly influences the chemical specie present. Phytoplankton produce reduced and methylated species [4,35] and release them into solution. The arsenite is gradually oxidized to arsenate at the same rate it is produced [21,33] while the methylated species appear to be stable for at least several months [3,23,33].

Some arsenic is incorporated into cellular components as both inorganic and organic compounds [35]. The nature of these compounds is uncertain although a number of arsenolipids have been isolated and identified such as: arsenocholine--[5,19]; trimethylarsoniumlactic acid--[10]; arsenobetaine--[14].

Zooplankton consume about 90% of the arsenic present in phytoplankton. Of the consumed phytoplankton, about 25% of the arsenic is excreted unchanged in the feces. Some of the excreted arsenic is regenerated as soluble arsenate while about 30% is lost to the sediments.

Higher members in the aquatic food chain consume the zooplankton but little of the arsenic is incorporated. The ingested arsenic is about equally inorganic and organic [35] and is rapidly excreted [16,27]. Biological magnification does not occur in marine food chains with most organisms containing 10-30 mg/kg [16,35,45].

The production of gaseous arsines in marine sediments and diffusion into the overlying water column are probably unimportant.

Residence times have been estimated for various components of the ecosystem (table 8). The times vary from 9 days for air, to 17 years for land plants, to 99.8 million years for oceanic sediments [24].

Table 8. Residence Time in Various Environmental Components[a]

	Year
Sediments	99,800,000
Ocean (dissolved)	9,400
Land	2,400
Terrestrial Biota	17
Oceanic Biota	0.07
Air (total)	0.03

[a]Mackenzie et al, [24]

In conclusion, man may alter the quantities of arsenic present in any component of the ecosystem in a localized area, but cannot change or stop the natural biological or chemical processes that occur. Little evidence is available to indicate that man's activities have altered the arsenic cycle on a global basis.

References

[1] Akins, M.B. and Lewis, R.J., Chemical distribution and gaseous evolution of arsenic-74 added to soils as DSMA-74 arsenic, Soil Sci. Soc. Amer. J., 40, [5], 655-658 (1976).

[2] Amasa, S.K., Arsenic pollution at Obuasi goldmine, town, and surrounding countryside, Environ. Health Perspect., 12, 131-135 (1975).

[3] Andreae, M.O., Arsenic speciation in seawater and interstitial waters: The influence of biological-chemical interactions on the chemistry of a trace element, Limnol. Oceanogr., 24, 440-452 (1979).

[4] Andreae, M.O. and Klumpp, D., Biosynthesis and release of organo-arsenic by marine algae, Environ. Sci. Tech., 13, [6], 738-741 (1979).

[5] Bottino, N.R., Cox, E.R., Irgolic, K.J., Maeda, S., et al, Arsenic uptake and metabolism by the alga tetraselmis chui, in Organometals and Organometalloids, F.E. Brinckman and J.M. Bellama, eds., pp.116-129, (American Chem. Soc., Wash., D.C., 1978).

[6] Braman, R.S., Arsenic in the environment, in Arsenical Pesticides, ACS Symp Ser No. 7, E.A. Woolson, ed., pp. 108-123, (American Chem. Soc., Wash., D.C., 1975).

[7] Carpenter, R., Peterson, M.L. and Jahnke, R.A., Sources, sinks, and cycling of arsenic in the Puget Sound region, in Estuarine Interactions, M.L. Wiley, ed., pp. 459-480, (Academic Press, New York, N.Y., 1978).

[8] Challenger, F., Biological methylation, Chem. Rev. 36, 315-361 (1945).

[9] Cheng, C.N. and Focht, D.D., Production of arsine and methylarsines in soil and in culture, Appl. Environ. Microbiol., 38, [3], 494-499 (1979).

[10] Cooney, R.V., Mumma, R.O. and Benson, A.A., Arsoniumphospholipid in algae, Proc. Nat. Acad. Sci., 75, [9], 4262-4264 (1978).

[11] Cox, D.P. and Alexander, M., Production of trimethylarsine gas from various arsenic compounds by three sewage fungi, Bull. Environ. Contam. Toxicol., 9, [2], 84-88 (1973).

[12] Crecelius, E.A., The geochemical cycle of arsenic in Lake Washington and its relation to other elements, Limnol. Oceanogr., 20, [3], 441-451 (1975).

[13] Crecelius, E.A. and Carpenter, R., Arsenic distribution in waters and sediments of the Puget Sound region, NSF-RANN. Trace. Contam. Abst., 1, [6], 615-625 (1974).

[14] Edmonds, J.S. and Francesconi, K.A., Methylated arsenic from marine fauna, Nature (London), 265-436 (1977).

[15] Ferguson, J.F. and Gavis, J., A review of the arsenic cycle in natural waters, Water Res., 6, 1259-1274 (1972).

[16] Fowler, S.W. and Unlu, M.Y., Factors affecting bioaccumulation and elimination of arsenic in the shrimp Lysmata seticaudata, Chemosphere, 9, 711-720 (1978).

[17] Hess, R.E. and Blancher, R.W., Arsenic stability in contaminated soils, Soil Sci. Soc. Amer. J., 40, [6], 847-852 (1976).

[18] Hocking, D., Kuchar, P., Plambeck, J.A. and Smith, R.A., The impact of gold smelter emissions on vegetation and soils of a sub-arctic forest-tundra transition ecosystem, J. Air Pollut. Control Assoc., 28, [2], 133-137 (1978).

[19] Irgolic, K.J., Woolson, E.A., Stockton, R.H., et al, Characterization of arsenic compounds formed by Daphnia magna and Tetraselmis chuii from inorganic arsenate, Environ. Health Perspect., 19, 61-66 (1977).

[20] Johnson, D.L., Bacterial reduction of arsenate in sea water, Nature (London), 240, 44-45 (1972).

[21] Johnson, D.L. and Pilson, M.E.Q., Spectrophotometric determination of arsenite, arsenate, and phosphate in natural waters, Anal. Chim. Acta., 58, 289-299 (1972).

[22] Jones, J.S. and Hatch, M.B., Spray residues and crop assimilation of arsenic and lead, Soil Sci., 60, 277-288 (1945).

[23] Lunde, G., Occurrence and transformation of arsenic in the marine environment, Environ. Health Perspect., 19, 47-52 (1977).

[24] Mackenzie, F.T., Lantzy, R.J., and Paterson, V., Global trace metal cycles and predictions, Math. Geol., 11, [2], 99-142 (1979).

[25] McBride, B.C. and Wolfe, R.S., Biosynthesis of dimethylarsine by methanobacterium, Biochemistry (Easton), 10, [23], 4312-4317 (1971).

[26] Miesch, A.T. and Huffman, Jr., C., Abundance and distribution of lead, zinc, cadmium, and arsenic in soils, in Helena Valley, Montana, Area Environmental Pollution Study, AP-91, 65-80 (USEPA, Office of Air Programs, Research Triangle Park, North Carolina, 1972).

[27] Penrose, W.R., Biosynthesis of organic arsenic compounds in Brown Trout, Salmo trutta, J. Fish Res. Bd. Can., 32, 2385-2390 (1975).

[28] Porter, E.K. and Peterson, P.J., Arsenic tolerance in grasses growing on mine waste, Environ. Pollut., 14, 255-265 (1977).

[29] Robinson, E.L., Arsenic in soil with five annual applications, Weed Sci., 23, [5], 341-343 (1975).

[30] Rocovich, S.E. and West, D.A., Arsenic tolerance in a population of the grass Andropogon scoparius Michx., Science 188, 263-264 (1975).

[31] Sandberg, G.R. and Allen, I.K., A proposed arsenic cycle in an agronomic ecosystem, in Arsenical Pesticides, ACS Symp Ser No. 7, E.A. Woolson, ed., pp. 124-147, (American Chem. Soc., Wash., D.C. 1975).

[32] Sandberg, G.R., Allen, I.K. and Dietz, Jr., E.A., Arsenic residues in soils treated with six annual applications of MSMA and cacodylic acid, The Ansul Co., Marinette, WI.

[33] Sanders, J.G., Interactions between arsenic species and marine algae, PhD. Thesis, Univ. North Carolina (1978).

[34] Sanders, J.G., Arsenic cycling in marine systems, Mar. Environ. Res. 3, 257-266 (1980).

[35] Sanders, J.G. and Windom, H.L., The uptake and reduction of arsenic species by marine algae, Estuar. Coast Mar. Sci., 10, [5], 555-568 (1980).

[36] Scanlon, D.H. and Duggan, J.C., Growth and element uptake of woody plants on fly ash, Environ. Sci. Tech., 13, [3], 311-315 (1979).

[37] Shariatpanahi, M., Anderson, A.C., Abdelghani, A.A., et al, Biotransformation of the pesticide sodium arsenate, J. Environ. Sci. Health Pt. B., B16, [1], 35-47 (1981).

[38] Temple, P.J., Linzon, S.N. and Chai, B.L., Contamination of vegetation and soil by arsenic emissions from secondary lead smelters, Environ. Pollut., 12, 311-320 (1977).

[39] Turekian, K.K., Nozaki, Y., and Benniger, L.K., Geochemistry of atmospheric radon and radon products, Ann. Rev. Earth Planet. Sci., 5, 227-255 (1977).

[40] Von Endt, D.W., Kearney, P.C., and Kaufman, D.D., Degradation of MSMA by soil microorganisms, J. Agr. Food Chem., 16, [1], 17-20 (1968).

[41] Walsh, L.M. and Keeney, D.R., Behavior and phytotoxicity of inorganic arsenicals in soils, in Arsenical Pesticides, ACS Symp Ser No. 7, E.A. Woolson, ed., pp. 35-52, (American Chem. Soc., Wash., D.C.,1975).

[42] Walsh, P.R., Duce, R.A., and Fasching, J.L., Considerations of the enrichment, sources, and flux of arsenic in the troposphere, J. Geophys. Res., 84, [C4], 1719-1726 (1979).

[43] Wedepohl, K.H., Arsenic, in Handbook of Geochemistry, II, 33, (Springer-Verlag, New York, N.Y., 1969).

[44] Wood, J.M., Metabolic cycles for toxic elements in aqueous systems, Rev. Intern. Oceanogr. Med., 31-32 (1973).

[45] Woolson, E.A., Bioaccumulation of arsenicals, in Arsenical Pesticides, ACS Symp Ser No. 7, E.A. Woolson, ed., (American Chem. Soc., Wash., D.C. 1975).

[46] Woolson, E.A., The persistence and chemical distribution of arsanilic acid in three soils, J. Agr. Food Chem., 23, [4], 677-681 (1975).

[47] Woolson, E.A., Fate of arsenicals in different environmental substrates, Environ. Health Perspect., 19, 73-81 (1977).

[48] Woolson, E.A., Generation of alkylarsines from soil, Weed Sci., 25, 412-416 (1977).

[49] Woolson, E.A., Axley, J.H., and Kearney, P.C., The chemistry and phytotoxicity of arsenic in soils: I. Contaminated field soils, Soil Sci. Soc. Amer. Proc. 35, [6], 938-943 (1971).

[50] Woolson, E.A., Axley, J.H., and Kearney, P.C., The chemistry and phytotoxicity of arsenic in soils: II. Effects of time and phosphorus, Soil Sci. Soc. Amer. Proc., 37, [2], 254-259 (1973).

DISCUSSION

R. D. Wauchope: You listed some of the highest values for plant tissue that I've ever seen. Is that internal or surface?

E. A. Woolson: That was from root uptake. The 6,640 (Table 5) was on the grass that had adapted itself to growing on high arsenic mine spoils.

R. D. Wauchope: This is interesting, because I haven't heard of adaptation of vegetation to arsenic levels. Where is that information?

E. A. Woolson: It was published in Science (Rochovich and West, 1975, Volume 188, pp. 263-264).

M. O. Andreae: I'd like to make two points about the possibility for biomethylation in anaerobic environments. The first is based on very recent data from an expedition into the Baltic Sea. A lot of processes which normally happen within sediments happen in the water column of the Baltic Sea because the water column is anoxic in some of the basins. Therefore, if processes were to occur in marine sediments which lead to biomethylation, we should very clearly see them in the Baltic because it's easy to sample and analyze, much more so than it is in the sedimentary pore waters. In fact, there's no evidence at all of methyls being present in the anoxic systems of the Baltic Sea. The second comment is on the source of the information that we have on biomethylation in sediments. These data are largely from experiments which use either samples of very highly contaminated environments or which enrich the cultures very highly with arsenic. This represents situations of high toxic stress which may cause the organisms to do things which they normally don't. In biochemical terms, this would be reflected both by the induction of enzymes and by population change towards arsenic metabolizing organisms. It's very, very dangerous to draw conclusions from highly enriched cultures to the processes that you'd expect in a natural environment.

E. A. Woolson: I would agree with that 100%.

REGISTERED ATTENDEES
ARSENIC SYMPOSIUM

National Bureau of Standards
Gaithersburg, Maryland

November 4-6, 1981

John R. Abernathy
Texas Agricultural Experiment Sta.
Rt. 3
Lubbock, TX 79401

John C. Alden
Woolfolk Chemical Works, Inc.
P.O. Box 938
Fort Valley, GA 31030

B. L. Allison, III
Koppers Co. Inc.
Forest Products Group
724 Koppers Bldg.
Pittsburgh, PA 15219

Robert Alvarez
Office of Standard Reference Matls.
National Bureau of Standards
Washington, DC 20234

Julian Andelman
Univ. of Pittsburgh
Industrial Health Services
130 De Soto St.
Pittsburgh, PA 15261

C. E. Anderson
Salsbury Laboratories, Inc.
2000 Rockford Road
Charles City, IA 50616

David Anderson
Corning Glass Works
Main Plant 21-1
Corning, NY 14831

Meinrat O. Andreae
Florida State Univ.
Dept. of Oceanography
Tallahassee, FL 32306

Richard Asmus
Mooney Chemical Company, Inc.
2301 Scranton Rd.
Cleveland, OH 44113

C. F. Baker
Nedlog Technology Group
13 Doering Way
Cranford, NJ 07016

William J. Baldwin
Koppers Company, Inc.
Forest Products Group
700 Koppers Building
Pittsburgh, PA 15219

Harold Barnes
Homesteak Mining Co.
1726 Cole Blvd.
Golden, CO 80401

Dave Barron
Bureau of Mines
4900 La Salle Rd.
Avondale, MD 20782

Richard John Bauer
Corning Glass Works
HP-ME2-(E-6)
Corning, NY 14830

Robert P. Beliles
U.S. Dept. Labor OSHA
200 Constitution Ave., NW
Washington, DC 20210

Gerry Berger
North American Minerals
2004 Hogback Rd.
Ann Arbor, MI 48104

Federico Bertolero
NCI - Frederick Cancer Research
Foundation
Bldg. 560, Rm. 32-93
Frederick, md 21701

Bharat Bhooshan
Consumer Product Safety Commission
5401 Westbard Avenue
Bethesda, MD 20207

Bill Booth
Sunshine Mining Company
P.O. Box 1080
Kellogg, ID 83837

Adriane P. Borgias
Pacific Gas and Electric
3400 Crow Canyon Rd.
San Ramon, CA 94583

Kenneth Boyer
FDA
200 C St., SW
Washington, DC 20204

Gail Brinkerhoff
OSHA
US Dept. of Labor
Washington, DC 20210

David Brooks
Office of Pesticides Program
Chrystal Mall #2
Arlington, VA

Richard Burau
Univ. of California - Davis
Dept. of Land, Air & Water Resource
Davis, CA 95616

Sam C. Carapella, Jr.
ASARCO Incorporated
901 Oak Tree Road
South Plainfield, NJ 07080

Quing Chen
Institute of Environmental Medicine
New York Univ. Medical Ctr.
Long Meadow Raod
Tuxedo, NY

Kenneth C. Chu
NTP
7910 Woodmont Ave.
Bethesda, MD 20205

J. E. Clemans
Western Electric Co.
P.O. Box 900
Princeton, NJ 08540

C. R. Coggins
Rentoikil Inc.
Felcourt
East Grinstead, ENGLAND

Jerry M. Cohen
ASARCO Incorporated
120 Broadway
New York, NY 10271

Anthony V. Colucci
A. V. Colucci and Associates, Inc.
15305 Calle Enrique, Suite D
Morgan Hill, CA 95037

Michael Z. Corrigan
Simpson Tohacher & Bartlett
1 Battery Park Plaza
New York, NY 10004

Lloyd J. Croesus
Whitmoyer Laboratories, Inc.
P.O. Box 18, 99 S. Fairland Ave.
Myerstown, PA 17067

John Cross
Boliden Intertrade Limited
Tubs Hill House, London Rd.
Sevenoaks Kent, TN131BL
UNITED KINGDOM

Dewey A. Cubit
Enviro Control, Inc.
11140 Rockville Pike
Rockville, MD 20852

Harry Day
US EPA Office of Pesticides Programs
Chrystal Mall #2, Rm. 809A
Arlington, VA 22210

Philippe De Nicolay
Penarroya
Tour Maine Montparnasse
Paris, FRANCE 75755

Denise Deschenes
Texas Eastern Transmission Corp.
P.O. Box 2521
Houston, TX 77001

Robert E. Donadio
OSHA
US Dept. of Labor
Washington, DC 20210

Ross Doughty
International Paper Co.
Box 797
Tuxedo Park, NY 10987

Patricia L. Dsida
Motor Vehicle Manufacturers
Assoc. of the US, Inc.
300 New Center Building
Detroit, MI 48202

Curtis E. Dungey
ASARCO Incorporated
120 Broadway
New York, NY 10271

Charles Dunne
ASARCO Inc.
120 Broadway
New York, NY 10271

Arnie Edelman
US EPA
401 M St., SW
Washington, DC 20460

Thomas M. Engel
Anaconda Copper Co.
555 17th St.
Denver, CO 80202

Philip E. Enterline
Univ. of Pittsburgh
Graduate Sch. of Pub. Health
130 De Soto St.
Pittsburgh, PA 15261

Louis D. Fitzgerald
ASARCO Incorporated
120 Broadway
New York, NY 10271

Robert J. Fensterheim
Chemical Manufacturers Assoc.
2501 M. Street, NW
Washington, DC 20037

George Fleming
Fleming Laboratories, Inc.
Box 34384
Charlotte, NC 28234

Lorne Forsyth
Cominco Ltd.
120 Adelaide St., W. Ste. 1700
Toronto, Ontario M5H 1T1

Charles Fraust
Western Electric
Dept. 220-555, Union Blvd.
Allentown, PA 18103

Arthur Furst
Univ. of San Francisco
Institute of Chemical Biology
San Francisco, CA 94117

Susan Gallerani
Diamond Shamrock
P.O. Box 9637
Houston, TX 77015

Thomas Gills
National Bureau of Standards
Washington, DC 20234

Leonard J. Goldwater
Duke University
Route 3, Box 197
Chapel Hill, NC 27514

Chuck Gordon
USDOL-OSHA
200 Constitution Ave., NW
Washington, DC 20210

Frank Gostomski
US EPA, Office of Water
Regulations & Standards
401 M St., SW, Rm. 2181 Mall
Washington, DC 20460

Barbara Goun
Center for Environmental Epidemiology
Univ. of Pittsburgh
Pittsburgh, PA 15261

Henri Hagon
Metallurgie Hoboken Overpelt
Greinea 3
Antwerp, BELGIUM

I. Harding-Barlow
Consultant
3717 Laguna Ave.
Palo Alto, CA 94306

Jorg W. Hensel
Mobay Chemical Corp.
Penn Lincoln Parkway West
Pittsburgh, PA 15205

John Herzfeld
BNA-Occupational Safety
Health Reporter
1231 25th St., NW
Washington, DC 20037

John D. Hite
Koppers Company, Inc.
1001 Koppers Building
Pittsburgh, PA 15215

Marian Holmes
Phelps Dodge Corp.
1025 Vermont Ave., #1201
Washington, DC 20005

Peter F. Holzberg
Maryland Casualty Co.
3910 Keswick Rd.
Baltimore, MD 21211

Ronald D. Hood
Dept. of Biology
The Univ. of Alabama
P.O. Box 1927
University, AL 35486

Joseph A. Hopkins, Jr.
US Dept. of Labor, OSHA
Office of Field Coordination
200 Constitution Ave., Rm. N3119
Washington, DC 20210

Diana Horton
Dept. of Agronomy
Univ. of Arkansas
Rt. 11, Box 83
Fayetteville, AR 70701

Kurt J. Irgolic
Texas A & M University
Chemistry Department
College Station, TX 77843

Wilma Ann Jancuk
Western Electric Co.
Engineering Research Ctr.
P.O. Box 900
Princeton, NJ 08540

Gunnar Jonsson
Boliden Metall AB
22 Sturegatan, Box 5508
Stockholm, Sweden S-11485

Bob Kayser
US EPA, Office of Toxics Integration
401 M St., SW, Rm. 435
Washington, DC 20460

George W. Keitt, Jr.
US EPA (OPP/OPTS)
4220 Linden St.
Fairfax, VA 22030

William H. Kirchhoff
National Bureau of Standards
Washington, DC 20234

Anders Kjellberg
Boliden Metall AB
22 Sturegatan, Box 5508
Stockholm, Sweden S-11485

Phyllis Kyner
U.S. Dept. of Labor - OSHA
2101 Ferry Ave., Suite 403
Camden, NJ 08104

Abbas Labbouf
Penn. Environmental Health
4614 Fifth Ave.
Pittsburgh, PA 15213

Frank L. Laird
Anaconda Copper Company
555 17th St., Rm. 1731
Denver, CO 80202

Robert M. Leach
Osmole Wood Preserving Co.
980 Ellicott Street
Buffalo, NY 14209

William H. Lederer
Koppers Company, Inc.
1201 Koppers Building
Pittsburgh, PA 15219

Anna Lee-Feldstein
Univ. of Michigan
School of Public Health
Ann Arbor, MI 48109

David A. Levine
Bernuth, Lembcke Co., Inc.
7600 West Tidwell, Suite 204
Houston, TX 77040

Roger Loebenstein
U.S. Bureau of Mines
2401 E St., NW
Washington, DC 20241

Jay H. Lubin
Environmental Epidemiology Branch/NCI
Landow Bldg., 3C09
Bethesda, MD 20205

J. P. McCarthy
Koppers Company, Inc.
1201 Koppers Building
Pittsburgh, PA 15219

Lottie McClendon
National Bureau of Standards
OEM
Washington, DC 20234

Charles Ray McClure
Consultant, Industrial Hygiene
510 Utterback Store Rd.
Great Falls, VA 22066

Glen A. McDonald
Equity Silver Mines Limited
Houston
British, Columbia

Paul McDaniel
Union Carbide Corp.
270 Park Ave., 22nd Floor
New York, NY 10017

T. C. McEntee
Thiokol/Ventron Division
150 Andover St.
Danvers, MA 01923

Steven L. Malish
Bell Laboratories
600 Mountain Ave.
Murray Hill NJ 07974

William L. Marcus
US EPA
Office of Dr. Water
401 M Street
Washington, DC 20460

John Martonik
US Dept. of Labor/Occupational Safety and Health Administration
200 Constitution Ave., NW
Washington, DC 20210

Allen A. Mattes
OSHA Office of Regulatory Analysis
200 Constitution Ave., NW, Rm. N3651
Washington, DC 20210

Noel B. Menesse
Diamond Shamrock Corporation
Purchasing Dept.
1100 Superior Ave.
Cleveland, OH 44114

A. C. Middleton
Koppers Company, Inc.
440 College Park Drive
Monroeville, PA 15146

John C. Middleton
United States Borax & Chem. Corp.
412 Crescent Way
Anaheim, CA 92801-6794

Andrew Miles
Radian Corp.
3024 Pickett Road
Durham, NC 27705

Chuck Mitchell
US EPA
Office of Health Research
401 M St., SW, Rm. 3812 Mall
Washington, DC 20460

G. Donald Munger
Diamond Shamrock Corporation
1100 Superior Ave.
Cleveland, OH 44114

Peter J. Neff
St. Joe International Co.
250 Park Ave.
New York, NY 10017

K. W. Nelson
ASARCO Incorporated
3422 South - 700 West
Salt Lake City, UT 84106

Norman R. Nelsen
Voluntary Purchasing Groups, Inc.
Two Shunpike Road
Madison, NJ 07940

Forrest H. Nielsen
USDA, ARS, Grand Forks Human
Nutrition Research Center
Box 7166, University Station
Grand Forks, ND 58202

John F. O'Callaghan
Leonard J. Buck and Co., Inc.
95 Madison Avenue
Morristown, NJ 07960

Ed Ohanian
US EPA, Office of Drinking Water
401 M St., SW, Rm. 1103 E.T.
Washington, DC 20460

Fred L. Omundson
OSMOSE
724 Maple Drive
Griffin, GA 30223

George W. Ozga
US Dept. of Labor - OSHA
200 Constitution Ave., NW
Washington, DC 20210

Edwin Parks
National Bureau of Standards
Washington, DC 20234

E. George Pazianos
PAZIANOS ASSOCIATES
211 9th St., NE
Washington, DC 20002

Stuart Anderson Peoples
Rt. 1, Box 2350
Davis, CA 95616

Warren Peters
US EPA, Office of Air Quality
Planning and Standards
Room 937
Research Triangle Park, NC 27711

Michel Pewinski
PENARROYA
Tour Montparnasse
Paris, France 75755

Richard T. Price
Diamond Shamrock Corporation
1149 Ellsworth Drive
Pasadena, TX 77501

Edward P. Radford
Univ. of Pittsbrugh
Graduate Sch. of Public Health
130 DeSoto Street, 517 Parran Hall
Pittsburgh, PA 15261

Amy Rispin
Office of Pesticides Program
Chrystal Mall #2
Arlington, VA 22210

D. A. Robbins
ASARCO Incorporated
3422 South 700 West
Salt Lake City, UT 84119

Gerald D. Roseberry
Pennwalt Corporation
3 Parkway
Philadelphia, PA 19102

Marie L. Roy
Ministry of Labour
400 University Ave.
Toronto, Ontario M7A 1T7
CANADA

Donald W. Rumsey
Enviro Control Inc.
The Dynamac Building
11140 Rockville Pike
Rockville, MD 20852

M. Jay Rupp, Jr.
Chessie Systems - 111
100 N. Charles St.
Baltimore, MD 21201

Emil Sandi
Health Protection Board
Health & Welfare Canada
Ottawa, Ontario
CANADA

Michael Sansonetti Jr.
United Mineral and Chemical Corp.
129 Hudson Street
New York, NY 10013

John B. Schaye
Powell Metals and Chemicals, Inc.
1122 Milford Ave.
Rockford, IL 61125

Erich Schmidt
Ministerio de Sanidad y Asitencia
Social D.M.S.A.
Edificio Sur Piso 2, Centro Simon Bolivar
Caracas, Venezuela 1010

Marvin A. Schneiderman
Clement Associates
1010 Wisconsin Ave., NW, Suite 660
Washington, DC 20007

Ronald F. Scholl
Hoge, Fenton, Jones & Appel, Inc.
P.O. Box 791
Monterey, CA 93940

Richard A. Schraufnagel
Arco Coal R & D
P.O. Box 2819
Dallas, TX 75221

Martin M. Schwartzberg
Pennwalt Corporation
3 Parkway
Philadelphia, PA 19102

Arthur Schwerdtle
Vineland Chemical Company
West Wheat Rd.
Vineland, NJ 08360

Jack F. Self
Chemical Operations Div. - Aerojet
Strategic Propulsion Co.
P.O. Box 15699C
Sacramento, CA 95813

Hasmukh Shah
Chemical Manufacturers Assoc.
2501 M St., NW
Washington, DC 20037

Simon Silver
Biology Department
Washington University
St. Louis, MO 63130

Vincent F. Simmon
Genex Corporation
12300 Washington Avenue
Rockville, MD 20852

Steven M. Skubik
Wausau Insurance Companies
10 Rooney Circle
W. Orange, NJ 07052

C. L. Smart
Mobil Oil Corp., Toxicology Div.
P.O. Box 1029
Princeton, NJ 08540

Francis J. Smith
Radian Corp.
7927 Jones Branch Dr.
McLean, VA 22102

Russell C. Smith
Salsbury Laboratories, Inc.
2000 Rockford Road
Charles City, IA 50616

J. Wanless Southwick
Utah State Dept. of Health
150 North West Temple
P.O. Box 2500
Salt Lake City, UT 84110

Howard G. Speck
C. Withington Co., Inc.
16 Pelham Parkway
Pelham Manor, NY 10803

Edward Stein
US Dept. of Labor/Occupational Safety
and Health Administration
200 Constitution Ave., NW
Washington, DC 20210

Patricia Stewart
OSHA
200 Constitution Ave., NW
Washington, DC 20210

Woodhall Stopford
Durham Medical Center
Durham, NC 27710

Langan W. Swent
Homestake Mining Co.
650 California St.
San Francisco, CA 94108

E. D. Switala
Owens-Corning Fiberglas Corp.
Fiberglas Tower
Toledo, OH 43659

Malcolm C. Thomas
Food & Drug Administration
5600 Fishers Lane
Rockville, MD 21701

Robert Toth
Pennwalt Corporation
1630 E. Shaw Ave., Suite 179
Fresno, CA 93710

J. Drake Watson
Pennwalt Corporation
900 1st Avenue
King of Prussia, PA 19406

R. Don Wauchope
USDA-ARS
Southern Weel Science Lab.
Stoneville, MS 38776

Carl S. Weiss
National Bureau of Standards
Bldg. 223, Rm. A329
Washington, DC 20234

John K. Whitley
Rentokil, Inc.
272 Prado North, 5600 Roswell Rd.
Atlanta, GA 30342

R. K. Willardson
Cominco American Incorporated
W. 818 Riverside
P.O. Box 3087
Spokane, WA 99220

Hans J. Woerner
Mineral Research Corp.
Rd. 8, Box 222
Concord, NC 28124

James R. Wolfe
Nedlog Corp.
Rd. 3, Box 668
Laramie, WY 82070

E. A. Woolson
USDA
Washington, DC

Hiroshi Yamauchi
Dept. Public Health
St. Marianna Univ. Sch. of Medicine
2095 Sugawo, Takatsu-ku
Kamasaki, JAPAN

Mary M. Yurachek
Occupational Safety and
Health Administration
200 Constitution Ave., NW, Rm. N3718
Washington, DC 20210

Virginia L. Zaratzian
USDA/FSIS/Science/RES
USDA ANNEX
300 12th St., SW
Washington, DC 20205

Ralph A. Zingaro
Texas A & M University
Dept. of Chemistry
College Station, TX 77843

Author Index

SUBJECT INDEX[1]

[1]The citations in this index refer to arsenic; and the term arsenic is used only when necessary for clarity.

C

F

G

R

T